Dietary Patterns, Physical Activity, and Lifestyle in the Onset, Prevention, and Management of Noncommunicable Diseases

Dietary Patterns, Physical Activity, and Lifestyle in the Onset, Prevention, and Management of Noncommunicable Diseases

Editors

William B. Grant
Ronan Lordan

MDPI

Basel • Beijing • Wuhan • Barcelona • Belgrade • Novi Sad • Cluj • Manchester

Editors
William B. Grant
Sunlight, Nutrition, and
Health Research Center
San Francisco, CA, USA

Ronan Lordan
University of Pennsylvania
Philadelphia, PA, USA

Editorial Office
MDPI
St. Alban-Anlage 66
4052 Basel, Switzerland

This is a reprint of articles from the Special Issue published online in the open access journal *Nutrients* (ISSN 2072-6643) (available at: https://www.mdpi.com/journal/nutrients/special_issues/dietary_lifestyle_noncommunicable_diseases).

For citation purposes, cite each article independently as indicated on the article page online and as indicated below:

Lastname, A.A.; Lastname, B.B. Article Title. *Journal Name* **Year**, *Volume Number*, Page Range.

ISBN 978-3-0365-9550-4 (Hbk)
ISBN 978-3-0365-9551-1 (PDF)
doi.org/10.3390/books978-3-0365-9551-1

Contents

About the Editors

William B. Grant

William B. Grant has a Ph.D. in physics from the University of California, Berkeley (1971). He is the director of the Sunlight, Nutrition and Health Research Center, a health-research organization founded in San Francisco in 2004. He has produced over 300 publications regarding vitamin D and over 50 regarding diet, including articles, reviews, editorials, and letters to the editor. His two main interests are the study of solar UV exposure and vitamin D supplementation for optimal health and the role of nutrition in reducing the risk of chronic diseases such as Alzheimer's disease and cancer.

Ronan Lordan

Dr. Ronan Lordan is a researcher of circadian biology and nutrition at the Institute for Translational Medicine and Therapeutics, Perelman School of Medicine, at the University of Pennsylvania. He obtained a Ph.D. in the field of nutrition and cardiovascular science and an MA in teaching, learning, and scholarship. Ronan has lectured in nutrition, genetics, and physiology at the University of Limerick. He has published several peer-reviewed articles and book chapters, and is an active reviewer and editor for several journals. His research interests include circadian biology, metabolism, and aging; dietary supplements; and discerning the mechanisms of platelet-activating factor-induced systemic inflammation in noncommunicable diseases.

Preface

Noncommunicable diseases are a significant burden on global health systems. According to the World Health Organization, the top ten causes of death globally in 2019 were, in order, ischemic heart disease (9 million deaths/year), stroke (6 million), chronic obstructive pulmonary disease (3 million), lower respiratory tract infections, neonatal conditions, trachea, bronchus, lung cancers, Alzheimer's disease and other dementias, diarrheal diseases, diabetes mellitus, and kidney diseases. Noncommunicable diseases account for the majority of the deaths worldwide. However, many of these are preventable if populations have access to and adopt healthy dietary patterns and lifestyle choices. The aim of this Special Issue is to attract cutting-edge research regarding dietary patterns, physical activity, and lifestyle in the onset, prevention, and management of noncommunicable disease.

William B. Grant and Ronan Lordan
Editors

nutrients

MDPI

Editorial

Dietary Patterns, Physical Activity, and Lifestyle in the Onset, Prevention, and Management of Noncommunicable Diseases

Ronan Lordan [1,*] and William B. Grant [2]

1 Institute for Translational Medicine and Therapeutics (ITMAT), Perelman School of Medicine, University of Pennsylvania, Philadelphia, PA 19104-5158, USA
2 Sunlight, Nutrition, and Health Research Center, P.O. Box 641603, San Francisco, CA 94164-1603, USA; williamgrant08@comcast.net
* Correspondence: ronan.lordan@pennmedicine.upenn.edu

Noncommunicable diseases (NCDs) are on the rise due to population growth and aging, which will cause a significant burden on global health systems [1]. Noncommunicable diseases account for most deaths worldwide. However, decades of research show that dietary and lifestyle factors play a significant role in the onset and progression of many noncommunicable diseases, including cardiovascular diseases (CVD) and cancers [2–5]. Furthermore, NCDs may be risk factors for communicable disease severity as is the case for the Coronavirus disease 2019 (COVID-19) [6,7], or indeed NCDs can increase the risk of developing other NCDs leading to individuals with many comorbidities. Therefore, there is an increased awareness in the development of programs addressing maladaptive dietary practices and lifestyles and increased research into how these factors may affect the onset, prevention, and management of NCDs. On the other hand, regular exercise is known to be beneficial for the prevention of NCDs and as an intervention for those who develop some NCDs such as CVD, type II diabetes mellitus, or obesity.

The aim of this Special Issue was to collate state-of-the-art original research articles and reviews addressing the role of dietary patterns, physical activity, and lifestyle in the onset, prevention, and management of NCDs. Several articles in the Special Issue investigated the role of dietary patterns in association with NCDs.

Hlaing-Hlaing et al. [8] investigated the association between the incidence of NCDs with diet quality measured using the alternative healthy eating index 2010, in an Australian longitudinal study of women's health. Their repeated cross-sectional multivariate logistic regression analysis over a 15-year period showed that baseline diet quality does not demonstrate early evidence of a protective effect against a future occurrence of NCDs or multimorbidity between the aged of 25–45 years. Further longitudinal and temporal analyses with sensitive markers for NCD risk factors are required.

Wang et al. [9] conducted a prospective cohort study that investigated the association between fruit intake and stroke with consideration for genetic disposition in a cohort of almost 35,000 individuals from the project Prediction for Atherosclerotic Cardiovascular Disease Risk in China (CHINA-PAR). This study reported that a greater amount of fruit consumption was associated with a lower risk of stroke (28–32%), whereby individuals who had a higher genetic risk of stroke may gain more benefits in terms of more stroke-free years and absolute risk reduction. These findings add to the body of evidence demonstrating the importance of following a healthy dietary pattern for the prevention of stroke [10].

A cross-sectional study of over 40,000 participants was conducted by Zhu, et al. [11], who analyzed the association between obesity and dyslipidaemia using data from the Shanghai suburban adult cohort and biobank study (SSACB). Their data shows that central obesity and to some extent general obesity were associated with lower levels of high-density lipoprotein cholesterol (HDL-C) and higher levels of triglycerides. Other factors such as gender, age, diabetes, and hypertension were also associated with various forms

Citation: Lordan, R.; Grant, W.B. Dietary Patterns, Physical Activity, and Lifestyle in the Onset, Prevention, and Management of Noncommunicable Diseases. *Nutrients* **2023**, *15*, 2540. https://doi.org/10.3390/nu15112540

Received: 11 May 2023
Accepted: 24 May 2023
Published: 30 May 2023

of dislipidaemia. Indeed, these findings indicate the importance of further research into effective interventions for obesity to reduce the risk of metabolic syndrome and potential cardiovascular diseases due to the high prevalence of dyslipidaemia.

On the other hand, young adults are a population less explored in nutrition research. Kurniawan et al. [12] conducted a study with 1-year follow-up to determine the effects of urbanization on urban or rural individuals and their metabolic profiles. Baseline urban (n = 106) and rural (n = 83) participants were young (16–25 years old) female Indonesian adults, some of which were examined in a 1-year follow-up (n = 81 and n = 66 participants, respectively). The urbanization of individuals who previously resided in rural areas appeared to be associated with less favorable changes in adipokine profiles and adiposity. These findings indicate that migrating from rural areas to urban areas may lead to maladaptive changes in lifestyle that affect health.

Understanding the mechanisms of how various lifestyle factors affect the onset and development of NCDs is important for developing prophylactics and therapeutics. Harishkumar et al. [13] reviewed the specific nutrients that may impart antithrombotic and antiatherosclerotic biological activity against proinflammatory signaling induced by platelet-activating factor (PAF) via its receptor (PAF-R). They report the in vitro, ex vivo, and nutritional evidence regarding the potential benefits of consuming various nutrients including vitamins, trace metals, polar lipids, and other micronutrients, which are constituents of healthy dietary patterns such as the Mediterranean diet. They focus on the potential cardiovascular benefits that may be derived from targeting the PAF signaling cascade and discuss the need for more research in the field, which has been echoed by others in the field studying the role of dietary components in PAF-induced atherosclerotic diseases [14].

Liu et al. [15] examined representative data from the youth risk behavior surveillance system (YRBSS) study database of over 73,000 adolescents in the United States. They showed that the prevalence of suicidal ideation and planning was higher among adolescents that did not meet optimum 24 h movement guidelines set forth by the YRBSS study or soft drink consumption $\geq$ 3 times/day was associated with increased suicidal ideation, planning, and attempts with/without medication. However, cautious interpretation is required as a causal relationship cannot be inferred from a cross-sectional design and most responses were self-reported introducing recall biases. However, the findings should give pause to policy makers generally about the importance of a healthy lifestyle pattern to prevent suicidal ideation, planning, and attempts [16,17].

Vitamin D research is also featured as a prominent element of this Special Issue. Grant and Boucher [18] reviewed how the seasonal varations of atomospheric temperature, humiduty, and solar radiation exposure affect seasonal variations of blood pressure, CVD rates, and respiratory viral infections. This review points to the importance of safe sun exposure in health and disease and the need for further public health research on the topic.

Two reviews in this Special Issue focused on the role of vitamin D-related risk factors for both maternal morbidity [19] and mortality [20]. Vitamin D deficiency has been previously associated with adverse pregnancy outcomes [21]. The first article conducted a systematic review of studies relating to vitamin D status during pregnancy and maternal outcomes [19]. The second article examined both morbidity and mortality as part of meta-analyses of the same topic. The authors show that vitamin D supplementation may be important for reducing the risk of pregnancy complications such as gestational diabetes, hypertension, preeclampsia, and others, and that low-circulating 25-hydroxyvitamin D [25(OH)D] concentrations during pregnancy were a risk factor for these complications [20]. These articles highlight the importance of vitamin D in the health of the mother and fetus during pregnancy.

The final two articles of the Special Issue address findings relating to the Coronavirus disease 2019 (COVID-19) pandemic and vitamin D. Vitamin D has been postulated as a potential prophylactic and therapeutic for COVID-19. Gupta, et al. [22] carried out case–control studies within two institutions of the University of California San Diego (UCSD) health system. They measured 25(OH)D concentrations in patients either prior to or post disease diagnosis within 180 days of diagnosis in January 2020. The authors reported that serum 25(OH)D status was not associated with an increased risk of COVID-19, but concentrations were reduced post infection. These findings provide further evidence that serum 25(OH)D concentrations measured after COVID-19 should not be used to determine the risk of COVID-19 with respect to 25(OH)D concentration but might be useful in predicting prognosis afterwards.

Post acute sequalae of COVID-19 (PASC) or more commonly known as long COVID-19 has emerged as a burdening health risk of SARS-CoV-2 infection. Barrea, et al. [23] conducted a review of the literature to determine if vitamin D levels play a role in long COVID. They surmise that vitamin D may play a role in long COVID due to its immunomulatory effects and its association with COVID-19 severity. A recent review by Moukayed [24] supports the findings of Barrea et al. [23] and suggests that higher doses of vitamin D supplementation may improve the health and survivorship of COVID-19 and long COVID-19 patients. However, further research is required to determine the role of vitamin D in long COVID aetiology and pathology.

This Special Issue has captured the breadth of research being undertaken regarding dietary patterns, physical activity, and lifestyle in the onset, prevention, and management of noncommunicable diseases.

Acknowledgments: W.B.G. and R.L. would like to thank all the contributing authors, reviewers, and editors to our Special Issue: Dietary Patterns, Physical Activity, and Lifestyle in the Onset, Prevention, and Management of Noncommunicable Diseases.

Conflicts of Interest: W.B.G.'s non-profit organization, Sunlight, Nutrition and Health Research Center, receives funding from Bio-Tech Pharmacal, Inc., R.L. has no conflict of interest to declare.

References

1. Di Cesare, M. Global trends of chronic non-communicable diseases risk factors. *Eur. J. Public Health* **2019**, *29*, ckz185.196. [CrossRef]
2. Yu, E.; Malik, V.S.; Hu, F.B. Cardiovascular disease prevention by diet modification: JACC Health Promotion Series. *J. Am. Coll. Cardiol.* **2018**, *72*, 914–926. [CrossRef] [PubMed]
3. Irigaray, P.; Newby, J.A.; Clapp, R.; Hardell, L.; Howard, V.; Montagnier, L.; Epstein, S.; Belpomme, D. Lifestyle-related factors and environmental agents causing cancer: An overview. *Biomed. Pharm.* **2007**, *61*, 640–658. [CrossRef]
4. Sinicrope, F.A. Increasing Incidence of Early-Onset Colorectal Cancer. *N. Engl. J. Med.* **2022**, *386*, 1547–1558. [CrossRef] [PubMed]
5. Lordan, R.; Tsoupras, A.; Zabetakis, I. The Potential Role of Dietary Platelet-Activating Factor Inhibitors in Cancer Prevention and Treatment. *Adv. Nutr.* **2019**, *10*, 148–164. [CrossRef]
6. Kluge, H.H.P.; Wickramasinghe, K.; Rippin, H.L.; Mendes, R.; Peters, D.H.; Kontsevaya, A.; Breda, J. Prevention and control of non-communicable diseases in the COVID-19 response. *Lancet* **2020**, *395*, 1678–1680. [CrossRef]
7. Bello, B.; Useh, U. COVID-19: Are non-communicable diseases risk factors for its severity? *Am. J. Health Promot.* **2021**, *35*, 720–729. [CrossRef]
8. Hlaing-Hlaing, H.; Dolja-Gore, X.; Tavener, M.; James, E.L.; Hure, A.J. Alternative Healthy Eating Index-2010 and Incident Non-Communicable Diseases: Findings from a 15-Year Follow Up of Women from the 1973–78 Cohort of the Australian Longitudinal Study on Women's Health. *Nutrients* **2022**, *14*, 4403. [CrossRef]
9. Wang, J.; Li, J.; Liu, F.; Huang, K.; Yang, X.; Liu, X.; Cao, J.; Chen, S.; Shen, C.; Yu, L.; et al. Genetic Predisposition, Fruit Intake and Incident Stroke: A Prospective Chinese Cohort Study. *Nutrients* **2022**, *14*, 5056. [CrossRef]
10. Niewada, M.; Michel, P. Lifestyle modification for stroke prevention: Facts and fiction. *Curr. Opin. Neurol.* **2016**, *29*, 9–13. [CrossRef]
11. Zhu, J.; Zhang, Y.; Wu, Y.; Xiang, Y.; Tong, X.; Yu, Y.; Qiu, Y.; Cui, S.; Zhao, Q.; Wang, N.; et al. Obesity and Dyslipidemia in Chinese Adults: A Cross-Sectional Study in Shanghai, China. *Nutrients* **2022**, *14*, 2321. [CrossRef]
12. Kurniawan, F.; Manurung, M.D.; Harbuwono, D.S.; Yunir, E.; Tsonaka, R.; Pradnjaparamita, T.; Vidiawati, D.; Anggunadi, A.; Soewondo, P.; Yazdanbakhsh, M.; et al. Urbanization and Unfavorable Changes in Metabolic Profiles: A Prospective Cohort Study of Indonesian Young Adults. *Nutrients* **2022**, *14*, 3326. [CrossRef]
13. Harishkumar, R.; Hans, S.; Stanton, J.E.; Grabrucker, A.M.; Lordan, R.; Zabetakis, I. Targeting the Platelet-Activating Factor Receptor (PAF-R): Antithrombotic and Anti-Atherosclerotic Nutrients. *Nutrients* **2022**, *14*, 4414. [CrossRef]
14. Antonopoulou, S.; Demopoulos, C.A. Protective Effect of Olive Oil Microconstituents in Atherosclerosis: Emphasis on PAF Implicated Atherosclerosis Theory. *Biomolecules* **2023**, *13*, 700. [CrossRef]
15. Liu, B.-P.; Jia, C.-X.; Li, S.-X. The Association of Soft Drink Consumption and the 24-Hour Movement Guidelines with Suicidality among Adolescents of the United States. *Nutrients* **2022**, *14*, 1870. [CrossRef]

16. Rodelli, M.; De Bourdeaudhuij, I.; Dumon, E.; Portzky, G.; DeSmet, A. Which healthy lifestyle factors are associated with a lower risk of suicidal ideation among adolescents faced with cyberbullying? *Prev. Med.* **2018**, *113*, 32–40. [CrossRef]
17. Berardelli, I.; Corigliano, V.; Hawkins, M.; Comparelli, A.; Erbuto, D.; Pompili, M. Lifestyle Interventions and Prevention of Suicide. *Front. Psychiatry* **2018**, *9*, 587. [CrossRef]
18. Grant, W.B.; Boucher, B.J. An Exploration of How Solar Radiation Affects the Seasonal Variation of Human Mortality Rates and the Seasonal Variation in Some Other Common Disorders. *Nutrients* **2022**, *14*, 2519. [CrossRef]
19. Morales-Suárez-Varela, M.; Uçar, N.; Peraita-Costa, I.; Huertas, M.F.; Soriano, J.M.; Llopis-Morales, A.; Grant, W.B. Vitamin D-Related Risk Factors for Maternal Morbidity during Pregnancy: A Systematic Review. *Nutrients* **2022**, *14*, 3166. [CrossRef]
20. Morales-Suárez-Varela, M.; Uçar, N.; Soriano, J.M.; Llopis-Morales, A.; Sanford, B.S.; Grant, W.B. Vitamin D-Related Risk Factors for Maternal Morbidity and Mortality during Pregnancy: Systematic Review and Meta-Analysis. *Nutrients* **2022**, *14*, 4124. [CrossRef]
21. Mulligan, M.L.; Felton, S.K.; Riek, A.E.; Bernal-Mizrachi, C. Implications of vitamin D deficiency in pregnancy and lactation. *Am. J. Obstet. Gynecol.* **2010**, *202*, 429.e1–429.e9. [CrossRef] [PubMed]
22. Gupta, D.; Menon, S.; Criqui, M.H.; Sun, B.K. Temporal Association of Reduced Serum Vitamin D with COVID-19 Infection: Two Single-Institution Case–Control Studies. *Nutrients* **2022**, *14*, 2757. [CrossRef] [PubMed]
23. Barrea, L.; Verde, L.; Grant, W.B.; Frias-Toral, E.; Sarno, G.; Vetrani, C.; Ceriani, F.; Garcia-Velasquez, E.; Contreras-Briceño, J.; Savastano, S.; et al. Vitamin D: A Role Also in Long COVID-19? *Nutrients* **2022**, *14*, 1625. [CrossRef] [PubMed]
24. Moukayed, M. A Narrative Review on the Potential Role of Vitamin D_3 in the Prevention, Protection, and Disease Mitigation of Acute and Long COVID-19. *Curr. Nutr. Rep.* **2023**, 1–9. [CrossRef]

nutrients

MDPI

Review

Vitamin D: A Role Also in Long COVID-19?

Luigi Barrea [1,2,†], Ludovica Verde [2,†], William B. Grant [3,†], Evelyn Frias-Toral [4], Gerardo Sarno [5], Claudia Vetrani [6], Florencia Ceriani [7], Eloisa Garcia-Velasquez [8], José Contreras-Briceño [9], Silvia Savastano [2,6], Annamaria Colao [2,6,10] and Giovanna Muscogiuri [2,6,10,*]

1 Dipartimento di Scienze Umanistiche, Università Telematica Pegaso, Via Porzio, Centro Direzionale, Isola F2, 80143 Napoli, Italy; luigi.barrea@unina.it
2 Centro Italiano per la Cura e il Benessere del Paziente con Obesità (C.I.B.O), Endocrinology Unit, Department of Clinical Medicine and Surgery, University Medical School of Naples, Via Sergio Pansini 5, 80131 Naples, Italy; lud.verde@studenti.unina.it (L.V.); ambulatoriocibo@gmail.com (S.S.); operafederico2@gmail.com (A.C.)
3 Sunlight, Nutrition, and Health Research Center, P.O. Box 641603, San Francisco, CA 94164-1603, USA; wbgrant@infionline.net
4 School of Medicine, Universidad Católica Santiago de Guayaquil, Av. Pdte. Carlos Julio Arosemena Tola, Guayaquil 090615, Ecuador; evelyn.frias@cu.ucsg.edu.ec
5 General Surgery and Kidney Transplantation Unit, San Giovanni di Dio e Ruggi d'Aragona University Hospital, 84131 Salerno, Italy; gsarno79@yahoo.it
6 Unit of Endocrinology, Dipartimento di Medicina Clinica e Chirurgia, Federico II University Medical School of Naples, Via Sergio Pansini 5, 80131 Naples, Italy; c.vetrani@libero.it
7 Nutrition School, Universidad de la Republica (UdelaR), Montevideo 11100, Uruguay; florceriani@gmail.com
8 Clinical Nutrition Service, Grupo Hospitalario Kennedy, Guayaquil 090615, Ecuador; eloisa.garciavelasquez@gmail.com
9 Nursing Program, Universidad Técnica Particular de Loja, San Cayetano Alto, Calle París, Loja 110107, Ecuador; jocontreras@utpl.edu.ec
10 Cattedra Unesco "Educazione Alla Salute e Allo Sviluppo Sostenibile", Federico II University, 80131 Naples, Italy
* Correspondence: giovanna.muscogiuri@unina.it; Tel.: +39-0817463779; Fax: +30-0817463688
† These authors contributed equally to this work.

Citation: Barrea, L.; Verde, L.; Grant, W.B.; Frias-Toral, E.; Sarno, G.; Vetrani, C.; Ceriani, F.; Garcia-Velasquez, E.; Contreras-Briceño, J.; Savastano, S.; et al. Vitamin D: A Role Also in Long COVID-19? *Nutrients* **2022**, *14*, 1625. https://doi.org/10.3390/nu14081625

Academic Editors: Carsten Carlberg and Connie Weaver

Received: 23 February 2022
Accepted: 1 April 2022
Published: 13 April 2022

Abstract: Coronavirus disease 2019 (COVID-19) has quickly become a global pandemic. Reports from different parts of the world indicate that a significant proportion of people who have recovered from COVID-19 are suffering from various health problems collectively referred to as "long COVID-19". Common symptoms include fatigue, shortness of breath, cough, joint pain, chest pain, muscle aches, headaches, and so on. Vitamin D is an immunomodulatory hormone with proven efficacy against various upper respiratory tract infections. Vitamin D can inhibit hyperinflammatory reactions and accelerate the healing process in the affected areas, especially in lung tissue. Moreover, vitamin D deficiency has been associated with the severity and mortality of COVID-19 cases, with a high prevalence of hypovitaminosis D found in patients with COVID-19 and acute respiratory failure. Thus, there are promising reasons to promote research into the effects of vitamin D supplementation in COVID-19 patients. However, no studies to date have found that vitamin D affects post-COVID-19 symptoms or biomarkers. Based on this scenario, this review aims to provide an up-to-date overview of the potential role of vitamin D in long COVID-19 and of the current literature on this topic.

Keywords: COVID-19; SARS-CoV-2; long COVID-19; vitamin D; inflammation

1. Introduction

Because of the recent advances in the pathophysiological mechanisms that occur with the novel coronavirus disease 2019 (COVID-19), there is growing interest to profoundly investigate the role of vitamin D and its deficiency to increase the susceptibility and negative results of COVID-19. Vitamin D is a secosteroid produced by the skin mainly due to exposure to sunlight in the form of cholecalciferol, and diet provides about 20% of the daily requirement of this vitamin [1].

Vitamin D deficiency has been defined as a serum concentration of 25-hydroxyvitamin D (25OHD) < 20 ng/mL (50 nmol/L) [2]. It is noteworthy to emphasize that it has been reported that its concentrations in women are lower than in men due to the proportion and distribution of fat tissue [3].

Goërtz et al. assessed 2113 patients with confirmed or suspected COVID-19 diagnosis and found that at least 87% of them continued with symptoms later than 60 days of the first symptom (of which 32% reported one or two symptoms and 55% reported three or more symptoms) [4]. The conditions (up to 60 days) following infection with COVID-19 are known as prolonged, long-lasting, post-acute, long-term, or chronic effects; among them, the most common symptoms are fatigue, dyspnea, and insomnia [5]. Vitamin D deficiency has also been related to all these symptoms [6]. Carpagnano et al. found a high prevalence of vitamin D deficiency in COVID-19 patients with acute respiratory failure [7]. It is important to emphasize that vitamin D deficiency is related to many other diseases and conditions that will increase the risk of developing a long-term COVID-19 [2,4,6,8]. In this respect, Savanelli et al. reported that vitamin D deficiency is the greatest predictor of the prevalence of dyslipidemia and hypertension in patients with coronary heart disease, suggesting the presence of both factors in cardiovascular risk in this group of patients [8]. Furthermore, according to a recent review, this virus may provoke a new onset of type 2 diabetes mellitus with undetermined clinical and metabolic components, providing a possible role for COVID-19 in developing type 2 diabetes mellitus [9].

For these reasons, vitamin D has been identified as one of the critical components for treating COVID-19 infection [6,10,11]. However, there is more to explain of how vitamin D works in prolonged COVID-19 patients. This review aims to provide a state of the art of vitamin D's role in long COVID-19 and the updated literature on this topic. The authors will present the impact of vitamin D deficiency on COVID-19 patients and those with long-lasting COVID-19.

2. Methods

In this review, data were summarized using a narrative approach, based on clinical expertise in interpretation of available evidence in the peer-reviewed journal literature. Both https://pubmed.ncbi.nlm.nih.gov/ (accessed on 15 September 2021) and https://scholar.google.com/ (accessed on 15 September 2021) were searched. Search terms included COVID-19, long COVID-19, mechanisms, risk, SARS-CoV-2, symptoms, and vitamin D. In addition, references regarding virus downregulation of vitamin D receptors were found at https://vitamindwiki.com/ (accessed on 15 September 2021). All studies evaluating vitamin D and long-COVID-19 were taken into consideration. However, the review only included (i) manuscripts in English; (ii) original articles; and (iii) prospective or retrospective observational (analytical or descriptive), experimental, or quasi-experimental studies. Non-original studies, including editorials and letters to the editor, were excluded except for one editorial that had a figure showing how vitamin D reduces risk of COVID-19. No limitations on the date of publication were imposed.

3. COVID-19 and Vitamin D

The world is still experiencing the pandemic of COVID-19 and its impacts. This pandemic has posed an immense threat to humans, and it is responsible for causing considerable morbidity and mortality worldwide [12]. At the pandemic's beginning, people needed to quarantine and change their lifestyle habits, such as food ingestion or exercise, and working from home became the new routine. All these modifications are associated with less time spent outside and, as a result, less sun exposure and reduced production of vitamin D [13].

Vitamin D deficiency is a known public worldwide health problem affecting over a billion people [14], and its consequences cannot be underestimated. Solid evidence in these subjects proposes that vitamin D has several roles besides bone and calcium metabolism [15]. Vitamin D deficiency has been related to a variety of diseases that involve infectious diseases, preeclampsia, cancers, dental caries and periodontitis, autoimmune disorders, cardiovascular disease (CVD), chronic inflammation, type 1 and 2 diabetes mellitus, and neurological disorders that significantly increases the risk of death from respiratory tract infections in otherwise healthy individuals [2,8,16–21]. Additionally, it has been reported that vitamin D deficiency in particular patient groups, such as those in intensive care units and kidney transplant recipients, had increased adverse outcomes and mortality rates [22].

Although having diseases such as those mentioned above is often considered to indicate increased risk of COVID-19, it is more likely that vitamin D deficiency accompanying the disease is the important factor, not the disease itself. An article on risk factors for COVID-19 on in-hospital mortality rates involving 66,646 inpatients in the U.S. included analyses with respect to comorbid diseases [23]. Interestingly, rates for patients with diabetes and hypertension were much higher for those with complicated disease than not, 25.2% vs. 16.5% for complicated diabetes and 28.9 vs. 14.8% for complicated hypertension. The differences for non-complicated diabetes and hypertension were not significant.

In clinical studies, low 25OHD concentrations were associated with acute respiratory tract infections, including influenza, and increased risk of community-acquired pneumonia [12]. Some retrospective studies have shown the association of 25OHD concentrations with the severity and mortality of COVID-19 cases [24].

Vitamin D has many mechanisms by which it can lower the risk of microbial infection and death. These mechanisms can be grouped into the physical barrier, natural cellular immunity, and adaptive immunity [25,26]. It is well-established that vitamin D defends the respiratory tract by maintaining strong junctions, exterminating covered viruses by cathelicidin and defensins, and lessening the generation of proinflammatory cytokines by the innate immune system that can reduce viral replication rates and diminish concentrations of proinflammatory cytokines. Therefore, vitamin D lowers the risk of a cytokine storm that can lead to pneumonia [10]. The vitamin D immunomodulation role can improve innate immunity through the secretion of antiviral peptides, which enhance mucosal resistance, influencing both tumor necrosis factor (TNF)-α and interferon-γ [12].

As mentioned before, vitamin D enhances innate cellular immunity by the induction of antimicrobial peptides, such as cathelicidin and defensins. Cathelicidins exhibit a direct antimicrobial activity against Gram-positive and Gram-negative bacteria, viruses, and fungi. The induction of cathelicidin and defensins can obstruct viral entrance into cells and repress viral replication [27]. The innate immune system generates both proinflammatory and anti-inflammatory cytokines in response to viral and bacterial infections, as observed in COVID-19 patients [28]. Vitamin D can also help reduce the production of proinflammatory Th1 cytokines, such as TNF-α and interferon-γ. Additionally, it decreases proinflammatory cytokines production and stimulates macrophages to generate anti-inflammatory cytokines [26].

Vitamin D stimulates autophagy by intensifying the expression of the light chain 3-autophagy marker. It is essential to mention that these actions are closely linked to apoptosis, which may aid viral replication. Therefore, vitamin D may be directly related to an adequate balance between autophagy and apoptosis to maximize these antiviral responses to infection [27].

It seems that severe acute respiratory syndrome coronavirus 2 (SARS-CoV-2) primarily uses the immune process during infection. It is accompanied by the release of many proinflammatory cytokines such as TNF-α, interleukin (IL)-6, and IL-1β, which are related to vascular hyperpermeability, lung injury, multiorgan failure, and COVID-19 severity, followed by hyper-reaction and cytokine storm in some patients, which can develop into a pathogenic process of acute respiratory disease syndrome [12]. The cytokine storm can lead to unusual activation of the adaptive immune pathway resulting from the alteration and disruption of the innate immune system, with proinflammatory cytokines and chemokines overflow. So, the improvement of immunity through better nutrition can be a substantial and vital factor to consider, and vitamin D shows a significant role in immune function [12] (Figure 1).

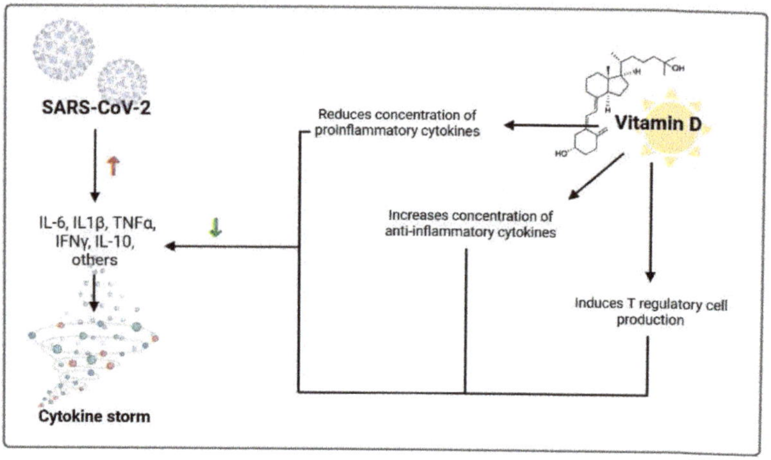

Figure 1. Mechanisms by which vitamin D could decrease the risk of cytokine storm. Red upward arrow indicates an increase while the green downward arrow indicates a decrease. Abbreviations: IL-6 = Interleukin-6; IL-1β = Interleukin-1β; TNFα = tumor necrosis factor α; INFγ = interferon γ; IL-10 = Interleukin-10.

COVID-19 has also been associated with cardiovascular sequelae. Myocardial injury has been reported, along with the elevation of cardiac biomarkers and electrocardiographic or echocardiographic changes, cardiomyopathy, arrhythmias, thrombotic complications, and cardiogenic shock [29] (Figure 2). Activation of the vitamin D receptor also modulates myocardial contractility, likely by regulating calcium flux and low 25OHD concentrations, indicating an increased risk of overall CVD and cardiovascular mortality [30].

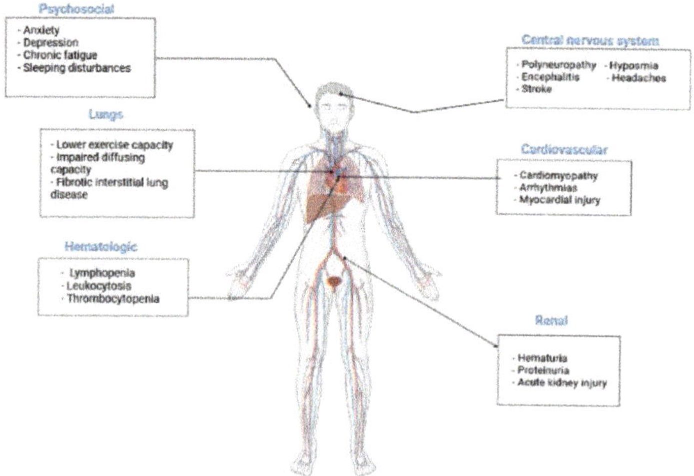

Figure 2. Potential long-term health consequences COVID-19. Some of the potential long-term manifestations are matched exercise capacity and carbon monoxide diffusing capacity, cardiovascular disorders, hematological manifestations and thrombotic complications, central nervous system, and psychosocial manifestations, as well as renal problems.

The large number of COVID-19 cases from the last winter is related to regular reports of the lowest 25OHD concentrations, whereas the number of cases diminished at the end of summer. These findings demonstrated that vitamin D's protective role in decreasing the risk of acquiring COVID-19 is associated with a seasonal condition. In addition, ultraviolet-A radiation induces the release of cutaneous photolabile nitric oxide (NO), which inhibits the replication of SARS-CoV2. Furthermore, NO impacts metabolic syndrome and CVD, both COVID-19 risk factors. Cherrie et al. demonstrated the relationship between ambient ultraviolet-A radiation and COVID-19 deaths (n = 62,219) across the USA. The authors concluded that lower COVID-19-specific mortality is associated with higher ambient ultraviolet-A radiation exposure [31]. In addition, case-fatality incidences grow with age and chronic conditions, both of which have been related to vitamin D deficiency [26,32].

The interplay between SARS-CoV-2 and angiotensin-converting enzyme angiotensin-converting enzyme 2 (ACE2) [33] is crucial for expressing diverse clinical and metabolic characteristics. ACE2 is the host receptor to access alveolar and intestinal epithelial cells [9]. It is known that vitamin D deficiency promotes the renin-angiotensin system, leading to chronic CVD and impaired lung function [12]. The expression of the ACE2 in different tissues is proportional to various symptoms of COVID-19, such as respiratory symptom issues, acute cardiac and kidney injuries, gastrointestinal and liver function abnormalities, and beta-cell damage [9].

It has been suggested that vitamin D deficiency and lack of vitamin D receptor (VDR) activation can aggravate this respiratory syndrome associated with SARS-CoV-2, as it triggers a wounding response in lung stellate cells [34]. It is essential to point out that multiple studies identify vitamin D as capable of repairing epithelial layers and damaged organs, and the use of this vitamin in various pathologies that induce inflammation [35], such as fibrosis, appears to show that vitamin D has antifibrotic properties [36–38]. At the same time, a 2021 review analyzed previous animal studies where it was demonstrated that vitamin D deficiency deepened the activation of the renin—angiotensin system and increased the TGF-β/SMAD signaling pathway, thus causing bleomycin-induced lung fibrosis. The same review associated vitamin D deficiency with an increased risk of pulmonary viral infection, since it may enrich type I interferon responses, critical actors of antiviral immunity [39]. Likewise, a remarkable link has been seen between the signaling of vitamin D, the VDR, and tissue barriers. These interactions are essential in the pathogenesis of various diseases such as cancer, atopic dermatitis, and inflammatory bowel diseases [40]. It is worth highlighting the importance of this vitamin in triggering the transforming growth factor (TGF)-β-signaling pathway, essential for the proper healing of skin wounds [41]. In this respect, similar results were obtained when analyzing the effects of VDR deactivation on wound healing in mouse corneas, showing that, if VDR is inactive, this affects the healing of the corneal epithelium [42].

In many studies, the immunomodulatory characteristics of vitamin D have been reported, as well as its significant role in the maintenance of the immune system correct homeostasis; well-designed randomized controlled trials are required to discover this fascinating and plausible role of vitamin D in protective immune responses against respiratory microbes and in preventing various types of acute respiratory tract infections including SARS-CoV-2 [12]. Despite the controversy on the effect of vitamin D status on COVID-19 infection, the authors believe that vitamin D deficiency is a modifiable risk factor of acute respiratory tract infections, so it has to be considered an inexpensive, safe, and readily-available supplement for these patients [43].

Many mechanisms have been proposed to explain how vitamin D reduces the risk of COVID-19 [27,44]. Table 1 presents a list of mechanisms proposed to explain how vitamin D can reduce the risk of COVID-19. Figures showing how the vitamin D mechanisms reduce the risk of COVID-19 can be found in several publications, e.g., [27,44–46].

Table 1. Proposed mechanisms whereby vitamin D reduces risk of COVID-19 (note, order of mechanisms should be carefully considered, perhaps placing more important ones near the beginning).

Effect	Mechanism	Reference
Inactivates viruses	Induction of cathelicidin	[47]
Reduces risk of cytokine storm	Reduces concentration of proinflammatory cytokines and increases concentration of anti-inflammatory cytokines	[24]
Reduces risk of cytokine storm	Induces T regulatory cell production	[27]
Reduces risk of pneumonia	Reduces risk of endothelial dysfunction	[48]
Increases the metabolic tolerance of the host to damage inflicted by the pathogen infection	Reduces matrix metalloproteinase-9 concentrations	[49]
Reduces free SARS-CoV-2 concentrations	Increases soluble ACE2 concentrations that can bind to SARS-CoV-2	[50]
Anti-viral effects	Balanced differentiation of effector CD8 and CD4 T cells	[51]
Reduces risk of myocarditis	Reduces concentration of catecholamines	[52]
Reduces risk of myocarditis	Inhibits RAS	[53]
Reduces risk of vascular dilation and permeability and hypotensin	Inhibits RAS-mediated bradykinin storm	[46]
Protects against the effects of histamines such as acute immune-mediated reactions [54], lung dysregulation [55], increase in Th2 and decrease in Th1 cytokines [56], and thus susceptibility to respiratory tract infections [57]	Preserves stability of mast cells, which can release histamine when activated.	[58]
Promotes adaptive immunity	Regulations of T cell proliferation	[27]
Neuroprotection	Reduces inflammation and oxidative stress	[59]
Protection against exacerbation by other viruses	Reduces risk of Epstein–Barr virus infection	[60]

Abbreviations: SARS-CoV-2 = severe acute respiratory syndrome coronavirus 2; ACE2 = angiotensin-converting enzyme 2; RAS = renin angiotensin system.

A recent meta-analysis with 23 studies (n = 2692) that evaluated the effect of vitamin D concentrations in COVID-19 patients reported that its deficiency appears to be related to an increased severity and mortality, but those studies did not indicate causality [61]. These determinations highlight the need for more randomized controlled trials to reach more solid conclusions.

On the other hand, an observational study based on 4.6 million community inhabitants supplemented with prescription cholecalciferol or calcifediol prescriptions in Barcelona who achieved 25OHD concentrations >30 ng/mL had about half the risk of SARS-CoV-2 infection, severe COVID-19, or COVID-19 mortality than those not treated [62]. An observational study of 4599 veterans enrolled in U.S. Veterans Affairs health care facilities who had a blood 25OHD test between 20 February and 8 November 2020 had a fully adjusted relative risk of 0.83 (95% CI, 0.72–0.96) for hospitalization for COVID-19 at higher 25OHD concentrations and fully adjusted relative risk of 0.65 (95% CI, 0.50–0.84) for mortality [63].

It is important to highlight that a clinical trial using high-dose cholecalciferol with 95 hospitalized COVID-19 patients in Turkey found a reduction by a factor of 2.14 (95% CI, 1.06–4.33) for mortality rate [64]. A pilot clinical trial in 76 Spanish patients hospitalized for COVID-19 found significantly lower admission to the Intensive Care Unit or mortality for COVID-19 patients treated with high-dose calcifediol shortly after entering the hospital [65]. Subsequent observational studies in Spain found similar results [66]. Thus, there is mounting evidence that higher 25OHD concentrations are associated with reduced risk of COVID-19.

Observational studies can be used to investigate the role of 25OHD concentrations in reducing the risk of SARS-CoV-2 infection and severity of COVID-19. Since having an acute inflammatory illness can lower 25OHD concentrations [67], measurement of 25OHD at the time of COVID-19 diagnosis is not as useful as measurement a few weeks to a few months prior to the disease. However, a recent study from Iran reported that, for 248 COVID-19 patients, 25OHD measured one year before COVID-19 gave similar odds ratios for incidence and death to 25OHD measured at time of admission for COVID-19 [68].

A study from Barcelona reviewed data on prescription supplementation with either cholecalciferol or calcifediol, achieved 25OHD concentrations, and risk of SARS-CoV-2 infection, severe COVID-19, or COVID-19 mortality in comparison with unsupplemented patients with 25OHD concentration <20 ng/mL [62]. Propensity score matching was used to generate appropriate unsupplemented patients. Patients taking cholecalciferol and achieving >30 ng/mL had a significantly lower risk of SARS-CoV-2 infection, severe COVID-19, and COVID-19 mortality. A subsequent article from Andalusia found that the risk of COVID-19 hospitalization was more strongly reduced for cholecalciferol or calcifediol prescriptions 15 days prior to hospitalization than 30 days prior [69].

Another recent article reported findings, from an observational study with U.S. veterans, about the association of vitamin D status and COVID-19-related hospitalization and mortality [63]. The 4599 participants were veterans treated by the U.S. Department of Veteran Affairs health care facilities who tested positive for SARS-CoV-2 and had a vitamin D blood test between 20 February and 8 November 2020, followed for up to 60 days. Vitamin D blood tests were used if obtained between 15 and 90 days prior to testing positive for SARS-CoV-2. Values for many factors were also measured and were used to determine whether they were independent predictors of hospitalization requiring treatment for COVID-19. These factors were included in the models for both hospitalization and mortality concerning 25OHD concentrations. Notably, in going from 15 to 60 ng/mL, hospitalization rates declined from 24.1 to 18.7% ($p = 0.009$), and mortality rates decreased from 10.4 to 5.7% ($p = 0.001$). This observational study appears to be very strong evidence for causality between 25OHD concentrations and severity of COVID-19.

4. Long COVID-19 and Vitamin D

The pandemic's beginning with COVID-19 was characterized by a great concern to contain the contagion of the disease. Almost two years after the health emergency was declared, the focus is the health of those who have survived the disease [70]. As of 18 February 2022, over 409 million confirmed cases and over 5.8 million deaths had been reported globally [71].

These data highlight the large number of people who had COVID-19 and have recovered; in some of them, the consequences will persist in the long term. It is estimated that one-third of patients have persisting symptoms for six months after contracting the infection [72]. Thus, there is an increased need to provide healthcare for long-term symptoms.

The risk factors for long COVID-19 differ somewhat from those for COVID-19 [73]. One study reported that having hypertension, obesity, a psychiatric condition, or an immunosuppressive condition was associated with increased risk of long COVID-19 [74]. On the other hand, long COVID-19 is more likely in women and the age group most affected is somewhat lower. One reason for middle age being a more important risk factor for long COVID-19 is that the risk of mortality increases rapidly with increasing age. A review of COVID-19 mortality rates for 66,646 inpatients in the U.S. admitted from April to June 2020 found increasing mortality rates with increasing age: 40–49 years, 5.8%; 50–59 years, 10.6%; 60–69 years, 18.0%; 70–79 years, 26.5%; and 80+ years, 34.4% [23]. For both COVID-19 mortality and long COVID-19, admission to an intensive care unit is a very important risk factor. Thus, the difference in age profile between mortality and long COVID-19 is that older COVID-19 patients are more likely to die. The reason for more women having long COVID-19 is also likely due to men having a higher mortality rate from COVID-19.

A study involving 4182 COVID-19 cases from Sweden, the UK, and the USA investigated the risk factors for long COVID-19 [75] A total of 558 participants reported symptoms lasting longer than 4 weeks, 189 > 8 weeks, and 95 > 12 weeks. Factors significantly associated with long COVID-19 were age (52 (43–59) years), asthma, heart disease, visit to a hospital, and number of symptoms.

A review of risk factors for long COVID-19 stated that several biomarkers were elevated including D-dimer, interleukin-6 (IL-6), C-reactive protein, procalcitonin, and neutrophils count [76]. A study conducted in western Mexico involving 22 vitamin D supplemented COVID-19 outpatients (mean 25OHD = 22.4 ng/mL) and 20 non-supplemented patients (mean 25OHD = 23.4 ng/mL) found that, although ferritin concentrations were significantly lower in supplemented patients, d-dimer concentrations were not significantly different [33]. A study in India also reported no significant effect on d-dimer concentrations with vitamin D supplementation of 69 COVID-19 patients [77]. Additionally, a high-dose vitamin D supplementation study conducted in Turkey involving 95 hospitalized COVID-19 patients found that increasing mean 25OHD concentration from 23 to 35 ng/mL had no significant effect on ferritin, d-dimer concentrations, but was associated with reduced fibrinogen concentrations [64]. As shown in Table 2, no mechanisms have been shown to reduce concentrations of biomarkers associated with long COVID-19.

Table 2. Evidence that vitamin D reduces concentrations of biomarkers associated with long COVID.

Biomarker	Approach	Finding	Reference
D-dimer, a coagulation biomarker	High-dose vitamin D supplementation on COVID-19 patients	No effect	[78]
Procalcitonin	Supplementation with 5000 IU/d vs. 1000 IU/d for 36 and 33 COVID-19 patients	No effect	[77]
Neutrophils count	Supplementation with 5000 IU/d vs. 1000 IU/d for 36 and 33 COVID-19 patients	Significant increase	[77]

Abbreviations: COVID-19 = coronavirus disease 2019; IU = international unit.

Epstein–Barr virus (EBV) reactivation appears to be a risk factor for severe COVID-19 and also appears to be associated with long COVID-19. A study in the UK involving 128 COVID-19 patients, 17 had EBV reactivation and more severe COVID-19 and adverse outcomes [79]. A study in Turkey found long COVID-19 in 56 of 185 COVID-19 patients and that 20 of 30 long COVID-19 patients were positive for EBV reactivation vs. 2 of 20 controls [80]. Vitamin D supplementation of 20,000 IU/week over 96 weeks was found to significantly reduce humoral immune responses to the latent EBV antigen EBNA1 in relapsing-remitting multiple sclerosis [81].

One of the reasons why vitamin D supplementation may be ineffective in treating long COVID-19 is that the SARS-CoV-2 virus can downregulate vitamin D receptors. This has been observed for cytomegalovirus infection [82,83], hepatitis B virus [84], and hepatitis C virus [85]. EBV has been found to block activation of gene expression through is EBNA-3-ptotein [86]. If downregulation is not complete, it might be that very high vitamin D doses would be able to have some effect, but not in the cells with VDRs downregulated.

Although many of the multi-organ manifestations of COVID-19 are known, the possible long-term implications remain unknown [87,88]. Given the recentness of the COVID-19 pandemic, it is not possible to estimate, by itself, the long-term effects. However, there are similar coronavirus events that have happened previously, such as SARS-CoV-1 and Middle East respiratory syndrome coronavirus [89]. Although the current COVID-19 has lower mortality rates than those mentioned above [90], the reports describe similarities, allowing us to know the possible long-term implications and thus take actions to minimize complications [89]. Some of the long-term manifestations of these other pandemics were: matched exercise capacity and carbon monoxide diffusing capacity, cardiovascular complications, hematological manifestations, thrombotic complications, central nervous system manifestations, and renal and gastrointestinal complications [89].

Wang et al. stated that COVID-19 could leave long-lasting consequences in at least three critical areas: pulmonary, neuronal, and neurodegenerative diseases such as Parkinson's disease, Alzheimer's disease, and multiple sclerosis [91].

A recent study based on survivors of COVID-19 found that 78 patients out of 100 who had recovered had abnormal cardiovascular findings on magnetic resonance imaging; also, 36 of them suffered dyspnea and unusual fatigue [92]. It should be noted that these consequences were not only observed in those patients who had a severe illness but also in those with mild and moderate presentations [43,93]. Many questions remain unanswered, but the variation in viral load and differential immune response can explain why some have long COVID-19 and others do not [94]. Ahearn-Ford et al. presented data that proposed that inflammatory cytokine pathways altered during infection could continue during convalescence [95].

A recent review stated that the effects on the central nervous system after an acute phase of COVID-19 could be perpetuated over time as a neuro-COVID-19. They reported 12 neurological deficits in long COVID-19, such as mental fog, tremors, confusion, and stiff limbs [96]. For their part, Logue et al. investigated the symptoms that persisted after COVID-19 infection in a longitudinal prospective cohort study with 234 patients. The results show that the most persistent manifestations were fatigue (13.6%) and loss of the sense of smell or taste (13.6%) [72].

The importance of vitamin D in long COVID-19 has recently been published [97]. This pro-hormone, fat-soluble is obtained to a greater extent through sun exposure, but there is also a lower contribution from diet [98]. Diet is an essential determinant of 25OHD concentrations. In particular, Crowe reported that 25OHD concentrations were higher in meat and fish eaters than in vegetarians and vegans, who exclude specific food sources of vitamin D from their diet [1]. More recently, a positive association has been reported between 25OHD concentrations and adherence to the Mediterranean diet, a nutritional pattern effective in preventing and treating obesity-related diseases due to the synergistic action of many nutrients with anti-inflammatory and antioxidant properties [99]. The best-known function of this vitamin is related to the normal mineralization of the bones since it contributes to the absorption of calcium in the intestine and the maintenance of adequate levels of calcium and phosphate in serum, having a fundamental role in the prevention of rickets in children and osteomalacia and osteoporosis in adults [100].

The possibility of vitamin D acting as an immunomodulator has generated great interest recently [101]. However, it has many other functions, including the modulation of cell growth, neuromuscular function, immune function, and a reduction in inflammation [100]. It is important to note that 25OHD concentrations can be decreased in the presence of acute inflammation.

Likewise, more studies are needed to understand better the health impact of the prolonged period of COVID-19 in these patients. Before the pandemic, it was already known that low 25OHD concentrations were associated with fatigue and muscle weakness in the general population. Townsend et al. investigated the relationship between 25OHD concentrations and fatigue and reduced exercise tolerance in 149 patients 79 days after COVID-19 [97]. They evaluated the participants using the Chalder Fatigue score, six-minute walk test, and the modified Borg scale. By applying multivariable linear and logistic regression models, they concluded that there was a correlation between vitamin D and persistent vitamin D fatigue and reduced exercise tolerance in this population of COVID-19 patients. It is important to note that this work only evaluated two of the ample diversity of long COVID-19 symptoms [97]. However, fatigue is the most common symptom of long COVID-19 and is seen in other viral infections [102].

Pizzini et al. studied, in a prospective, multicenter study on long-term sequelae after suffering COVID-19 in 109 patients, the associations of 25OHD concentrations with the presentation of COVID-19 [103]. It was observed that a high proportion of patients presented alteration of vitamin D metabolism eight weeks after diagnosis. Patients with severe COVID-19, most likely due to prolonged hospitalization, showed a disturbing parathyroid-vitamin-D axis within their recovery phase. However, low 25OHD concentrations were not related to the burden of persistent symptoms, concluding that although vitamin D deficiency is common among COVID-19 patients, it was not associated with long-term disease outcomes [103]. Due to the novelty of the disease and the different reported results, it is essential to continue with more studies to evaluate the possible effect of vitamin D in the long post-COVID-19 period.

Overall, the COVID-19 pathology is still characterized by cytokine storm, resulting in endothelial inflammation, microvascular thrombosis, and multiple organ failure [104]. Hyperinflammation is a critical component of severe COVID-19, which is associated with poor outcomes underneath the cytokine storm umbrella term [105]. Thus, an important way to minimize or avoid long COVID-19 is to raise 25OHD concentrations before SARS-CoV-2 infection or COVID-19.

Another way to reduce the risk of long COVID-19 is to aggressively treat SARS-CoV-2 infection and COVID-19 as soon as possible after symptoms are manifest. In one study, raising serum 25OHD concentrations to a mean value near 35 ng/mL in a few days to two weeks for hospitalized COVID-19 patients significantly reduced mortality rates but did not seem to affect symptoms [64]. On the other hand, treating hospitalized COVID-19 patients with high-dose calcifediol has been found to significantly reduce admission to the Intensive Care Unit and death rates [65].

It should be noted that most of the research results are based on COVID-19 variants that are no longer dominant such as the Delta variant. The Omicron variant is associated with 30–45% lower attendance for emergency care and 50–70% lower hospital admission rate in the UK than the Delta variant was [106]. A preliminary report from South Africa also indicates that Omicron COVID-19 is much less severe than Delta COVID-19 [107]. Oxygen therapy use was 75% lower, mechanical ventilation use was nearly 90% lower, admission to intensive care units was about 40% lower, length of stay was 70% lower, and death rates were 90% lower. Although the mean age of Omicron COVID-19 patients (36 years vs. 59 years for Delta COVID-19 patients) explains some of the differences, it implies that those who survived Omicron COVID-19 are much less likely to experience serious long COVID-19. However, it is too soon to determine whether this will be the case.

5. Conclusions

The number of COVID-19 diagnosed cases and recovered patients continues to rise with each new SARS-CoV-2 variant, although slowing in summer and fall due to various effects of sunlight. The latter is a particular group that demands special healthcare services since a third of those patients will have persisting symptoms six months or more after recovering from the disease. They will be affected by long-term sequelae causing pulmonary, neuronal, and neurodegenerative diseases.

The functions of vitamin D in bone and calcium metabolism are well-known; however, this pandemic has reinforced the known immunomodulatory effects of this vitamin. Its deficiency has been linked to infectious diseases, some types of cancers, CVD, and chronic inflammation, among others. Additionally, it is related to the severity and mortality of COVID-19 cases, finding a high prevalence of vitamin D deficiency in patients with COVID-19 with acute respiratory failure.

Although more evidence is needed on the effect of vitamin D in COVID-19 (acute and long-term phases), the fundamental role of this vitamin on immune function is evident. So, it appears to be an inexpensive and safe supplement to add as part of COVID-19 treatment.

Given the recent nature of the pandemic and the few studies on prolonged COVID-19 and vitamin D, well-randomized controlled trials are necessary to better understand the role of vitamin D in the protective immune response against prolonged COVID-19.

Author Contributions: Conceptualization, L.B., L.V. and G.M.; methodology, G.M. and L.B.; validation, G.M.; investigation and data curation, C.V., E.F.-T., F.C., E.G.-V. and J.C.-B.; writing, original draft preparation, C.V., E.F.-T., F.C., E.G.-V. and J.C.-B.; writing—review and editing, G.M., L.V., L.B. and W.B.G.; visualization, A.C., W.B.G., G.S. and S.S.; supervision, A.C. and S.S. All authors have read and agreed to the published version of the manuscript.

Funding: No funding was received for the preparation of this review.

Institutional Review Board Statement: Not applicable.

Informed Consent Statement: Not applicable.

Acknowledgments: The assistance of the staff is gratefully appreciated.

Conflicts of Interest: The authors do not have any financial or other conflicts of interest to declare.

References

1. Crowe, F.L.; Steur, M.; Allen, N.E.; Appleby, P.N.; Travis, R.C.; Key, T.J. Plasma concentrations of 25-hydroxyvitamin D in meat eaters, fish eaters, vegetarians and vegans: Results from the EPIC–Oxford study. *Public Health Nutr.* **2011**, *14*, 340–346. [CrossRef] [PubMed]
2. Barrea, L.; Frias-Toral, E.; Pugliese, G.; Garcia-Velasquez, E.; Carignano, M.D.L.A.; Savastano, S.; Colao, A.; Muscogiuri, G. Vitamin D in obesity and obesity-related diseases: An overview. *Minerva Endocrinol.* **2021**, *46*, 177–192. [CrossRef] [PubMed]
3. Muscogiuri, G.; Barrea, L.; Di Somma, C.; Laudisio, D.; Salzano, C.; Pugliese, G.; De Alteriis, G.; Colao, A.; Savastano, S. Sex Differences of Vitamin D Status across BMI Classes: An Observational Prospective Cohort Study. *Nutrients* **2019**, *11*, 3034. [CrossRef]
4. Goërtz, Y.M.J.; Van Herck, M.; Delbressine, J.M.; Vaes, A.W.; Meys, R.; Machado, F.V.C.; Houben-Wilke, S.; Burtin, C.; Posthuma, R.; Franssen, F.M.E.; et al. Persistent symptoms 3 months after a SARS-CoV-2 infection: The post-COVID-19 syndrome? *ERJ Open Res.* **2020**, *6*, 00542–02020. [CrossRef]
5. Callard, F.; Perego, E. How and why patients made Long COVID. *Soc. Sci. Med.* **2020**, *268*, 113426. [CrossRef]
6. Pereira, M.; Dantas Damascena, A.D.; Galvão Azevedo, L.M.G.; de Almeida Oliveira, T.D.A.; da Mota Santana, J.D.M. Vitamin D deficiency aggravates COVID-19: Systematic review and meta-analysis. *Crit. Rev. Food Sci. Nutr.* **2020**, *62*, 1308–1316. [CrossRef] [PubMed]
7. Carpagnano, G.E.; Di Lecce, V.; Quaranta, V.N.; Zito, A.; Buonamico, E.; Capozza, E.; Palumbo, A.; Di Gioia, G.; Valerio, V.N.; Resta, O. Vitamin D deficiency as a predictor of poor prognosis in patients with acute respiratory failure due to COVID-19. *J. Endocrinol. Investig.* **2020**, *44*, 765–771. [CrossRef]
8. Savanelli, M.C.; Scarano, E.; Muscogiuri, G.; Barrea, L.; Vuolo, L.; Rubino, M.; Savastano, S.; Colao, A.; Di Somma, C. Cardiovascular risk in adult hypopituitaric patients with growth hormone deficiency: Is there a role for vitamin D? *Endocrine* **2015**, *52*, 111–119. [CrossRef]
9. Mahrooz, A.; Muscogiuri, G.; Buzzetti, R.; Maddaloni, E. The complex combination of COVID-19 and diabetes: Pleiotropic changes in glucose metabolism. *Endocrine* **2021**, *72*, 317–325. [CrossRef]
10. Damascena, A.D.; Azevedo, L.M.G.; Oliveira, T.A.; Santana, J.D.M.; Pereira, M. Addendum to vitamin D deficiency aggravates COVID-19: Systematic review and meta-analysis. *Crit. Rev. Food Sci. Nutr.* **2021**, 1–6. [CrossRef]
11. Giustina, A. Hypovitaminosis D and the endocrine phenotype of COVID-19. *Endocrine* **2021**, *72*, 1–11. [CrossRef] [PubMed]
12. Ali, N. Role of vitamin D in preventing of COVID-19 infection, progression and severity. *J. Infect. Public Health* **2020**, *13*, 1373–1380. [CrossRef] [PubMed]
13. Muscogiuri, G.; Barrea, L.; Savastano, S.; Colao, A. Nutritional recommendations for COVID-19 quarantine. *Eur. J. Clin. Nutr.* **2020**, *74*, 850–851. [CrossRef] [PubMed]
14. Lips, P.; de Jongh, R.T.; van Schoor, N.M. Trends in Vitamin D Status Around the World. *JBMR Plus* **2021**, *5*, e10585. [CrossRef] [PubMed]
15. Altieri, B.; Grant, W.B.; Della Casa, S.; Orio, F.; Pontecorvi, A.; Colao, A.; Sarno, G.; Muscogiuri, G. Vitamin D and pancreas: The role of sunshine vitamin in the pathogenesis of diabetes mellitus and pancreatic cancer. *Crit. Rev. Food Sci. Nutr.* **2017**, *57*, 3472–3488. [CrossRef] [PubMed]
16. Barrea, L.; Muscogiuri, G.; Annunziata, G.; Laudisio, D.; De Alteriis, G.; Tenore, G.C.; Colao, A.; Savastano, S. A New Light on Vitamin D in Obesity: A Novel Association with Trimethylamine-N-Oxide (TMAO). *Nutrients* **2019**, *11*, 1310. [CrossRef]
17. Barrea, L.; Muscogiuri, G.; Laudisio, D.; Di Somma, C.; Salzano, C.; Pugliese, G.; De Alteriis, G.; Colao, A.; Savastano, S. Phase Angle: A Possible Biomarker to Quantify Inflammation in Subjects with Obesity and 25(OH)D Deficiency. *Nutrients* **2019**, *11*, 1747. [CrossRef]
18. Holick, M.F. The vitamin D deficiency pandemic: Approaches for diagnosis, treatment and prevention. *Rev. Endocr. Metab. Disord.* **2017**, *18*, 153–165. [CrossRef]
19. Miao, Z.; Wang, S.; Wang, Y.; Guo, L.; Zhang, J.; Liu, Y.; Yang, Q. A Potential Linking between Vitamin D and Adipose Metabolic Disorders. *Can. J. Gastroenterol. Hepatol.* **2020**, *2020*, 2656321. [CrossRef]

20. Muscogiuri, G.; Barrea, L.; Altieri, B.; Di Somma, C.; Bhattoa, H.P.; Laudisio, D.; Duval, G.T.; Pugliese, G.; Annweiler, C.; Orio, F.; et al. Calcium and Vitamin D Supplementation. Myths and Realities with Regard to Cardiovascular Risk. *Curr. Vasc. Pharmacol.* **2019**, *17*, 610–617. [CrossRef]
21. Grant, W.B.; Al Anouti, F.; Boucher, B.J.; Dursun, E.; Gezen-Ak, D.; Jude, E.B.; Karonova, T.; Pludowski, P. A Narrative Review of the Evidence for Variations in Serum 25-Hydroxyvitamin D Concentration Thresholds for Optimal Health. *Nutrients* **2022**, *14*, 639. [CrossRef] [PubMed]
22. Sarno, G.; Russo, E.; Ferrara, A.; Cerbone, V.; Villa, R. COVID-19 Infection in Kidney Transplant Recipients in Italy: Management Issues in a Kidney Transplant Center. *Exp. Clin. Transplant.* **2021**, *19*, 284–286. [CrossRef]
23. E Goodman, K.; Magder, L.S.; Baghdadi, J.D.; Pineles, L.; Levine, A.R.; Perencevich, E.N.; Harris, A.D. Impact of Sex and Metabolic Comorbidities on Coronavirus Disease 2019 (COVID-19) Mortality Risk Across Age Groups: 66 646 Inpatients Across 613 U.S. Hospitals. *Clin. Infect. Dis.* **2020**, *73*, e4113–e4123. [CrossRef]
24. Daneshkhah, A.; Agrawal, V.; Eshein, A.; Subramanian, H.; Roy, H.K.; Backman, V. Evidence for possible association of vitamin D status with cytokine storm and unregulated inflammation in COVID-19 patients. *Aging Clin. Exp. Res.* **2020**, *32*, 2141–2158. [CrossRef] [PubMed]
25. Barrea, L.; Muscogiuri, G.; Frias-Toral, E.; Laudisio, D.; Pugliese, G.; Castellucci, B.; Garcia-Velasquez, E.; Savastano, S.; Colao, A. Nutrition and immune system: From the Mediterranean diet to dietary supplementary through the microbiota. *Crit. Rev. Food Sci. Nutr.* **2020**, *61*, 3066–3090. [CrossRef] [PubMed]
26. Grant, W.B.; Lahore, H.; McDonnell, S.L.; Baggerly, C.A.; French, C.B.; Aliano, J.L.; Bhattoa, H.P. Evidence that Vitamin D Supplementation Could Reduce Risk of Influenza and COVID-19 Infections and Deaths. *Nutrients* **2020**, *12*, 988. [CrossRef] [PubMed]
27. Bilezikian, J.P.; Bikle, D.; Hewison, M.; Lazaretti-Castro, M.; Formenti, A.M.; Gupta, A.; Madhavan, M.V.; Nair, N.; Babalyan, V.; Hutchings, N.; et al. Mechanisms in Endocrinology: Vitamin D and COVID-19. *Eur. J. Endocrinol.* **2020**, *183*, R133–R147. [CrossRef] [PubMed]
28. Huang, C.; Wang, Y.; Li, X.; Ren, L.; Zhao, J.; Hu, Y.; Zhang, L.; Fan, G.; Xu, J.; Gu, X.; et al. Clinical features of patients infected with 2019 novel coronavirus in Wuhan, China. *Lancet* **2020**, *395*, 497–506. [CrossRef]
29. Salabei, J.K.; Asnake, Z.T.; Ismail, Z.H.; Charles, K.; Stanger, G.-T.; Abdullahi, A.H.; Abraham, A.T.; Okonoboh, P. COVID-19 and the Cardiovascular System: An Update. *Am. J. Med. Sci.* **2022**. [CrossRef]
30. Acharya, P.; Dalia, T.; Ranka, S.; Sethi, P.; A Oni, O.; Safarova, M.S.; Parashara, D.; Gupta, K.; Barua, R.S. The Effects of Vitamin D Supplementation and 25-hydroxyvitamin D Levels on The Risk of MI and Mortality. *J. Endocr. Soc.* **2021**, *5*, bvab124. [CrossRef]
31. Cherrie, M.; Clemens, T.; Colandrea, C.; Feng, Z.; Webb, D.; Weller, R.; Dibben, C. Ultraviolet A radiation and COVID-19 deaths in the USA with replication studies in England and Italy*. *Br. J. Dermatol.* **2021**, *185*, 363–370. [CrossRef] [PubMed]
32. Sarno, G.; Montalti, R. COVID-19 infection in kidney transplant recipients: Endocrine and metabolic issues. *Minerva Endocrinol.* **2021**, *46*, 293–295. [CrossRef] [PubMed]
33. Sánchez-Zuno, G.; González-Estevez, G.; Matuz-Flores, M.; Macedo-Ojeda, G.; Hernández-Bello, J.; Mora-Mora, J.; Pérez-Guerrero, E.; García-Chagollán, M.; Vega-Magaña, N.; Turrubiates-Hernández, F.; et al. Vitamin D Levels in COVID-19 Outpatients from Western Mexico: Clinical Correlation and Effect of Its Supplementation. *J. Clin. Med.* **2021**, *10*, 2378. [CrossRef]
34. Evans, R.M.; Lippman, S.M. Shining Light on the COVID-19 Pandemic: A Vitamin D Receptor Checkpoint in Defense of Unregulated Wound Healing. *Cell Metab.* **2020**, *32*, 704–709. [CrossRef] [PubMed]
35. Ahmad, S.; Arora, S.; Khan, S.; Mohsin, M.; Mohan, A.; Manda, K.; Syed, M.A. Vitamin D and its therapeutic relevance in pulmonary diseases. *J. Nutr. Biochem.* **2020**, *90*, 108571. [CrossRef]
36. Ramirez, A.M.; Wongtrakool, C.; Welch, T.; Steinmeyer, A.; Zügel, U.; Roman, J. Vitamin D inhibition of pro-fibrotic effects of transforming growth factor β1 in lung fibroblasts and epithelial cells. *J. Steroid Biochem. Mol. Biol.* **2010**, *118*, 142–150. [CrossRef] [PubMed]
37. Monti, N.; Cucina, A. Fibrosis: A Role for Vitamin D. Organisms. *J. Biol. Sci.* **2020**, *4*, 26–41. [CrossRef]
38. Gurtner, G.C.; Werner, S.; Barrandon, Y.; Longaker, M.T. Wound repair and regeneration. *Nature* **2008**, *453*, 314–321. [CrossRef]
39. Lucchini, A.C.; Gachanja, N.N.; Rossi, A.G.; Dorward, D.A.; Lucas, C.D. Epithelial Cells and Inflammation in Pulmonary Wound Repair. *Cells* **2021**, *10*, 339. [CrossRef]
40. Zhang, Y.-G.; Wu, S.; Sun, J. Vitamin D, vitamin D receptor and tissue barriers. *Tissue Barriers* **2013**, *1*, e23118. [CrossRef]
41. Rieger, S.; Zhao, H.; Martin, P.; Abe, K.; Lisse, T.S. The role of nuclear hormone receptors in cutaneous wound repair. *Cell Biochem. Funct.* **2014**, *33*, 1–13. [CrossRef] [PubMed]
42. Elizondo, R.A.; Yin, Z.; Lu, X.; Watsky, M.A. Effect of Vitamin D Receptor Knockout on Cornea Epithelium Wound Healing and Tight Junctions. *Investig. Opthalmology Vis. Sci.* **2014**, *55*, 5245–5251. [CrossRef] [PubMed]
43. Zemb, P.; Bergman, P.; Camargo, C.A., Jr.; Cavalier, E.; Cormier, C.; Courbebaisse, M.; Hollis, B.; Joulia, F.; Minisola, S.; Pilz, S.; et al. Vitamin D deficiency and the COVID-19 pandemic. *J. Glob. Antimicrob. Resist.* **2020**, *22*, 133–134. [CrossRef]
44. Mercola, J.; Grant, W.B.; Wagner, C.L. Evidence Regarding Vitamin D and Risk of COVID-19 and Its Severity. *Nutrients* **2020**, *12*, 3361. [CrossRef]
45. Cutolo, M.; Paolino, S.; Smith, V. Evidences for a protective role of vitamin D in COVID-19. *RMD Open* **2020**, *6*, e001454. [CrossRef] [PubMed]

46. Garvin, M.R.; Alvarez, C.; Miller, J.I.; Prates, E.T.; Walker, A.M.; Amos, B.K.; Mast, A.E.; Justice, A.; Aronow, B.; Jacobson, D.A. A mechanistic model and therapeutic interventions for COVID-19 involving a RAS-mediated bradykinin storm. *eLife* **2020**, *9*, e59177. [CrossRef]
47. White, J.H. Emerging Roles of Vitamin D-Induced Antimicrobial Peptides in Antiviral Innate Immunity. *Nutrients* **2022**, *14*, 284. [CrossRef]
48. Zhang, J.; McCullough, P.A.; Tecson, K.M. Vitamin D deficiency in association with endothelial dysfunction: Implications for patients withCOVID-19. *Rev. Cardiovasc. Med.* **2020**, *21*, 339–344. [CrossRef]
49. Hardy, E.; Fernandez-Patron, C. Targeting MMP-Regulation of Inflammation to Increase Metabolic Tolerance to COVID-19 Pathologies: A Hypothesis. *Biomolecules* **2021**, *11*, 390. [CrossRef]
50. Martínez-Moreno, J.; Hernandez, J.C.; Urcuqui-Inchima, S. Effect of high doses of vitamin D supplementation on dengue virus replication, Toll-like receptor expression, and cytokine profiles on dendritic cells. *Mol. Cell. Biochem.* **2019**, *464*, 169–180. [CrossRef]
51. Kalia, V.; Studzinski, G.P.; Sarkar, S. Role of vitamin D in regulating COVID-19 severity—An immunological perspective. *J. Leukoc. Biol.* **2021**, *110*, 809–819. [CrossRef] [PubMed]
52. Tønnesen, R.; Schwarz, P.; Hovind, P.; Jensen, L.T. Modulation of the sympathetic nervous system in youngsters by vitamin-D supplementation. *Physiol. Rep.* **2018**, *6*, e13635. [CrossRef]
53. Ben-Eltriki, M.; Hopefl, R.; Wright, J.M.; Deb, S. Association between Vitamin D Status and Risk of Developing Severe COVID-19 Infection: A Meta-Analysis of Observational Studies. *J. Am. Coll. Nutr.* **2021**, 1–11. [CrossRef]
54. Peddapalli, A.; Gehani, M.; Kalle, A.; Peddapalli, S.; Peter, A.; Sharad, S. Demystifying Excess Immune Response in COVID-19 to Reposition an Orphan Drug for Down-Regulation of NF-κB: A Systematic Review. *Viruses* **2021**, *13*, 378. [CrossRef] [PubMed]
55. Movat, H.Z. The role of histamine and other mediators in microvascular changes in acute inflammation. *Can. J. Physiol. Pharmacol.* **1987**, *65*, 451–457. [CrossRef] [PubMed]
56. Branco, A.C.C.C.; Yoshikawa, F.S.Y.; Pietrobon, A.J.; Sato, M.N. Role of Histamine in Modulating the Immune Response and Inflammation. *Mediat. Inflamm.* **2018**, *2018*, 9524075. [CrossRef]
57. Chen, G.; Wu, D.; Guo, W.; Cao, Y.; Huang, D.; Wang, H.; Wang, T.; Zhang, X.; Chen, H.; Yu, H.; et al. Clinical and immunological features of severe and moderate coronavirus disease 2019. *J. Clin. Investig.* **2020**, *130*, 2620–2629. [CrossRef] [PubMed]
58. Kakavas, S.; Karayiannis, D.; Mastora, Z. The Complex Interplay between Immunonutrition, Mast Cells, and Histamine Signaling in COVID-19. *Nutrients* **2021**, *13*, 3458. [CrossRef]
59. Menéndez, S.G.; Giménez, V.M.M.; Holick, M.F.; Barrantes, F.J.; Manucha, W. COVID-19 and neurological sequelae: Vitamin D as a possible neuroprotective and/or neuroreparative agent. *Life Sci.* **2022**, *297*, 120464. [CrossRef]
60. Hanwell, H.E.; Banwell, B. Assessment of evidence for a protective role of vitamin D in multiple sclerosis. *Biochim. Biophys. Acta (BBA)-Mol. Basis Dis.* **2011**, *1812*, 202–212. [CrossRef]
61. Oscanoa, T.J.; Amado, J.; Vidal, X.; Laird, E.; A Ghashut, R.; Romero-Ortuno, R. The relationship between the severity and mortality of SARS-CoV-2 infection and 25-hydroxyvitamin D concentration—a metaanalysis. *Adv. Respir. Med.* **2021**, *89*, 145–157. [CrossRef] [PubMed]
62. Oristrell, J.; Oliva, J.C.; Casado, E.; Subirana, I.; Domínguez, D.; Toloba, A.; Balado, A.; Grau, M. Vitamin D supplementation and COVID-19 risk: A population-based, cohort study. *J. Endocrinol. Investig.* **2021**, *45*, 167–179. [CrossRef] [PubMed]
63. Seal, K.H.; Bertenthal, D.; Carey, E.; Grunfeld, C.; Bikle, D.D.; Lu, C.M. Association of Vitamin D Status and COVID-19-Related Hospitalization and Mortality. *J. Gen. Intern. Med.* **2022**, *37*, 853–861. [CrossRef] [PubMed]
64. Gönen, M.S.; Alaylıoğlu, M.; Durcan, E.; Özdemir, Y.; Şahin, S.; Konukoğlu, D.; Nohut, O.K.; Ürkmez, S.; Küçükece, B.; Balkan, I.I.; et al. Rapid and Effective Vitamin D Supplementation May Present Better Clinical Outcomes in COVID-19 (SARS-CoV-2) Patients by Altering Serum INOS1, IL1B, IFNg, Cathelicidin-LL37, and ICAM1. *Nutrients* **2021**, *13*, 4047. [CrossRef]
65. Entrenas Castillo, M.E.; Entrenas Costa, L.M.E.; Vaquero Barrios, J.M.V.; Alcalá Díaz, J.F.A.; López Miranda, J.L.; Bouillon, R.; Quesada Gomez, J.M.Q. Effect of calcifediol treatment and best available therapy versus best available therapy on intensive care unit admission and mortality among patients hospitalized for COVID-19: A pilot randomized clinical study. *J. Steroid Biochem. Mol. Biol.* **2020**, *203*, 105751. [CrossRef]
66. Bouillon, R.; Quesada-Gomez, J.M. Vitamin D Endocrine System and COVID-19. *JBMR Plus* **2021**, *5*, e10576. [CrossRef]
67. Smolders, J.; van den Ouweland, J.; Geven, C.; Pickkers, P.; Kox, M. Letter to the Editor: Vitamin D deficiency in COVID-19: Mixing up cause and consequence. *Metabolism* **2020**, *115*, 154434. [CrossRef]
68. Fatemi, A.; Ardehali, S.H.; Eslamian, G.; Noormohammadi, M.; Malek, S. Association of vitamin D deficiency with COVID-19 severity and mortality in Iranian people: A prospective observational study. *Acute Crit. Care* **2021**, *36*, 300–307. [CrossRef]
69. Loucera, C.; Peña-Chilet, M.; Esteban-Medina, M.; Muñoyerro-Muñiz, D.; Villegas, R.; Lopez-Miranda, J.; Rodriguez-Baño, J.; Túnez, I.; Bouillon, R.; Dopazo, J.; et al. Real world evidence of calcifediol or vitamin D prescription and mortality rate of COVID-19 in a retrospective cohort of hospitalized Andalusian patients. *Sci. Rep.* **2021**, *11*, 1–12. [CrossRef]
70. Klok, F.A.; Boon, G.J.; Barco, S.; Endres, M.; Geelhoed, J.M.; Knauss, S.; Rezek, S.A.; Spruit, M.A.; Vehreschild, J.; Siegerink, B. The Post-COVID-19 Functional Status scale: A tool to measure functional status over time after COVID-19. *Eur. Respir. J.* **2020**, *56*, 2001494. [CrossRef]
71. World Health Organization. Weekly Epidemiological Update on COVID-19. Available online: https://www.who.int/publications/m/item/weekly-epidemiological-update-on-COVID-19---15-february-2022 (accessed on 1 March 2022).

72. Logue, J.K.; Franko, N.M.; McCulloch, D.J.; McDonald, D.; Magedson, A.; Wolf, C.R.; Chu, H.Y. Sequelae in Adults at 6 Months After COVID-19 Infection. *JAMA Netw. Open* **2021**, *4*, e210830. [CrossRef] [PubMed]
73. Crook, H.; Raza, S.; Nowell, J.; Young, M.; Edison, P. Long COVID—mechanisms, risk factors, and management. *BMJ* **2021**, *374*, n1648. [CrossRef] [PubMed]
74. Tenforde, M.W.; Kim, S.S.; Lindsell, C.J.; Rose, E.B.; Shapiro, N.I.; Files, D.C.; Gibbs, K.W.; Erickson, H.L.; Steingrub, J.S.; Smithline, H.A.; et al. Symptom Duration and Risk Factors for Delayed Return to Usual Health Among Outpatients with COVID-19 in a Multistate Health Care Systems Network—United States, March–June 2020. *MMWR. Morb. Mortal. Wkly. Rep.* **2020**, *69*, 993–998. [CrossRef] [PubMed]
75. Sudre, C.H.; Murray, B.; Varsavsky, T.; Graham, M.S.; Penfold, R.S.; Bowyer, R.C.; Pujol, J.C.; Klaser, K.; Antonelli, M.; Canas, L.S.; et al. Attributes and predictors of long COVID. *Nat. Med.* **2021**, *27*, 626–631. [CrossRef]
76. Yong, S.J. Long COVID or post-COVID-19 syndrome: Putative pathophysiology, risk factors, and treatments. *Infect. Dis.* **2021**, *53*, 737–754. [CrossRef] [PubMed]
77. Sabico, S.; Enani, M.A.; Sheshah, E.; Aljohani, N.J.; Aldisi, D.A.; Alotaibi, N.H.; Alshingetti, N.; Alomar, S.Y.; Alnaami, A.M.; Amer, O.E.; et al. Effects of a 2-Week 5000 IU versus 1000 IU Vitamin D3 Supplementation on Recovery of Symptoms in Patients with Mild to Moderate COVID-19: A Randomized Clinical Trial. *Nutrients* **2021**, *13*, 2170. [CrossRef]
78. Rastogi, A.; Bhansali, A.; Khare, N.; Suri, V.; Yaddanapudi, N.; Sachdeva, N.; Puri, G.D.; Malhotra, P. Short term, high-dose vitamin D supplementation for COVID-19 disease: A randomised, placebo-controlled, study (SHADE study). *Postgrad. Med. J.* **2020**. [CrossRef]
79. Xie, Y.; Cao, S.; Dong, H.; Lv, H.; Teng, X.; Zhang, J.; Wang, T.; Zhang, X.; Qin, Y.; Chai, Y.; et al. Clinical characteristics and outcomes of critically ill patients with acute COVID-19 with Epstein-Barr virus reactivation. *BMC Infect. Dis.* **2021**, *21*, 1–8. [CrossRef]
80. Gold, J.; Okyay, R.; Licht, W.; Hurley, D. Investigation of Long COVID Prevalence and Its Relationship to Epstein-Barr Virus Reactivation. *Pathogens* **2021**, *10*, 763. [CrossRef]
81. Røsjø, E.; Lossius, A.; Abdelmagid, N.; Lindstrøm, J.C.; Kampman, M.T.; Jørgensen, L.; Sundström, P.; Olsson, T.; Steffensen, L.H.; Torkildsen, Ø.; et al. Effect of high-dose vitamin D3 supplementation on antibody responses against Epstein–Barr virus in relapsing-remitting multiple sclerosis. *Mult. Scler. J.* **2016**, *23*, 395–402. [CrossRef]
82. Rieder, F.J.; Gröschel, C.; Kastner, M.-T.; Kosulin, K.; Laengle, J.; Zadnikar, R.; Marculescu, R.; Schneider, M.; Lion, T.; Bergmann, M.; et al. Human cytomegalovirus infection downregulates vitamin-D receptor in mammalian cells. *J. Steroid Biochem. Mol. Biol.* **2016**, *165*, 356–362. [CrossRef] [PubMed]
83. Robak, O.; Kastner, M.-T.; Stecher, C.; Schneider, M.; Andreas, M.; Greinix, H.; Kallay, E.; Honsig, C.; Steininger, C. Cytomegalovirus Infection Downregulates Vitamin D Receptor in Patients Undergoing Hematopoietic Stem Cell Transplantation. *Transplantation* **2020**, *105*, 1595–1602. [CrossRef]
84. Gotlieb, N.; Tachlytski, I.; Lapidot, Y.; Sultan, M.; Safran, M.; Ben-Ari, Z. Hepatitis B virus downregulates vitamin D receptor levels in hepatoma cell lines, thereby preventing vitamin D-dependent inhibition of viral transcription and production. *Mol. Med.* **2018**, *24*, 53. [CrossRef]
85. Abdel-Mohsen, M.A.; El-Braky, A.A.-A.; Ghazal, A.A.E.-R.; Shamseya, M.M. Autophagy, apoptosis, vitamin D, and vitamin D receptor in hepatocellular carcinoma associated with hepatitis C virus. *Medicine* **2018**, *97*, e0172. [CrossRef] [PubMed]
86. Yenamandra, S.P.; Hellman, U.; Kempkes, B.; Darekar, S.D.; Petermann, S.; Sculley, T.; Klein, G.; Kashuba, E. Epstein-Barr virus encoded EBNA-3 binds to vitamin D receptor and blocks activation of its target genes. *Cell Mol. Life Sci.* **2010**, *67*, 4249–4256. [CrossRef] [PubMed]
87. Gupta, A.; Madhavan, M.V.; Sehgal, K.; Nair, N.; Mahajan, S.; Sehrawat, T.S.; Bikdeli, B.; Ahluwalia, N.; Ausiello, J.C.; Wan, E.Y.; et al. Extrapulmonary manifestations of COVID-19. *Nat. Med.* **2020**, *26*, 1017–1032. [CrossRef]
88. Wadman, M.; Couzin-Frankel, J.; Kaiser, J.; Matacic, C. A rampage through the body. *Science* **2020**, *368*, 356–360. [CrossRef]
89. Higgins, V.; Sohaei, D.; Diamandis, E.P.; Prassas, I. COVID-19: From an acute to chronic disease? Potential long-term health consequences. *Crit. Rev. Clin. Lab. Sci.* **2020**, *58*, 297–310. [CrossRef]
90. Wu, A.; Peng, Y.; Huang, B.; Ding, X.; Wang, X.; Niu, P.; Meng, J.; Zhu, Z.; Zhang, Z.; Wang, J.; et al. Genome Composition and Divergence of the Novel Coronavirus (2019-nCoV) Originating in China. *Cell Host Microbe* **2020**, *27*, 325–328. [CrossRef]
91. Wang, F.; Kream, R.M.; Stefano, G.B. Long-Term Respiratory and Neurological Sequelae of COVID-19. *Med. Sci. Monit.* **2020**, *26*, e928996. [CrossRef]
92. Puntmann, V.O.; Carerj, M.L.; Wieters, I.; Fahim, M.; Arendt, C.; Hoffmann, J.; Shchendrygina, A.; Escher, F.; Vasa-Nicotera, M.; Zeiher, A.M.; et al. Outcomes of Cardiovascular Magnetic Resonance Imaging in Patients Recently Recovered From Coronavirus Disease 2019 (COVID-19). *JAMA Cardiol.* **2020**, *5*, 1265. [CrossRef]
93. Yelin, D.; Wirtheim, E.; Vetter, P.; Kalil, A.C.; Bruchfeld, J.; Runold, M.; Guaraldi, G.; Mussini, C.; Gudiol, C.; Pujol, M.; et al. Long-term consequences of COVID-19: Research needs. *Lancet Infect. Dis.* **2020**, *20*, 1115–1117. [CrossRef]
94. Baig, A.M. Deleterious Outcomes in Long-Hauler COVID-19: The Effects of SARS-CoV-2 on the CNS in Chronic COVID Syndrome. *ACS Chem. Neurosci.* **2020**, *11*, 4017–4020. [CrossRef]
95. Ahearn-Ford, S.; Lunjani, N.; McSharry, B.; MacSharry, J.; Fanning, L.; Murphy, G.; Everard, C.; Barry, A.; McGreal, A.; al Lawati, S.M.; et al. Long-term disruption of cytokine signalling networks is evident in patients who required hospitalization for SARS-CoV-2 infection. *Allergy* **2021**, *76*, 2910–2913. [CrossRef] [PubMed]

96. Marshall, M. The lasting misery of coronavirus long-haulers. *Nature* **2020**, *585*, 339–341. [CrossRef] [PubMed]
97. Townsend, L.; Dyer, A.; McCluskey, P.; O'Brien, K.; Dowds, J.; Laird, E.; Bannan, C.; Bourke, N.; Cheallaigh, C.N.; Byrne, D.; et al. Investigating the Relationship between Vitamin D and Persistent Symptoms Following SARS-CoV-2 Infection. *Nutrients* **2021**, *13*, 2430. [CrossRef] [PubMed]
98. Muscogiuri, G.; Barrea, L.; Scannapieco, M.; Di Somma, C.; Scacchi, M.; Aimaretti, G.; Savastano, S.; Colao, A.; Marzullo, P. The lullaby of the sun: The role of vitamin D in sleep disturbance. *Sleep Med.* **2018**, *54*, 262–265. [CrossRef]
99. Barrea, L.; Muscogiuri, G.; Laudisio, D.; Pugliese, G.; De Alteriis, G.; Colao, A.; Savastano, S. Influence of the Mediterranean Diet on 25-Hydroxyvitamin D Levels in Adults. *Nutrients* **2020**, *12*, 1439. [CrossRef]
100. Georgakopoulou, V.E.; Mantzouranis, K.; Damaskos, C.; Karakou, E.; Melemeni, D.; Mermigkis, D.; Petsinis, G.; Sklapani, P.; Trakas, N.; Tsiafaki, X. Correlation Between Serum Levels of 25-Hydroxyvitamin D and Severity of Community-Acquired Pneumonia in Hospitalized Patients Assessed by Pneumonia Severity Index: An Observational Descriptive Study. *Cureus* **2020**, *12*, e8947. [CrossRef]
101. Wu, D.; Lewis, E.D.; Pae, M.; Meydani, S.N. Nutritional Modulation of Immune Function: Analysis of Evidence, Mechanisms, and Clinical Relevance. *Front. Immunol.* **2019**, *9*, 3160. [CrossRef]
102. Raveendran, A.; Misra, A. Post COVID-19 Syndrome ("Long COVID") and Diabetes: Challenges in Diagnosis and Management. *Diabetes Metab. Syndr. Clin. Res. Rev.* **2021**, *15*, 102235. [CrossRef]
103. Pizzini, A.; Aichner, M.; Sahanic, S.; Böhm, A.; Egger, A.; Hoermann, G.; Kurz, K.; Widmann, G.; Bellmann-Weiler, R.; Weiss, G.; et al. Impact of Vitamin D Deficiency on COVID-19—A Prospective Analysis from the CovILD Registry. *Nutrients* **2020**, *12*, 2775. [CrossRef] [PubMed]
104. Andrade, B.S.; Siqueira, S.; Soares, W.D.A.; Rangel, F.D.S.; Santos, N.; Freitas, A.D.S.; da Silveira, P.R.; Tiwari, S.; Alzahrani, K.; Góes-Neto, A.; et al. Long-COVID and Post-COVID Health Complications: An Up-to-Date Review on Clinical Conditions and Their Possible Molecular Mechanisms. *Viruses* **2021**, *13*, 700. [CrossRef] [PubMed]
105. Mehta, P.; Fajgenbaum, D.C. Is severe COVID-19 a cytokine storm syndrome: A hyperinflammatory debate. *Curr. Opin. Rheumatol.* **2021**, *33*, 419–430. [CrossRef]
106. Mahase, E. COVID-19: Hospital admission 50–70% less likely with omicron than delta, but transmission a major concern. *BMJ* **2021**, *375*. [CrossRef] [PubMed]
107. Maslo, C.; Friedland, R.; Toubkin, M.; Laubscher, A.; Akaloo, T.; Kama, B. Characteristics and Outcomes of Hospitalized Patients in South Africa During the COVID-19 Omicron Wave Compared with Previous Waves. *JAMA* **2022**, *327*, 583. [CrossRef]

nutrients

MDPI

Article

The Association of Soft Drink Consumption and the 24-Hour Movement Guidelines with Suicidality among Adolescents of the United States

Bao-Peng Liu [1,2], Cun-Xian Jia [1,2,*] and Shi-Xue Li [3,*]

1 Department of Epidemiology, School of Public Health, Cheeloo College of Medicine, Shandong University, Jinan 250012, China; liubaopeng@sdu.edu.cn
2 Center for Suicide Prevention Research, Shandong University, Jinan 250012, China
3 Centre for Health Management and Policy Research, School of Public Health, Cheeloo College of Medicine, Shandong University, Jinan 250012, China
* Correspondence: jiacunxian@sdu.edu.cn (C.-X.J.); shixueli@sdu.edu.cn (S.-X.L.)

Abstract: Background: Evidence is lacking for the association of the behaviors of the 24 h movement guidelines including sleep duration, physical activity, screen time, and soft drink consumption with suicidality among adolescents. Methods: Data were extracted from a national representative sample of Youth Risk Behavior Surveys (YRBS) in the United States from 2011 to 2019. Binary logistic regression models with complex sampling designs were used to explore the association of the recommendations of the 24 h movement guidelines and soft drink consumption with suicidality. Results: The total prevalence of suicidal ideation, suicide plan, suicide attempt, and suicide attempt with medical treatment was higher among adolescents who did not meet all the recommendations in the 24 h movement guidelines and had a higher level of soft drink consumption. Totally, not meeting all the recommendations of the 24 h movement guidelines was significantly associated with an increased risk of suicidal ideation (OR: 1.69, 95% CI: 1.30–2.19) and suicide plan (OR: 1.76, 95% CI: 1.34–2.33) compared with adolescents who meet all the recommendations. Soft drink consumption of $\geq$3 times/day was associated with an increased risk of suicidality including suicidal ideation, suicide plan, suicide attempt, and suicide attempt with medical treatment, regardless of sex. Soft drink consumption of $\geq$3 times/day was significantly associated with an increased risk of suicide attempt and suicide attempt with medical treatment, regardless of whether the recommendations of physical activity, screen time, and sleep duration were met. Conclusion: Age-appropriate sleep duration, no more than 2 h of screen time per day, at least 1 h of physical activity per day as contained in the 24 h movement guidelines and less than one soft drink consumption per day are good targets to prevent involvement in suicidality. More actions for intervening in the movement and dietary behaviors among adolescents are needed to maintain physical and mental health.

Keywords: 24 h movement guidelines; soft drink; suicidality; adolescent

Citation: Liu, B.-P.; Jia, C.-X.; Li, S.-X. The Association of Soft Drink Consumption and the 24-Hour Movement Guidelines with Suicidality among Adolescents of the United States. *Nutrients* **2022**, *14*, 1870. https://doi.org/10.3390/nu14091870

Academic Editors: William B. Grant and Maria Luz Fernandez

Received: 20 February 2022
Accepted: 26 April 2022
Published: 29 April 2022

1. Introduction

Suicide among adolescents brings a great burden of diseases worldwide and psychological pressure to the family [1,2]. Previous reports had identified the imperative of suicide, which is the fourth leading cause of death among 15–29 years old worldwide and the second leading cause of death among 10–34 years old in the United States (U.S.) [1,3,4]. Plenty of studies have materially identified recognized risk factors of suicide among adolescents such as depression [5], acute stressful events, chronic adversity in early life, familial, and genetic factors, and so on [2,5,6]. However, a perspective on lifestyle including dietary behaviors and physical activity should also be paid enough attention in adolescents, which are the important factors for physical and mental health [7–10].

In 2018, the World Health Organization (WHO) released some initiatives and guidelines about physical activity and set a goal to reduce physical inactivity by 15% by 2030 [11,12]. The guideline from the WHO recommends children and adolescents should do at least an average of 60 min per day of moderate to vigorous intensity, mostly aerobic, physical activity, and limit the amount of time spent being sedentary, particularly the amount of recreational screen time but without a precise threshold [12–14]. The Canadian 24 h movement guidelines for children and adolescents, which were released in 2016, are an integrative goal of physical activity, screen time, and sleep duration [15]. The detailed content of the 24 h movement guidelines was that an accumulation of at least 60 min per day of moderate to vigorous physical activity, uninterrupted 9–11 h of sleep per night for those aged 5–13 years and 8–10 h per night for those aged 14–17 years, and no more than 2 h per day of recreational screen time [15]. These guidelines adhere to the criteria of the WHO, specifically the threshold of sedentary behaviors, and add the recommendations of sleep duration, which agrees with the recommendations from the National Sleep Foundation [16] in children and adolescents. The integrative index could better reflect the effect of movement among children and adolescents compared with a single indicator and give us more evidence for protecting children and adolescents from adverse outcomes.

The prevalence of meeting all the recommendations of the 24 h movement guidelines reported in the previous studies is between 1.0% and 9.4% among children and adolescents [9,17–22]. What is more, the males and younger adolescents were reported to have a higher prevalence of meeting all the recommendations contained in the guidelines according to previous studies [9,18,20,23]. Previous studies have reported that meeting all the recommendations of the 24 h movement guidelines are associated with obesity or being overweight [20], global cognition [22], and mental health such as internalizing and externalizing behaviors [10], impulsivity [21], psychological distress [24], depressive symptoms [25,26], and anxiety [25]. Moreover, only one study, performed by Sampasa-Kanyinga et al. using the data from the Ontario Student Drug Use and Health Survey [9], reported the association of the 24 h movement guidelines and suicidality including suicide ideation, and suicide attempt by sex and age among children and adolescents. However, this study did not report the overall association of meeting the recommendations of the 24 h movement guidelines with suicidal ideation and suicide attempt, and lack of the association of the recommendations of the 24 h movement guidelines and suicide plan or suicide attempt with medical treatment, which are an important index of suicidality.

Soft drinks, especially the consumption of sweetened beverages, were found to be highly correlated to loneliness [27], sedentary behaviors [28], physical status such as unhealthy weight status [29,30] and early menarche [31], aggressive behaviors [32–34] and mental health [35–39], which are reported to be associated with suicidality among adolescents. Although previous studies also reported that soft drinks and sweetened drinks are directly associated with an increased risk of suicidality [32,40–42], most of the recent studies are from non-US and low- and middle-income countries [40–42]. Moreover, Solnick et al. used the national data of the Youth Risk Behavior Survey (YRBS) of the U.S. in 2009 to explore the associations among soft drinks, aggression, and suicidality [32]. However, this study did not examine the dose–response association and the evidence for the association in the recent 10 years is limited to the U.S.

More attention should be paid to the interactive association of movement and dietary behaviors with suicidality and mental health. A previous study used the data of YRBS in 2019 to explore the association of sleep duration, screen time, physical activity, and dietary behaviors (not including soft drinks) with suicidality [43]. Another study also using the data of YRBS in 2019 and latent class analysis tried to build a new variable of lifestyle including all the variables of the 24 h movement guidelines and dietary behaviors and explore their association with suicidality [44]. However, there are no studies to explore the association between soft drinks and suicidality by different recommendations of the 24 h movement guidelines, namely sleep duration, screen time, physical activity, and integrative index. In addition, the interactive association of not meeting the recommendations of the

24 h movement guidelines and more consumption of soft drinks with suicidality is also rarely reported.

This study used the data from YRBS of the U.S. from 2011 to 2019 and aimed to (1) document the weighted prevalence of suicidality including suicidal ideation, suicide plan, suicide attempt, and suicide attempt with medical treatment in total and by the level of soft drink consumption or different recommendations of the 24 h movement guidelines; (2) document the prevalence of meeting all, two, or one of the recommendations of the 24 h movement guidelines; (3) explore the weighted association of the 24 h movement guidelines and soft drink consumption with suicidality; (4) report the association of soft drink consumption and suicidality by different recommendations of the 24 h movement guidelines among adolescents of the U.S.

2. Methods

2.1. Design and Participants

The Youth Risk Behavior Surveillance System (YRBSS), developed in 1990 by the Centers for Disease Control and Prevention (CDC) in the U.S., aimed to monitor health-risk behaviors during childhood and early adolescence. YRBS, which was conducted every two years with different participants, was a national school-based survey of representative samples of 9th through to 12th-grade students. Employing a three-stage cluster sample design, YRBS included public and private schools in the 50 states and the District of Columbia. The first-stage sampling frame, namely, primary sampling units (PSUs), consisted of large-sized counties or groups of smaller, adjacent counties. In the second stage of sampling, selected schools from PSUs and one or two entire classes in each chosen school and in each of the grades 9–12 were randomly selected in the final stage of sampling. A weight based on sex, race/ethnicity, and school grade is applied to each record to adjust for student nonresponse and oversampling of Black and Hispanic students. The protocol of national YRBS was approved by the institutional review board of CDC and is publicly available. A self-administered computer-scannable questionnaire with anonymity was used with the voluntary procedure and parental permission. YRBS was a repeated cross-sectional database and reflected the status of high school in the U.S. More details about YRBS can be seen at the website [45] and previously published studies about YRBS [46,47]. In consideration of data integrity (the data on physical activity began in 2011), this study included the data of five recent 10-year surveys (2011, 2013, 2015, 2017, and 2019). The sample size of the five surveys was 15,425, 13,583, 15,624, 14,765, and 13,677, respectively, and a total of 73,074 adolescents were examined eventually in this study.

2.2. Independent Variables

2.2.1. Soft Drink Consumption

Soft drink consumption was measured by the question: *During the past 7 days, how many times did you drink a can, bottle, or glass of soda or pop, such as Coke, Pepsi, or Sprite? (do not count diet soda or diet pop)?* Response options included not drinking soda or pop during the past 7 days, drinking 1 to 3 times during the past 7 days, 4 to 6 times during the past 7 days, 1 time per day, 2 times per day, 3 times per day, 4 or more times per day. These were categorized into none, <1 time per day, 1–2 times per day, and 3 times or above per day in this study.

2.2.2. The Recommendations of the 24 h Movement Guidelines

The recommendations of the 24 h movement guidelines included physical activity, screen time, and sleep duration. Physical activity was measured by the question: *During the past 7 days, how many days were you physically active for a total of at least 60 min per day? (Add up all the time you spent in any kind of physical activity that increased your heart rate and made you breathe hard some of the time).* Responses were dichotomized into 7 days (every day) and lower than 7 days. Screen time was extracted from two questions: *On an average school day how many hours do you (1) watch TV and (2) play video or computer games or use a computer for something that is not schoolwork?* After summing the time of the two questions, responses were dichotomized into above 2 h and 2 h or below. Sleep duration was measured by the question: *On the average school night, how many hours of sleep do you get?* Responses were dichotomized into adherence to the recommendations and not according to the guidelines (9–11 h per night for 11–13-year-olds; 8–10 h per night for 14–17-year-olds, or 7–9 h per night for those $\geq$18 years of age) [16,48].

Eventually, meeting the recommendations of the 24 h movement guidelines was assessed by two new variables: (1) meeting all the three criteria or not, and (2) meeting all the three criteria, meeting physical activity and screen time, meeting physical activity and sleep duration, meeting screen time and sleep duration, meeting physical activity only, meeting screen time only, meeting sleep duration only, and meeting none of the three criteria. The first variable was used to assess the prevalence of meeting all the recommendations of the 24 h movement guidelines and the association with suicidality. The second variable was used to check the distribution of meeting all and part recommendations of the 24 h movement guidelines.

2.3. Dependent Variables

Suicidality, namely, suicidal ideation, suicide plan, suicide attempt, and suicide attempt with medical treatment were the dependent variables in this study. Suicidal ideation was measured by the question: *During the past 12 months, did you ever seriously consider attempting suicide?* A suicide plan was measured by the question: *During the past 12 months, did you ever make a plan about how you would attempt suicide?* Responses for suicidal ideation and suicide plan were dichotomized into yes and no. Suicide attempt was measured by the question: *During the past 12 months, how many times did you actually attempt suicide?* Responses were dichotomized into none and 1 time or above. Suicide attempt with medical treatment was measured by the question: *If you attempted suicide during the past 12 months, did any attempt result in an injury, poisoning, or overdose that had to be treated by a doctor or nurse?* Responses were dichotomized into yes and no.

2.4. Covariates

2.4.1. Demographic Factors

The demographic factors in this study included age, sex, race, and year of the survey. Age was categorized into 14 years old or below, 15 years old, 16 years old, 17 years old, and 18 years old or above. Race was ascertained with two questions. The first question was "Are you Hispanic or Latino?" and the second question was "What is your race?". If the adolescents responded "yes" to the first question, they were identified as "Hispanic/Latino". Otherwise, the second question would be asked with the response options of "White", "Black or African American" and "others" (American Indian or Alaska Native, Asian, Native Hawaiian, or Other Pacific Islander). The year of the survey was used as a multinominal variable in this study.

2.4.2. Weight Status

Age- and sex-specific Body Mass Index (BMI) was used to determine normal or underweight, overweight or obese in this study. The participants were considered overweight when the BMI percentile was at or above the 85th percentile and obese when the BMI percentile was at or above the 95th percentile for BMI by age and sex. The program and technical documentation for calculating and discriminating weight status could be seen on the website [49] and a previous study [50].

2.4.3. Dietary Behaviors

Dietary behaviors in this study included vegetable, fruit, milk, and breakfast consumption. The responses of vegetables and fruit were dichotomized into one or more times per day and less than one time per day. The responses to milk consumption were dichotomized into one or more glasses per day and less than one glass per day. Breakfast consumption was categorized into daily and not daily. The question's wording and detailed responses can be seen in Table S1.

2.4.4. Depressive Symptoms

Depressive symptoms were measured by the question: *During the past 12 months, did you ever feel so sad or hopeless almost every day for two weeks or more in a row that you stopped doing some usual activities?* The responses to this question were yes or no. This question is valid for depressive symptoms according to a previous study [44].

More details of the questions and responses associated with covariates, independent variables, and dependent variables can be seen in Table S1.

2.5. Statistical Analysis

The software of R version 4.1.0 was used to perform all the analyses in this study. A series of analyses related to complex sampling design was used to get valid point estimates and corresponding confidence intervals. The weighted prevalence of suicidality in total or by the recommendations of the 24 h movement guidelines and soft drink consumption was reported in this study. Pearson Chi-squared statistics with the second-order correction of the Rao–Scott Chi-square test [51] were used to explore the differences in the weighted prevalence of suicidality by the recommendations of the 24 h movement guidelines and soft drink consumption. The *p*-values for the differences were computed with a Satterthwaite approximation to the distribution and with denominator degrees of freedom as recommended by Thomas and Rao [52]. The confidence intervals of weighted prevalence were estimated by the methods proposed by Korn and Graubard [53]. Venn diagrams, which could display weight percentage clearly, were used to show the distributions of meeting the recommendations of the 24 h movement guidelines. Binary logistic regression models with a complex sampling design were used to show the association of meeting all the recommendations of the 24 h movement guidelines and soft drink consumption with suicidality, including suicidal ideation, suicide plan, suicide attempt, and suicide attempt with medical treatment after adjusting age, sex, race, survey year, weight status, depressive symptoms, and dietary behaviors including milk, fruit, vegetable, and breakfast consumption. Simultaneously, the association between soft drink consumption and suicidality by different recommendations of the 24 h movement guidelines was explored in this study.

Sensitivity analysis of missing data by multiple imputations by chained equations (MICE) was used to explore the stability of the associations among soft drink consumption, 24 h movement guidelines, and suicidality [46,54]. Sensitivity analysis of the association among the 24 h movement guidelines, soft drink consumption, and suicidality by omitting weight status and depressive symptoms was also performed in consideration of its confounding effect on the association. In addition, E-values were utilized to assess the sensitivity of potential unmeasured confounding results [55]. E-values for each exposure were calculated using an online calculator (website: www.evalue-calculator.com, accessed on 6 April 2022) with reporting the estimates and limits of corresponding 95% CI [56].

3. Results

3.1. Characteristics of Included Participants

Among 73,074 included participants, 87.9% of high-school students were 15-years-old or above. The ratio of boy/girl was 0.99:1 (36,108/36,497, others are missing). The proportions of White, Black/African American, Hispanic/Latino were 43.0%, 16.8%, and 27.1%. A total of 14.7% and 13.2% of the participants were overweight and obese. More details on the distribution of age, sex, race, and weight status, and unweighted proportions of dietary behaviors (soft drink, vegetable, fruit, milk, and breakfast consumption), the recommendations of the 24 h movement guidelines, depressive symptoms, and suicidality could be seen in Table S2.

3.2. The Weighted Prevalence of Suicidality by the Recommendations of the 24 h Movement Guidelines and Levels of Soft Drink Consumption

As shown in Table 1, the total prevalence of suicidal ideation, suicide plan, suicide attempt and suicide attempt with medical treatment was 17.3% (16.8–17.8%), 14.0% (13.5–14.5%), 8.1% (7.7–8.5%), and 2.6% (2.4–2.8%), respectively. The prevalence of suicidality in the group of meeting all the recommendations of the 24 h movement guidelines is significantly lower than in those not meeting all the recommendations. A lower prevalence of suicidality could also be seen in other recommendations, namely appropriate sleep duration, screen time ≤2 h/day, or physical activity ≥1 h/day.

Table 1. The weighted prevalence of meeting the recommendations of the 24 h movement guidelines and suicidality by levels of soft drink consumption among adolescents of the U.S.

Variables	Suicidality, % (95% CI)			
	Suicidal Ideation	**Suicide Plan**	**Suicide Attempt**	**Suicide Attempt with Medical Treatment**
Total	17.3 (16.8–17.8)	14.0 (13.5–14.5)	8.1 (7.7–8.5)	2.6 (2.4–2.8)
The recommendations of the 24 h movement guidelines				
Appropriate sleep duration [a]	11.4 (10.8–12.0)	9.5 (9.0–10.1)	5.2 (4.7–5.7)	1.5 (1.3–1.8)
Inappropriate sleep duration [a]	20.2 (19.6–20.8)	16.3 (15.6–16.9)	9.2 (8.7–9.7)	2.9 (2.6–3.2)
p for difference	<0.001	<0.001	<0.001	<0.001
Screen time ≤2 h/day	14.7 (13.8–15.6)	11.9 (11.0–12.8)	7.1 (6.3–7.9)	2.4 (2.0–2.8)
Screen time >2 h/day	18.1 (17.5–18.6)	14.7 (14.2–15.3)	8.3 (7.9–8.6)	2.5 (2.3–2.7)
p for difference	<0.001	<0.001	0.005	0.496
Physical activity ≥1 h/day	13.0 (12.3–13.8)	10.9 (10.2–11.6)	6.3 (5.8–6.8)	2.1 (1.8–2.3)
Physical activity <1 h/day	18.8 (18.2–19.4)	15.2 (14.6–15.8)	8.7 (8.2–9.2)	2.7 (2.4–3.0)
p for difference	<0.001	<0.001	<0.001	<0.001
Meeting all the recommendations	6.8 (5.4–8.2)	4.7 (3.6–5.9)	3.5 (2.3–4.7)	1.2 (0.3–2.0)
Not meeting all the recommendations	17.8 (17.2–18.3)	14.4 (13.9–15.0)	8.0 (7.6–8.5)	2.5 (2.3–2.7)
p for difference	<0.001	<0.001	<0.001	0.036
Soft drink consumption				
None	15.3 (14.5–16.2)	12.5 (11.7–13.3)	6.6 (6.0–7.2)	2.1 (1.7–2.4)
<1 time/day	16.8 (16.1–17.4)	13.5 (12.9–14.2)	7.3 (6.8–7.7)	2.1 (1.8–2.3)
1–2 times/day	18.5 (17.3–19.7)	14.4 (13.3–15.5)	8.5 (7.7–9.4)	2.8 (2.3–3.3)
≥3 time/day	24.5 (22.9–26.2)	20.9 (19.2–22.7)	16.0 (14.5–17.5)	6.4 (5.3–7.5)
p for difference	<0.001	<0.001	<0.001	<0.001

[a] Appropriate sleep duration means 9–11 h/day for adolescents aged 11–13, 8–10 h/day for adolescents aged 14–17, and 7–9 h/day for adolescents aged above 18 years. CI: confidence interval.

The prevalence of suicidal ideation, suicide plan, suicide attempt, and suicide attempt with medical treatment associated with soft drink consumption of ≥3 time/day was 24.5% (22.9–26.2%), 20.9% (19.2–22.7%), 16.0% (14.5–17.5%) and 6.4% (5.3–7.5%). There was a significant difference in the prevalence of suicidal ideation, suicide plan, suicide attempt, and suicide attempt with medical treatment across different levels of soft drink consumption (all $p < 0.001$). As the frequency of soft drink consumption increased, the prevalence of suicidal ideation, suicide plan, suicide attempt, and suicide attempt with medical treatment increased.

3.3. *The Weighted Prevalence of Meeting the Recommendations of the 24 h Movement Guidelines*

The prevalence of meeting the relative recommendations of the 24 h movement guidelines can be seen in Figure 1. The prevalence of meeting all the recommendations contained in the guidelines was 3.1% in total, 4.3% for boys, and 1.9% for girls. Venn diagrams, shown in Figure 1, also gave us some findings on meeting two recommendations of the 24 h movement guidelines. The prevalence of only meeting the recommendations of sleep duration and screen time was 5.0%, 4.2%, and 5.9% in total, for boys, and for girls, respectively. The prevalence of only meeting the recommendations of sleep duration and physical activity was 6.9%, 10.1%, and 3.7% in total, for boys, and for girls, respectively. The prevalence of only meeting the recommendations of physical activity and screen time was 4.8%, 5.8%, and 3.9% in total, for boys, and for girls, respectively.

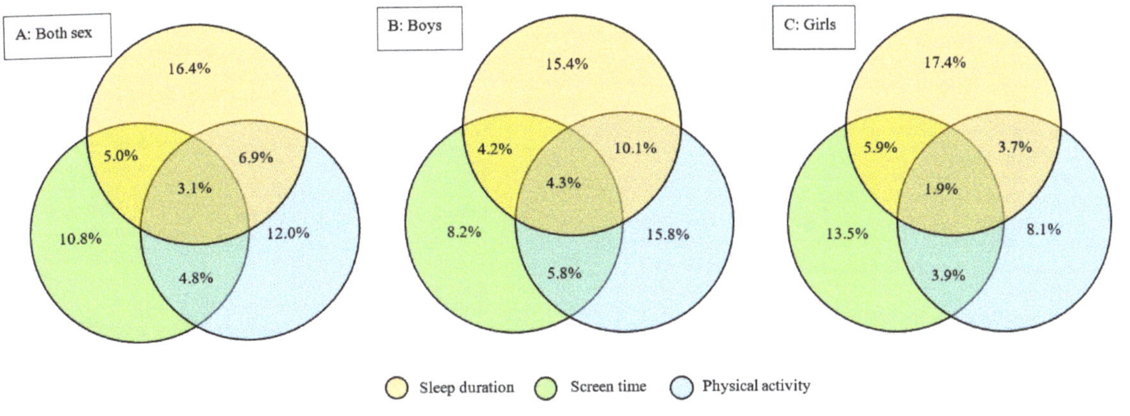

Figure 1. Venn diagrams showing the weighted prevalence of meeting all and part recommendations of the 24 h movement guidelines in total and by sex among the adolescents of the U.S.

3.4. *The Association of the 24 h Movement Guidelines and Soft Drink Consumption with Suicidality*

Totally, not meeting all the recommendations of the 24 h movement guidelines was significantly associated with an increased risk of suicidal ideation (OR: 1.69, 95% CI: 1.30–2.19), and suicide plan (OR: 1.76, 95% CI: 1.34–2.32), compared with adolescents who meet all the recommendations. However, the association between meeting all the recommendations of the 24 h movement guidelines and suicide attempt (OR: 1.12, 95% CI: 0.74–1.68), and suicide attempt with medical treatment (OR: 1.04, 95% CI: 0.49–2.23) was not statistically significant. In the group of boys, similar results compared with the total estimates were found to be with a higher risk of suicide ideation (OR: 2.18, 95% CI: 1.51–3.13) and suicide plan (OR: 2.28, 95% CI: 1.56–3.34) associated with not meeting all the recommendations of the 24 h movement guidelines. The association of meeting all the recommendations of the 24 h movement guidelines with suicide attempt and suicide attempt with medical treatment was also not statistically significant among the boys. Additionally, the association between meeting all the recommendations and suicidality was not found to be statistically significant among the girls.

Soft drink consumption of 1–2 times/day was only found to be associated with an increased risk of suicidal ideation (OR: 1.15, 95% CI: 1.02–1.30) and suicide attempt (OR: 1.21, 95% CI: 1.04–1.41) in total, and suicide attempt (OR: 1.32, 95% CI: 1.09–1.59) and suicide attempt with medical treatment among the girls (OR: 1.47, 95% CI: 1.05–2.05).

Soft drink consumption of $\geq$3 times/day was associated with an increased risk of suicidality including suicidal ideation, suicide plan, suicide attempt, and suicide attempt with medical treatment whether in overall estimates or subgroup analysis by sex. Moreover, there was a linear dose–response relationship for soft drink consumption associated with an increased risk of suicidality among adolescents regardless of sex. More details can be seen in Table 2.

Table 2. The association of the 24 h movement guidelines and soft drink consumption with suicidality among adolescents of the U.S.

Variables	Suicidal Ideation, OR (95% CI) [a]	Suicide Plan, OR (95% CI) [a]	Suicide Attempt, OR (95% CI) [a]	Suicide attempt with medical treatment, OR (95% CI) [a]
		Total		
		24 h movement guidelines		
Meeting all the recommendations	Reference	Reference	Reference	Reference
Not meeting all the recommendations	1.69 (1.30–2.19) ***	1.76 (1.34–2.32) ***	1.12 (0.74–1.68)	1.04 (0.49–2.23)
Soft drink consumption				
None	Reference	Reference	Reference	Reference
<1 time/day	1.03 (0.94–1.12)	1.03 (0.95–1.13)	1.03 (0.92–1.16)	0.96 (0.77–1.21)
1–2 times/day	1.15 (1.02–1.30) *	1.06 (0.93–1.20)	1.21 (1.04–1.41) *	1.25 (0.93–1.67)
≥3 times/day	1.37 (1.20–1.55) ***	1.45 (1.26–1.68) ***	2.20 (1.86–2.61) ***	2.49 (1.90–3.27) ***
p for trend	<0.001	<0.001	<0.001	<0.001
		Boy		
		24 h movement guidelines		
Meeting all the recommendations	Reference	Reference	Reference	Reference
Not meeting all the recommendations	2.18 (1.51–3.13) ***	2.28 (1.56–3.34) ***	1.68 (0.87–3.23)	0.80 (0.27–2.37)
Soft drink consumption				
None	Reference	Reference	Reference	Reference
<1 time/day	0.97 (0.83–1.13)	1.01 (0.87–1.18)	0.79 (0.61–1.02)	0.58 (0.38–0.89) *
1–2 times/day	1.11 (0.91–1.36)	1.01 (0.83–1.24)	0.99 (0.75–1.32)	0.87 (0.53–1.43)
≥3 times/day	1.40 (1.15–1.71) ***	1.52 (1.22–1.90) ***	2.09 (1.59–2.76) ***	2.58 (1.60–4.16) ***
p for trend	<0.001	0.001	<0.001	<0.001
		Girl		
		24 h movement guidelines		
Meeting all the recommendations	Reference	Reference	Reference	Reference
Not meeting all the recommendations	1.30 (0.89–1.90)	1.36 (0.90–2.07)	0.78 (0.44–1.35)	1.53 (0.71–3.30)
Soft drink consumption				
None	Reference	Reference	Reference	Reference
<1 time/day	1.04 (0.94–1.17)	1.03 (0.93–1.15)	1.15 (0.99–1.33)	1.17 (0.92–1.49)
1–2 times/day	1.16 (0.99–1.36)	1.09 (0.91–1.29)	1.32 (1.09–1.59) **	1.47 (1.05–2.05) *
≥3 times/day	1.29 (1.08–1.55) **	1.36 (1.13–1.63) **	2.13 (1.73–2.62) ***	2.19 (1.64–2.94) ***
p for trend	0.003	0.004	<0.001	<0.001

[a] All the estimates in these tables were adjusted for age, sex, race, survey year, weight status, depressive symptoms, and dietary behaviors including milk, fruit, vegetable, and breakfast consumption. Sex was not adjusted in the stratified models. OR: odds ratio, CI: confidence interval, *** $p < 0.001$, ** $p < 0.01$, * $p < 0.05$.

3.5. Subgroup Analyses of the Association between Soft Drink Consumption and Suicidality by Different Recommendations of the 24 h Movement Guidelines

Several findings emerged in the subgroup analyses. Firstly, the association between soft drink consumption and suicidality, regardless of suicidal ideation, suicide plan, suicide attempt, and suicide attempt with medical treatment was statistically significant among adolescents who were not meeting all the recommendations of the 24 h movement guidelines. The association was not found in adolescents who were meeting all the recommendations. Secondly, there was a significant interaction of not meeting the recommendation of screen time and soft drink consumption of <1 time/day on suicide plan and suicide attempt (Table S3). Thirdly, soft drink consumption of ≥3 times/day was significantly associated with an increased risk of suicide attempt and suicide attempt with medical treatment regardless of whether the adolescent was meeting the recommendations of physical activity, screen time, and sleep duration. Fourthly, the association between soft drink consumption and suicidal ideation and suicide plan was not statistically significant among adolescents who were not meeting all the recommendations of sleep duration, regardless of the frequency. More details of subgroup analyses can be seen in Figure 2.

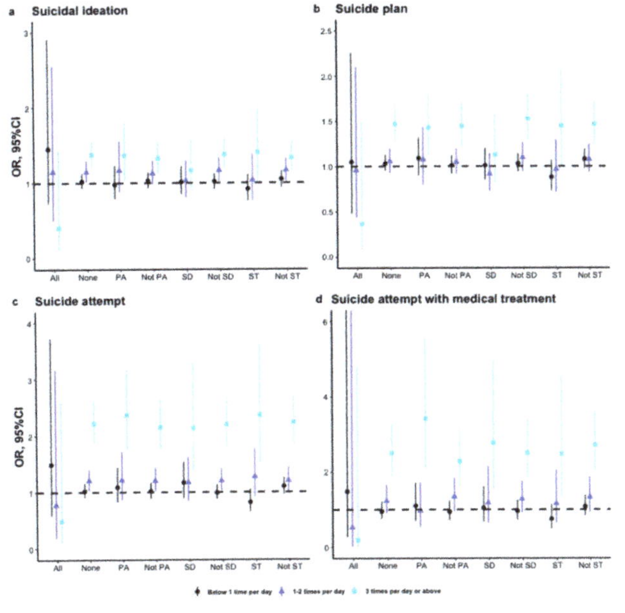

Figure 2. The association between soft drink consumption and suicidality by different recommendations of the 24 h movement guidelines among adolescents in the U.S. (OR: odds ratios, CI: confidence interval; All: meeting all the recommendations of the 24 h movement guidelines; None: meeting none of the recommendations of the 24 h movement guidelines; PA: meeting the recommendations of physical activity; Not PA: not meeting the recommendations of physical activity; SD: meeting the recommendations of sleep duration; Not PA: not meeting the recommendations of sleep duration; ST: meeting the recommendations of screen time; Not PA: not meeting the recommendations of screen time. The estimates of meeting all the recommended behaviors and not were adjusted for age, sex, race, survey year, weight status, depressive symptoms, and dietary behaviors including milk, fruit, vegetable, and breakfast consumption. The estimates of PA and not PA were adjusted for age, sex, race, survey year, weight status, depressive symptoms, and dietary behaviors including milk, fruit, vegetable, breakfast consumption, sleep duration, and screen time. The estimates of SD and not SD were adjusted for age, sex, race, survey year, weight status, depressive symptoms, and dietary behaviors including milk, fruit, vegetable, breakfast consumption, sleep duration, and physical activity. The estimates of ST and not ST were adjusted for age, sex, race, survey year, weight status, depressive symptoms, and dietary behaviors including milk, fruit, vegetable, breakfast consumption, sleep duration, and physical activity.).

3.6. Sensitivity Analysis

Multiple imputations by chained equations (MICE) were performed to explore the effect of missing data on the association among the recommendations of the 24 h movement guidelines, soft drink consumption, and suicidality. The estimates associated with the risks were slightly changed and revealed that the estimates were stable.

In addition, a sensitivity analysis (Table S4) of the association among the 24 h movement guidelines, soft drink consumption, and suicidality by omitting weight status and depressive symptoms was also performed. Although the effects were enhanced and lower levels of soft drink consumption were statistically associated with increased risk of suicidality, the association of the 24 h movement guidelines and soft drink consumption with suicidality was similar to previous estimates.

The E-values (Table S5) were relatively large, particularly for the association with three times per day or above of soft drink consumption. Our findings show that any unobserved confounder could be adequate to fully explain away these effect estimates and to move the CIs to null, while a weak confounder could not do so.

4. Discussions

4.1. Recommendations of the 24 h Movement Guidelines and Suicidality

To our knowledge, this is the first study to report the prevalence of meeting the recommendations of the 24 h movement guidelines taking advantage of the integrated index of physical activity, screen time, and age-appropriate sleep duration in the study of YRBS. Although Zhu et al. [20], using data from the 2016–2017 National Survey of Children's Health (NSCH) of the U.S., reported a higher prevalence (9.4%) of meeting all the recommendations of the 24 h movement guidelines, this study reported a comparable prevalence (3.1%, 95% CI: 2.8–3.4%) with previous studies [9,17–19,21,22]. Similar to most previous studies [9,18,20,23], the boys had a higher prevalence of meeting all the recommendations contained in the guidelines in this study. Despite all this, children and adolescents worldwide were reported to have a low prevalence of meeting all the recommendations of the 24 h movement guidelines.

Previous studies usually explored the association between one variable in an adolescent's lifestyle such as sedentary behaviors [57,58], screen time [59], sleep duration [60], physical activity [61], and suicidality. This study used the integrated index, namely the 24 h movement guidelines, which could reflect the movement of adolescents effectively to explore the association with suicidality. The findings in this study were consistent with a previous study from Canada [9], reporting that not meeting all the recommendations of the 24 h movement guidelines could significantly increase the risk of suicidal ideation and suicide attempt only among the boys. Moreover, our study added some evidence that the associations for suicide plan and suicide attempt with medical treatment and the total estimates for the associations among adolescents.

This study reported that there was no statistically significant association between meeting the 24 h movement guidelines and suicidality, regardless of suicidal ideation, suicide plan, suicide attempt, and suicide attempt with medical treatment among the girls, which was somewhat consistent with a previous study [9]. Several mechanisms may explain the differences by sex. Firstly, the girls have a lower prevalence of alcohol use [62], which might mediate the associations between meeting movement guidelines and suicidality [63]. Furthermore, the girls have a lower prevalence of suicide attempt [6] and a lower prevalence of meeting the recommendations of the 24 h movement guidelines [9,18,20,23], which might not have enough statistical power to detect the associations.

4.2. Soft Drink Consumption and Suicidality

The proportion of consuming no soft drink being 23.5% in this study is similar to previous studies [34,40,41]. Although different cut-offs for the levels of soft drink consumption were used in previous studies, only a higher frequency of soft drink consumption, namely above one time per day was associated with increased risk of suicidality including suicidal ideation, suicide plan, and suicide attempt in the previous studies [32,40–42]. This study also added some new evidence for the association between high levels of soft drink consumption and suicide attempt with medical treatment. Moreover, this study also had similar conclusions that the association of soft drink consumption with suicide attempt would be not changed in the subgroup analysis of sex with a previous study [42]. What is more, a significant association, regardless of sex, was also found in other behaviors of suicidality including suicidal ideation, suicide plan, and suicide attempt with medical treatment.

Although a previous study using the data of YRBS in 2009 reported soft drink consumption daily or above was associated with suicidality [32], this study added to the evidence of the recent 10 years for the linear dose–response relationship among adolescents in the U.S. The estimated risk in this study was fully adjusted by the dietary behaviors and depressive symptoms, which were not performed in previous studies. The mechanism from soft drink consumption to suicidality could be explained by mental problems. Many previous studies have examined that soft drink consumption was associated with depressive symptoms [35–39], which is highly related to suicidality among adolescents. Moreover, a high-sugar diet in adolescents was highly related to neuroinflammation, depressive-like behavior [64], stress-driven, emotional and addictive behaviors [65], which might be related to suicidality. In addition, previous studies reported that soft drink consumption was related to obesity or being overweight [29,30], which might cause inflammation among depressed patients [66]. This path could also be a reason for suicidality among adolescents in consideration of the positive effect of being overweight, inflammation, and depression on suicidality [67,68]. A high-sugar diet could increase anxiety-like and depressive-like behavior [69,70], decrease cognitive performance [71], and chronic psychological stress, development of metabolic syndrome (MetS), and behavioral impairment [72] among mice. This evidence from animals could provide some mediating paths from a high-sugar diet to suicidality.

4.3. *Interactive Association of the 24 h Movement Guidelines and Soft Drink Consumption with Suicidality*

Previous studies also tried to explore the combined association of lifestyle including dietary behaviors and movement behaviors with suicidality by the methods of exploring the individual association for included variables or using latent class analysis [39,40], limited studies focus on the interactive association with suicidality. It is worth noting that a significant association between any level of soft drink consumption and suicidality was not found in the group that met all the recommendations of the 24 h movement guidelines in this study. It might be explained that the negative effect of soft drinks could be decreased when adolescents have good habits of movement. In other groups including those not meeting all the recommendations and meeting the recommendations of sleep duration, screen time, and physical activity contained in the 24 h movement guidelines, the higher level of soft drink consumption, namely above two times per day, was significantly associated with suicidality. It is worth noting the important role of controlling the consumption of soft drinks and keeping suitable movement among adolescents.

4.4. *Strengths and Limitations*

This study had some strengths. Firstly, this study used national school-based data from representative samples of 9th through to 12th-grade students to emphasize a linear dose–response relationship between soft drink consumption and suicidality, namely suicidal ideation, suicide plan, suicide attempt, and suicide attempt with medical treatment. In addition, this is the first study using YRBS to report the recommendations of the 24 h movement guidelines and explore their association with suicidality. Furthermore, this is also the first study to explore the association of soft drink consumption and suicidality with different recommendations of the 24 h movement guidelines.

Some limitations are worth mentioning in this study. Firstly, a causal relationship is not able to be confirmed given the cross-sectional design. More cohort studies are needed to explore the relationship in the future. Secondly, all the questions were self-reported, and recall bias and information bias were unavoidable. Movement variables such as sleep duration, physical activity, and screen time were not measured by wearable devices, which might lead to information bias. Thirdly, soft drinks in this study did not include energy drinks, which were not collected in YRBS, which might bring some effect on the association. Fourthly, owing to the design of YRBS focusing on schools, findings are not suitable for extending to the entire population of adolescents. Lastly, some important socioeconomic factors such as family income, occupation of parents, and dietary habits of parents were not included in this database, which could be associated with confounding factors. Although the database included some factors such as alcohol and cigarette use, it was not able to be included as the covariates owing to the limited sample size in the subgroup of girls or meeting all the recommendations of the 24 h movement guidelines.

5. Conclusions

The present study supported the evidence that not meeting the recommendations of the 24 h movement guidelines and a high level of soft drink consumption could increase the risk of suicidality. It is implicated that the association of soft drink consumption with suicidality was not statistically significant when adolescents meet all the recommendations of the 24 h movement guidelines. These findings emphasize the importance of age-appropriate sleep duration, limited screen time ($\leq$2 h/day), and appropriate physical activity ($\geq$1 h/day) contained in the 24 h movement guidelines, and less consumption of soft drinks for preventing suicidality among adolescents. Relevant departments, schools, and families should formulate corresponding measures to ensure these beneficial behaviors and prevent at-risk adolescents from adverse behaviors.

Supplementary Materials: The following supporting information can be downloaded at: https://www.mdpi.com/article/10.3390/nu14091870/s1, Table S1: Question-wording and details for included variables; Table S2: Characteristics of included variables among youth risk behavior surveys by survey year (2011–2019); Table S3: Interactive association of soft drink consumption and the recommendations of the 24 h movement guidelines with suicidality; Table S4: Sensitivity analysis of the association among 24-h movement guidelines, soft drink consumption and suicidality by omitting weight status and depressive symptoms; Table S5: E-value analysis for the association among soft drink consumption, 24-h movement guideline, and suicidality.

Author Contributions: B.-P.L.: Conceptualization, Formal analysis, Funding acquisition, Validation, Visualization, Writing—original draft, Writing—review and editing; S.-X.L.: Conceptualization, Project administration, Supervision, Writing—review and editing; C.-X.J.: Conceptualization, Funding acquisition, Project administration, Supervision, Writing—review and editing. All authors have read and agreed to the published version of this manuscript.

Funding: This work was supported by the National Natural Science Foundation of China (NSFC) (No: 82103954; 30972527; 81573233].

Institutional Review Board Statement: The study was conducted in accordance with the Declaration of Helsinki. The study was approved by the institutional review board of CDC and is publicly available.

Informed Consent Statement: Informed consent was obtained from all subjects involved in the study.

Data Availability Statement: The data can be downloaded from https://www.cdc.gov/healthyyouth/data/yrbs/data.htm/.

Conflicts of Interest: The authors declare no conflict of interest.

References

1. World Health Organization. Suicide Worldwide in 2019. Available online: https://www.who.int/publications/i/item/9789240026643 (accessed on 1 October 2021).
2. Turecki, G.; Brent, D.A. Suicide and suicidal behaviour. *Lancet* **2016**, *387*, 1227–1239. [CrossRef]

3. National Center for Injury Prevention and Control. 10 Leading Causes of Death by Age Group, United States. Available online: https://www.cdc.gov/injury/images/lc-charts/leading_causes_of_death_by_age_group_2018_1100w850h.jpg (accessed on 1 October 2021).
4. May, A.M.; Czyz, E.K.; West, B.T. Differentiating Adolescent Suicide Attempters and Ideators: A Classification Tree Analysis of Risk Behaviors. *J. Adolesc. Health* **2020**, *67*, 837–850. [CrossRef] [PubMed]
5. Thapar, A.; Collishaw, S.; Pine, D.S.; Thapar, A.K. Depression in adolescence. *Lancet* **2012**, *379*, 1056–1067. [CrossRef]
6. Hawton, K.; Saunders, K.E.; O'Connor, R.C. Self-harm and suicide in adolescents. *Lancet* **2012**, *379*, 2373–2382. [CrossRef]
7. Khalid, S.; Williams, C.M.; Reynolds, S.A. Is there an association between diet and depression in children and adolescents? A systematic review. *Br. J. Nutr.* **2016**, *116*, 2097–2108. [CrossRef] [PubMed]
8. Cunha, C.M.; Costa, P.R.F.; de Oliveira, L.P.M.; Queiroz, V.A.O.; Pitangueira, J.C.D.; Oliveira, A.M. Dietary patterns and cardiometabolic risk factors among adolescents: Systematic review and meta-analysis. *Br. J. Nutr.* **2018**, *119*, 859–879. [CrossRef] [PubMed]
9. Sampasa-Kanyinga, H.; Chaput, J.P.; Goldfield, G.S.; Janssen, I.; Wang, J.; Hamilton, H.A.; Colman, I. 24-hour movement guidelines and suicidality among adolescents. *J. Affect. Disord.* **2020**, *274*, 372–380. [CrossRef]
10. Sampasa-Kanyinga, H.; Colman, I.; Goldfield, G.S.; Janssen, I.; Wang, J.; Tremblay, M.S.; Barnes, J.D.; Walsh, J.J.; Chaput, J.P. 24-Hour Movement Behaviors and Internalizing and Externalizing Behaviors Among Youth. *J. Adolesc. Health* **2021**, *68*, 969–977. [CrossRef]
11. World Health Organization. Global Action Plan on Physical Activity 2018–2030: More Active People for a Healthier World: At-a-Glance. Available online: https://apps.who.int/iris/handle/10665/272721 (accessed on 1 October 2021).
12. World Health Organization. WHO Guidelines on Physical Activity and Sedentary Behaviour. Available online: https://apps.who.int/iris/handle/10665/336656 (accessed on 1 October 2021).
13. Bull, F.C.; Al-Ansari, S.S.; Biddle, S.; Borodulin, K.; Buman, M.P.; Cardon, G.; Carty, C.; Chaput, J.P.; Chastin, S.; Chou, R.; et al. World Health Organization 2020 guidelines on physical activity and sedentary behaviour. *Br. J. Sports Med.* **2020**, *54*, 1451–1462. [CrossRef]
14. Chaput, J.P.; Willumsen, J.; Bull, F.; Chou, R.; Ekelund, U.; Firth, J.; Jago, R.; Ortega, F.B.; Katzmarzyk, P.T. 2020 WHO guidelines on physical activity and sedentary behaviour for children and adolescents aged 5-17 years: Summary of the evidence. *Int. J. Behav. Nutr. Phys. Act.* **2020**, *17*, 141. [CrossRef]
15. Tremblay, M.S.; Carson, V.; Chaput, J.P.; Connor Gorber, S.; Dinh, T.; Duggan, M.; Faulkner, G.; Gray, C.E.; Gruber, R.; Janson, K.; et al. Canadian 24-Hour Movement Guidelines for Children and Youth: An Integration of Physical Activity, Sedentary Behaviour, and Sleep. *Appl. Physiol. Nutr. Metab.* **2016**, *41*, S311–S327. [CrossRef] [PubMed]
16. Hirshkowitz, M.; Whiton, K.; Albert, S.M.; Alessi, C.; Bruni, O.; DonCarlos, L.; Hazen, N.; Herman, J.; Katz, E.S.; Kheirandish-Gozal, L.; et al. National Sleep Foundation's sleep time duration recommendations: Methodology and results summary. *Sleep Health* **2015**, *1*, 40–43. [CrossRef]
17. Janssen, I.; Roberts, K.C.; Thompson, W. Adherence to the 24-Hour Movement Guidelines among 10- to 17-year-old Canadians. *Health Promot. Chronic Dis. Prev. Can.* **2017**, *37*, 369–375. [CrossRef] [PubMed]
18. Roberts, K.C.; Yao, X.; Carson, V.; Chaput, J.P.; Janssen, I.; Tremblay, M.S. Meeting the Canadian 24-Hour Movement Guidelines for Children and Youth. *Health Rep.* **2017**, *28*, 3–7.
19. Shi, Y.; Huang, W.Y.; Sit, C.H.; Wong, S.H. Compliance With 24-Hour Movement Guidelines in Hong Kong Adolescents: Associations With Weight Status. *J. Phys. Act. Health* **2020**, *17*, 287–292. [CrossRef] [PubMed]
20. Zhu, X.; Healy, S.; Haegele, J.A.; Patterson, F. Twenty-Four-Hour Movement Guidelines and Body Weight in Youth. *J. Pediatr.* **2020**, *218*, 204–209. [CrossRef] [PubMed]
21. Guerrero, M.D.; Barnes, J.D.; Walsh, J.J.; Chaput, J.P.; Tremblay, M.S.; Goldfield, G.S. 24-Hour Movement Behaviors and Impulsivity. *Pediatrics* **2019**, *144*, e20190187. [CrossRef]
22. Walsh, J.J.; Barnes, J.D.; Cameron, J.D.; Goldfield, G.S.; Chaput, J.P.; Gunnell, K.E.; Ledoux, A.A.; Zemek, R.L.; Tremblay, M.S. Associations between 24 hour movement behaviours and global cognition in US children: A cross-sectional observational study. *Lancet Child Adolesc. Health* **2018**, *2*, 783–791. [CrossRef]
23. Sampasa-Kanyinga, H.; Colman, I.; Goldfield, G.S.; Janssen, I.; Wang, J.; Hamilton, H.A.; Chaput, J.P. Associations between the Canadian 24 h movement guidelines and different types of bullying involvement among adolescents. *Child Abuse Negl.* **2020**, *108*, 104638. [CrossRef]
24. Sampasa-Kanyinga, H.; Chaput, J.P.; Goldfield, G.S.; Janssen, I.; Wang, J.; Hamilton, H.A.; Ferro, M.A.; Colman, I. The Canadian 24-Hour Movement Guidelines and Psychological Distress among Adolescents: Les Directives canadiennes en matière de mouvement sur 24 heures et la détresse psychologique chez les adolescents. *Can. J. Psychiatry* **2021**, *66*, 624–633. [CrossRef]
25. Lu, S.; Cheval, B.; Yu, Q.; Hossain, M.M.; Chen, S.T.; Taylor, A.; Bao, R.; Doig, S.; Li, J.; Wang, T.; et al. Associations of 24-Hour Movement Behavior with Depressive Symptoms and Anxiety in Children: Cross-Sectional Findings from a Chinese Sample. *Healthcare* **2021**, *9*, 1532. [CrossRef] [PubMed]
26. Sampasa-Kanyinga, H.; Colman, I.; Goldfield, G.S.; Janssen, I.; Wang, J.; Podinic, I.; Tremblay, M.S.; Saunders, T.J.; Sampson, M.; Chaput, J.P. Combinations of physical activity, sedentary time, and sleep duration and their associations with depressive symptoms and other mental health problems in children and adolescents: A systematic review. *Int. J. Behav. Nutr. Phys. Act.* **2020**, *17*, 72. [CrossRef] [PubMed]

27. Pengpid, S.; Peltzer, K. Loneliness is associated with poor mental health, social-environmental factors, and health risk behaviours among national samples of in-school adolescents in four Caribbean countries. *Psychol. Health Med.* **2022**, *27*, 559–570. [CrossRef] [PubMed]
28. Ashdown-Franks, G.; Vancampfort, D.; Firth, J.; Smith, L.; Sabiston, C.M.; Stubbs, B.; Koyanagi, A. Association of leisure-time sedentary behavior with fast food and carbonated soft drink consumption among 133,555 adolescents aged 12-15 years in 44 low- and middle-income countries. *Int. J. Behav. Nutr. Phys. Act.* **2019**, *16*, 35. [CrossRef]
29. Chaves, O.C.; Velasquez-Melendez, G.; Costa, D.; Caiaffa, W.T. Soft drink consumption and body mass index in Brazilian adolescents: National Adolescent Student Health Survey. *Rev. Bras. Epidemiol.* **2018**, *21*, e180010. [CrossRef]
30. Iannotti, R.J.; Wang, J. Patterns of physical activity, sedentary behavior, and diet in U.S. adolescents. *J. Adolesc. Health* **2013**, *53*, 280–286. [CrossRef]
31. Mueller, N.T.; Jacobs, D.R., Jr.; MacLehose, R.F.; Demerath, E.W.; Kelly, S.P.; Dreyfus, J.G.; Pereira, M.A. Consumption of caffeinated and artificially sweetened soft drinks is associated with risk of early menarche. *Am. J. Clin. Nutr.* **2015**, *102*, 648–654. [CrossRef]
32. Solnick, S.J.; Hemenway, D. Soft drinks, aggression and suicidal behaviour in US high school students. *Int. J. Inj. Contr. Saf. Promot.* **2014**, *21*, 266–273. [CrossRef]
33. Holubcikova, J.; Kolarcik, P.; Madarasova Geckova, A.; Reijneveld, S.A.; van Dijk, J.P. The mediating effect of daily nervousness and irritability on the relationship between soft drink consumption and aggressive behaviour among adolescents. *Int. J. Public Health* **2015**, *60*, 699–706. [CrossRef]
34. Shi, Z.; Malki, A.; Abdel-Salam, A.G.; Liu, J.; Zayed, H. Association between Soft Drink Consumption and Aggressive Behaviour among a Quarter Million Adolescents from 64 Countries Based on the Global School-Based Student Health Survey (GSHS). *Nutrients* **2020**, *12*, 694. [CrossRef]
35. Mrug, S.; Jones, L.C.; Elliott, M.N.; Tortolero, S.R.; Peskin, M.F.; Schuster, M.A. Soft Drink Consumption and Mental Health in Adolescents: A Longitudinal Examination. *J. Adolesc. Health* **2021**, *68*, 155–160. [CrossRef]
36. Park, S.; Rim, S.J.; Lee, J.H. Associations between dietary behaviours and perceived physical and mental health status among Korean adolescents. *Nutr. Diet* **2018**, *75*, 488–493. [CrossRef] [PubMed]
37. Sim, E.; Sohn, W.; Choi, E.S.; Noh, H. Sugar-sweetened beverage consumption frequency in Korean adolescents: Based on the 2015 Youth Risk Behavior Web-Based Survey. *Int. Dent. J.* **2019**, *69*, 376–382. [CrossRef] [PubMed]
38. Pabayo, R.; Dias, J.; Hemenway, D.; Molnar, B.E. Sweetened beverage consumption is a risk factor for depressive symptoms among adolescents living in Boston, Massachusetts, USA. *Public Health Nutr.* **2016**, *19*, 3062–3069. [CrossRef] [PubMed]
39. Hong, S.A.; Peltzer, K. Dietary behaviour, psychological well-being and mental distress among adolescents in Korea. *Child Adolesc. Psychiatry Ment. Health* **2017**, *11*, 56. [CrossRef]
40. Pengpid, S.; Peltzer, K. High Carbonated Soft Drink Intake is Associated with Health Risk Behavior and Poor Mental Health among School-Going Adolescents in Six Southeast Asian Countries. *Int. J. Environ. Res. Public Health* **2019**, *17*, 132. [CrossRef]
41. Pan, X.; Zhang, C.; Shi, Z. Soft drink and sweet food consumption and suicidal behaviours among Chinese adolescents. *Acta Paediatr.* **2011**, *100*, e215–e222. [CrossRef]
42. Jacob, L.; Stubbs, B.; Koyanagi, A. Consumption of carbonated soft drinks and suicide attempts among 105,061 adolescents aged 12-15 years from 6 high-income, 22 middle-income, and 4 low-income countries. *Clin. Nutr.* **2020**, *39*, 886–892. [CrossRef]
43. Li, X.; Chi, G.; Taylor, A.; Chen, S.T.; Memon, A.R.; Zhang, Y.; Song, Y.; Li, J.; Luo, X.; Zou, L. Lifestyle Behaviors and Suicide-Related Behaviors in Adolescents: Cross-Sectional Study Using the 2019 YRBS Data. *Front. Public Health* **2021**, *9*, 766972. [CrossRef]
44. Xiao, Y.; Romanelli, M.; Lindsey, M.A. A latent class analysis of health lifestyles and suicidal behaviors among US adolescents. *J. Affect. Disord.* **2019**, *255*, 116–126. [CrossRef]
45. Centers for disease control and prevention of United States. Youth Risk Behavior Surveillance System Data & Document. Available online: https://www.cdc.gov/healthyyouth/data/yrbs/data.htm, (accessed on 1 October 2021).
46. Baiden, P.; LaBrenz, C.A.; Asiedua-Baiden, G.; Muehlenkamp, J.J. Examining the intersection of race/ethnicity and sexual orientation on suicidal ideation and suicide attempt among adolescents: Findings from the 2017 Youth Risk Behavior Survey. *J. Psychiatr. Res.* **2020**, *125*, 13–20. [CrossRef] [PubMed]
47. Jacobs, W.; Nabors, L.; Mahabee-Gittens, M.E.; Merianos, A.L. E-cigarette and marijuana use and the attainment of obesity prevention guidelines among U.S. adolescents. *Prev. Med. Rep.* **2021**, *23*, 101445. [CrossRef]
48. Ross, R.; Chaput, J.P.; Giangregorio, L.M.; Janssen, I.; Saunders, T.J.; Kho, M.E.; Poitras, V.J.; Tomasone, J.R.; El-Kotob, R.; McLaughlin, E.C.; et al. Canadian 24-Hour Movement Guidelines for Adults aged 18–64 years and Adults aged 65 years or older: An integration of physical activity, sedentary behaviour, and sleep. *Appl. Physiol. Nutr. Metab.* **2020**, *45*, S57–S102. [CrossRef] [PubMed]
49. Centers for Disease Control and Prevention of United States. Growth Chart Training. Available online: https://www.cdc.gov/nccdphp/dnpao/growthcharts/resources/sas.htm (accessed on 1 October 2021).
50. Fleary, S.A. Combined Patterns of Risk for Problem and Obesogenic Behaviors in Adolescents: A Latent Class Analysis Approach. *J. Sch. Health* **2017**, *87*, 182–193. [CrossRef] [PubMed]
51. Rao, J.; Scott, A. On Chi-squared Tests For Multiway Contigency Tables with Proportions Estimated From Survey Data. *Ann. Stat.* **1984**, *12*, 46–60. [CrossRef]

52. Thomas, D.; Rao, J. Small-sample comparison of level and power for simple goodness-of-fit statistics under cluster sampling. *JASA* **1987**, *82*, 630–636. [CrossRef]
53. Korn, E.; Graubard, B. Confidence Intervals For Proportions With Small Expected Number of Positive Counts Estimated From Survey Data. *Surv. Methodol.* **1998**, *23*, 193–201.
54. van Buuren, S.; Groothuis-Oudshoorn, K. mice: Multivariate Imputation by Chained Equations in R. *J. Stat. Softw.* **2011**, *45*, 1–67. [CrossRef]
55. VanderWeele, T.J.; Ding, P. Sensitivity Analysis in Observational Research: Introducing the E-Value. *Ann. Intern. Med.* **2017**, *167*, 268–274. [CrossRef]
56. Mathur, M.B.; Ding, P.; Riddell, C.A.; VanderWeele, T.J. Web Site and R Package for Computing E-values. *Epidemiology* **2018**, *29*, e45–e47. [CrossRef]
57. Ma, C.; Li, Z.; Tong, Y.; Zhao, M.; Magnussen, C.G.; Xi, B. Leisure sedentary time and suicide risk among young adolescents: Data from 54 low- and middle-income countries. *J. Affect. Disord.* **2022**, *298*, 457–463. [CrossRef] [PubMed]
58. Vancampfort, D.; Stubbs, B.; Mugisha, J.; Firth, J.; Van Damme, T.; Smith, L.; Koyanagi, A. Leisure-time sedentary behavior and suicide attempt among 126,392 adolescents in 43 countries. *J. Affect. Disord.* **2019**, *250*, 346–353. [CrossRef] [PubMed]
59. Coyne, S.M.; Hurst, J.L.; Dyer, W.J.; Hunt, Q.; Schvanaveldt, E.; Brown, S.; Jones, G. Suicide Risk in Emerging Adulthood: Associations with Screen Time over 10 years. *J. Youth Adolesc.* **2021**, *50*, 2324–2338. [CrossRef] [PubMed]
60. Chiu, H.Y.; Lee, H.C.; Chen, P.Y.; Lai, Y.F.; Tu, Y.K. Associations between sleep duration and suicidality in adolescents: A systematic review and dose-response meta-analysis. *Sleep Med. Rev.* **2018**, *42*, 119–126. [CrossRef]
61. Felez-Nobrega, M.; Haro, J.M.; Vancampfort, D.; Koyanagi, A. Sex difference in the association between physical activity and suicide attempts among adolescents from 48 countries: A global perspective. *J. Affect. Disord.* **2020**, *266*, 311–318. [CrossRef]
62. Kivimäki, P.; Kekkonen, V.; Valtonen, H.; Tolmunen, T.; Honkalampi, K.; Tacke, U.; Hintikka, J.; Lehto, S.M.; Laukkanen, E. Alcohol use among adolescents, aggressive behaviour, and internalizing problems. *J. Adolesc.* **2014**, *37*, 945–951. [CrossRef]
63. Borges, G.; Bagge, C.L.; Cherpitel, C.J.; Conner, K.R.; Orozco, R.; Rossow, I. A meta-analysis of acute use of alcohol and the risk of suicide attempt. *Psychol. Med.* **2017**, *47*, 949–957. [CrossRef]
64. Harrell, C.S.; Zainaldin, C.; McFarlane, D.; Hyer, M.M.; Stein, D.; Sayeed, I.; Neigh, G.N. High-fructose diet during adolescent development increases neuroinflammation and depressive-like behavior without exacerbating outcomes after stroke. *Brain Behav. Immun.* **2018**, *73*, 340–351. [CrossRef]
65. Jacques, A.; Chaaya, N.; Beecher, K.; Ali, S.A.; Belmer, A.; Bartlett, S. The impact of sugar consumption on stress driven, emotional and addictive behaviors. *Neurosci. Biobehav. Rev.* **2019**, *103*, 178–199. [CrossRef]
66. Shelton, R.C.; Miller, A.H. Inflammation in depression: Is adiposity a cause? *Dialogues Clin. Neurosci.* **2011**, *13*, 41–53. [CrossRef]
67. Brundin, L.; Bryleva, E.Y.; Thirtamara Rajamani, K. Role of Inflammation in Suicide: From Mechanisms to Treatment. *Neuropsychopharmacology* **2017**, *42*, 271–283. [CrossRef] [PubMed]
68. Liu, B.-P.; Fang, H.-J.; Jia, C.-X. The Serial Mediation of the Association between Breakfast Skipping and Suicidality by Weight Status and Depressive Symptoms: Findings from the National Youth Risk Behavior Surveys of the United States. *Nutrients* **2022**, *14*, 956. [CrossRef] [PubMed]
69. Harrell, C.S.; Burgado, J.; Kelly, S.D.; Johnson, Z.P.; Neigh, G.N. High-fructose diet during periadolescent development increases depressive-like behavior and remodels the hypothalamic transcriptome in male rats. *Psychoneuroendocrinology* **2015**, *62*, 252–264. [CrossRef] [PubMed]
70. Chakraborti, A.; Graham, C.; Chehade, S.; Vashi, B.; Umfress, A.; Kurup, P.; Vickers, B.; Chen, H.A.; Telange, R.; Berryhill, T.; et al. High Fructose Corn Syrup-Moderate Fat Diet Potentiates Anxio-Depressive Behavior and Alters Ventral Striatal Neuronal Signaling. *Front. Neurosci.* **2021**, *15*, 669410. [CrossRef]
71. Cisternas, P.; Salazar, P.; Serrano, F.G.; Montecinos-Oliva, C.; Arredondo, S.B.; Varela-Nallar, L.; Barja, S.; Vio, C.P.; Gomez-Pinilla, F.; Inestrosa, N.C. Fructose consumption reduces hippocampal synaptic plasticity underlying cognitive performance. *Biochim. Biophys. Acta* **2015**, *1852*, 2379–2390. [CrossRef]
72. de Sousa Rodrigues, M.E.; Bekhbat, M.; Houser, M.C.; Chang, J.; Walker, D.I.; Jones, D.P.; Oller do Nascimento, C.M.P.; Barnum, C.J.; Tansey, M.G. Chronic psychological stress and high-fat high-fructose diet disrupt metabolic and inflammatory gene networks in the brain, liver, and gut and promote behavioral deficits in mice. *Brain Behav. Immun.* **2017**, *59*, 158–172. [CrossRef]

nutrients

MDPI

Article

Obesity and Dyslipidemia in Chinese Adults: A Cross-Sectional Study in Shanghai, China

Junjie Zhu [1,2,†], Yue Zhang [3,†], Yiling Wu [4], Yu Xiang [1], Xin Tong [1], Yuting Yu [1], Yun Qiu [1], Shuheng Cui [1], Qi Zhao [1], Na Wang [1], Yonggen Jiang [4,*] and Genming Zhao [1,5,*]

1 Key Laboratory of Public Health Safety of Ministry of Education, Department of Epidemiology, School of Public Health, Fudan University, Shanghai 200032, China; zhujunjie233@163.com (J.Z.); xiangyuhk2016@163.com (Y.X.); danmoweiliangtx@gmail.com (X.T.); 17211020011@fudan.edu.cn (Y.Y.); qiuyun2018@fudan.edu.cn (Y.Q.); cuishuheng1995@outlook.com (S.C.); zhaoqi@shmu.edu.cn (Q.Z.); na.wang@fudan.edu.cn (N.W.)

2 Department of Epidemiology and Health Statistics, School of Public Health, Dali University, Dali 671000, China

3 Department of Epidemiology, School of Public Health, Shanxi Medical University, Taiyuan 030001, China; 17111020009@fudan.edu.cn

4 Songjiang District Center for Disease Prevention and Control, Shanghai 201600, China; aries2119@163.com

5 Shanghai Institute of Infectious Disease and Biosecurity, School of Life Science, Fudan University, Shanghai 200032, China

* Correspondence: sjjkzx1106@126.com (Y.J.); gmzhao@shmu.edu.cn (G.Z.)

† These authors contributed equally to this work.

Abstract: This study examined the association of obesity and dyslipidemia according to body measurements among Chinese adults in Shanghai, a place in the process of rapid urbanization. Using the baseline data of the Shanghai Suburban Adult Cohort and Biobank study (SSACB), the subjects completed questionnaires and physical examinations, and fasting blood was collected for biochemical assays. We estimated the odds ratios (OR) and 95% confidence interval (CI) by multivariable logistic regression. The prevalence was 12.9% and 28.8% in both general and central obesity, respectively. Compared with the non-obese, the general or central obesity participants had a higher level of TC, TG, LDL-C and lower level of HDL-C. The OR (95%CI) for dyslipidemia was 1.79 (1.69–1.91) and 1.91 (1.83–2.00) in general or central obesity, respectively. Positive associations were also observed between obesity and high TC, high LDL-C, low HDL-C and high TG, with the adjusted OR ranging from 1.11 to 2.00. Significant modifying effect of gender, age, hypertension, and diabetes were found in the association of obesity and different forms of dyslipidemia. The findings of our study indicated that participants with obesity, including general or central obesity, have a higher prevalence of dyslipidemia and gender, age, hypertension, and diabetes might be potential modifiers of the association. More effective attention and interventions should be directed to managing body weight to reduce the prevalence of dyslipidemia.

Keywords: dyslipidemia; body measurements; general obesity; central obesity; Chinese adults

Citation: Zhu, J.; Zhang, Y.; Wu, Y.; Xiang, Y.; Tong, X.; Yu, Y.; Qiu, Y.; Cui, S.; Zhao, Q.; Wang, N.; et al. Obesity and Dyslipidemia in Chinese Adults: A Cross-Sectional Study in Shanghai, China. *Nutrients* **2022**, *14*, 2321. https://doi.org/10.3390/nu14112321

Academic Editors: William B. Grant and Ronan Lordan

Received: 21 April 2022
Accepted: 29 May 2022
Published: 31 May 2022

1. Introduction

Dyslipidemia is a common metabolism abnormality involving plasma lipids and lipoproteins, categorized by elevated levels of total cholesterol (TC), triglycerides (TG) and low-density lipoprotein cholesterol (LDL-C), and/or decreased levels of high-density lipoprotein cholesterol (HDL-C). It is a dominant cause of morbidity, mortality, and one of the primary independent modifiable factors for cardiovascular diseases [1–3], diabetes [4] and stroke [5] in most countries. With rapid socioeconomic growth, improved standard of living and changes in lifestyles, dyslipidemia has been estimated to be about to rise dramatically worldwide in absolute terms [6,7]. In China, dyslipidemia has attracted much attention in recent years, and is inadequately treated and uncontrolled [8].

Globally, the number of obese people has raised seriously, and obesity has turned into one of the most vital public health threats in the last decades. The data in prevalent overweight/obesity show a rise from 13.4% to 26.4% and a rise from 18.6% to 37.4% in prevalent abdominal obesity in Chinese adults [9]. Several studies have examined the prevalence and risk factors of dyslipidemia in China [10–12]. These studies have showed that general obesity significantly influences lipid level; meanwhile, the association of obesity with lipid abnormalities depends not only on general obesity, but on central obesity [13]. Considering the increasing prevalence of dyslipidemia and its health burden, there should be a greater focus on the association of obesity with dyslipidemia. However, these studies were conducted in either urban or rural China. The Songjiang and Jiading District of Shanghai is a suburban area with rapid urbanization, and has experienced huge economic development. There has been a significant change in lifestyles in this region, such as westernized diets, sedentary work, and decreased physical activity, all of which are recognized as main risk factors for dyslipidemia [14–16]. Furthermore, little research has explored the modification of gender, age, hypertension and diabetes on the association of obesity with different types of dyslipidemia. This study aims to assess the association of obesity with dyslipidemia according to anthropometric indices, and to analyze the potential effect modification on these associations in a Shanghai natural cohort who live in a rapidly urbanizing suburban area.

2. Materials and Methods

2.1. Study Design and Population

This study was based on the Shanghai Suburban Adult Cohort and Biobank study (SSACB), which is an ongoing large-scale natural cohort study to identify risk factors for chronic noncommunicable diseases in Chinese adults. The SSACB has been described in great details previously [17]. Briefly, recruitment was conducted through multistage cluster sampling. Seven study sites were selected according to their economic status and population, including four communities from Songjiang (Xinqiao, Sheshan, Maogang, and Zhongshan) and three communities from Jiading (Huangdu. Anting, and Huating). From each community, one third of the villages or committees were randomly selected. According to participant willingness, all residents had lived in Shanghai for at least five years; those aged 20 to 74 years were included, while those with disabilities, terminal illnesses, perceptual impairments, or pregnant or lactating women were excluded. During the period April 2016 to October 2017, 44,887 participants were recruited and interviewed. Furthermore, participants who had malignant neoplasms, liver cirrhosis and thyroid diseases ($n = 3106$), extreme values of body mass index (BMI) or waist circumference (WC) ($n = 490$), unreasonable values for energy intake ($n = 721$) or physical activity ($n = 164$) were excluded. Finally, 40,406 subjects were involved in this analysis.

2.2. Physical Examination and Biochemical Assays

Anthropometric measurements (height, weight, WC) were taken two times by licensed physicians in the communities. Blood pressure was measured two times at five minute intervals using a digital sphygmomanometer, calculated by an average of the two measurements. Serum samples (2 mL) were collected into serum separation tubes, on an empty stomach and in the morning. After collection of the serum fraction, it was stored at −80 °C for no longer than 6 h in a freezer and then transported to the DiAn medical laboratory. Assays of serum TC, LDL-C, HDL-C, and TG were performed using enzyme colorimetry. Glycated hemoglobin (HbA1c) was ascertained using high performance liquid chromatography. Fasting plasma glucose (FPG) was determined using the glycokinase method.

2.3. Diagnostic Criteria

BMI was calculated as weight divided by standing height squared (kg/m^2), and more than 28 kg/m^2 was described as general obesity; central obesity is described as having a WC of equal or greater than 90 cm in males and equal or greater than 85 cm in females [18]. We excluded participants with a BMI less than 15 or more than 40 kg/m^2 or WC less than 50 or more than 150 cm from analysis [19]. The diagnostic criteria for dyslipidemia were: TC $\geq$ 6.22 mmol/L, or LDL-C $\geq$ 4.14 mmol/L, or HDL-C < 1.04 mmol/L, or TG $\geq$ 2.26 mmol/L, or a self-reported history of hyperlipidemia [20]. Hypertension was defined as a self-reported history of hypertension, or a documented history of hypertension in the medical record, or having a resting systolic blood pressure (SBP) $\geq$ 140 mmHg and/or a diastolic blood pressure (DBP) $\geq$ 90 mmHg [21]. Diabetes was defined by current ADA criteria: fasting plasma glucose (FPG) $\geq$ 7.0 mmol/L, or glycosylated hemoglobin (HbA1c) concentration $\geq$ 6.5%, or a self-reported history of diabetes, or a documented history in the medical record [22].

2.4. Assessment of Covariates

Structured questionnaires were used to collect the following variables: age, gender, marital status, education level, alcohol consumption, smoking, physical activity, and China Healthy Diet Index (CHDI). The subjects were separated into two categories: <60 and $\geq$60 years old. Marital status was recorded in two categories: married, or other (unmarried, divorced, widowed, or separated). Education level was recorded as three categories by years of schooling: $\leq$6 years, 7–12 years, and $\geq$12 years. Both smoking and alcohol drinking were separated into two categories: never or ever. Physical activity was expressed as the metabolic equivalent of task (MET)-hours/week and durations over 16 h/day were considered implausible [23]. Overall diet quality was assessed by using the CHDI established by the Chinese Center for Diseases Prevention and Control, which has been described previously in detail [24].

2.5. Statistical Analysis

Baseline characteristics of all participants were compared according to whether or not they were generally obese or centrally obese. Variables with continuous measurements were presented as mean $\pm$ standard deviation (SD) or median and interquartile range (IQR), and categorical variables as frequency (n) and proportion (%). The Kolmogorov-Smirnov test was used to determine if the data were normally distributed. Student's t test or Mann-Whitney U test were conducted to compare the differences of continuous data, and χ^2 tests for categorical data. The odds ratio (OR) and 95% confidence intervals (CI) for obesity with different types of dyslipidemia were assessed by using multivariate Logistic regression models. A variety of variables were adjusted, including gender, age, marital status, education level, physical activity, alcohol consumption, smoking, diabetes, hypertension, and CHDI. We tested the potential effect modification by adding multiplicative interaction terms in the multivariable logistic regression models and the interaction terms with $p < 0.05$ were considered statistically significant. Stratified analyses were conducted according to age (<60 and $\geq$60 years), gender, hypertension (yes, no), and diabetes (yes, no), which were potential effect modifiers for the associations. All data analyses were carried out using SAS version 9.4 (Institute Inc., Cary, NC, USA). All p-values were 2-tailed, and an alpha-level of 0.05 was considered statistically significant.

3. Results

3.1. Baseline Characteristics

A total of 40,406 participants included 16,793 males (41.6%) and 23,613 females (58.4%) in our study. The average age was 56 $\pm$ 11 years-old, which was higher for general or central obesity subjects than for non-obesity (all $p < 0.001$). The basic characteristics of participants according to general and central obesity are demonstrated in Table 1. The prevalence of general obesity and central obesity were 12.9% and 28.8%, respectively. According to BMI categories, the prevalent obesity in males was higher compared with that in females (13.8% vs. 12.2%), while 29.8% of males and 28.1% of females had central obesity. Those who were exposed to a relatively low level of education were inclined to suffer a higher prevalence of obesity. Participants with obesity had significantly higher prevalence of diabetes and hypertension, compared with the participants without obesity. The mean values of TC, TG, LDL-C, and FPG increased significantly, and the HDL-C level decreased ($p < 0.001$) in both general obesity and central obesity group.

Table 1. Basic characteristics of the participants based on general and central obesity.

Characteristics	All Subjects (n = 40,406)	BMI		p	WC		p
		Non-General Obesity (n = 35,207)	General Obesity [a] (n = 5199)		Non-Central Obesity (n = 28,761)	Central Obesity [b] (n = 11,645)	
Age (years)	56 ± 11	56 ± 11	57 ± 10	<0.001	55 ± 11	59 ± 10	<0.001
Gender				<0.001			<0.001
Male	16,793 (41.6)	14,471 (41.1)	2322 (44.7)		11,789 (41.0)	5004 (43.0)	
Female	23,613 (58.4)	20,736 (58.9)	2877 (55.3)		16,972 (59.0)	6641 (57.0)	
Education level				<0.001			<0.001
0–6 years	16,557 (41.0)	14,072 (40.0)	2485 (47.8)		10,823 (37.6)	5734 (49.2)	
7–12 years	21,017 (52.0)	18,561 (52.7)	2456 (47.2)		15,561 (54.1)	5456 (46.9)	
>12 years	2832 (7.0)	2574 (7.3)	258 (5.0)		2377 (8.3)	455 (3.9)	
Marriage				0.73			0.61
Married	37,584 (93.0)	32,754 (93.0)	4830 (92.9)		26,764 (93.1)	10,820 (92.9)	
Other	2822 (7.0)	2453 (7.0)	369 (7.1)		1997 (6.9)	825 (7.1)	
Physical activity				0.38			<0.001
Low	13,237 (32.8)	11,574 (32.9)	1663 (32.0)		9587 (33.3)	3650 (31.3)	
Moderate	13,771 (34.1)	11,994 (34.1)	1777 (34.2)		9840 (34.2)	3931 (33.8)	
High	13,398 (33.2)	11,639 (33.1)	1759 (33.8)		9334 (32.5)	4064 (34.9)	
Smoking				0.14			0.003
Ever	10,082 (25.0)	8742 (24.8)	1340 (25.8)		7059 (24.5)	3023 (26)	
Never	30,324 (75.1)	26,465 (75.2)	3859 (74.2)		21,702 (75.5)	8622 (74)	
Alcohol drinking				0.002			<0.001
Ever	5563 (13.8)	4774 (13.6)	789 (15.2)		3758 (13.1)	1805 (15.5)	
Never	34,843 (86.2)	30,433 (86.4)	4410 (84.8)		25,003 (86.9)	9840 (84.5)	
Diabetes				<0.001			<0.001
Yes	5738 (14.2)	4420 (12.6)	1318 (25.4)		3154 (11.0)	2584 (22.2)	
No	34,668 (85.8)	30,787 (87.5)	3881 (74.7)		25,607 (89)	9061 (77.8)	
Hypertension				<0.001			<0.001
Yes	19,696 (48.8)	15,948 (45.3)	3748 (72.1)		11,227 (45.0)	7102 (70.2)	
No	20,710 (51.3)	19,259 (54.7)	1451 (27.9)		16,769 (58.3)	3941 (33.8)	
CHDI	70.05 ± 9.29	70.2 ± 9.31	69.04 ± 9.13	<0.001	70.51 ± 9.28	68.91 ± 9.22	<0.001
HBA1c (%)	4.9 (4.4, 5.5)	4.8 (4.4, 5.4)	5.1 (4.5, 5.9)	<0.001	4.8 (4.4, 5.4)	5.0 (4.4, 5.8)	<0.001
FPG (mmol/L)	5.81 ± 0.85	5.77 ± 0.81	6.09 ± 1.00	<0.001	5.72 ± 0.78	6.03 ± 0.97	<0.001
TC (mmol/L)	4.91 ± 0.94	4.90 ± 0.93	4.99 ± 0.99	<0.001	4.88 ± 0.92	4.99 ± 0.97	<0.001
TG (mmol/L)	1.70 ± 1.26	1.64 ± 1.2	2.11 ± 1.55	<0.001	1.56 ± 1.12	2.03 ± 1.49	<0.001
HDL-C(mmol/L)	1.4 ± 0.35	1.42 ± 0.35	1.26 ± 0.31	<0.001	1.44 ± 0.35	1.29 ± 0.32	<0.001
LDL-C (mmol/L)	2.74 ± 0.83	2.74 ± 0.83	2.78 ± 0.88	0.004	2.73 ± 0.82	2.78 ± 0.87	<0.001

[a] General obesity defined as BMI $\geq$ 28.0 kg/m^2, [b] central obesity defined as WC $\geq$ 90 cm in males and $\geq$85 cm in females.

3.2. Prevalence of Different Forms of Dyslipidemia

The prevalence of different forms of dyslipidemia is shown by BMI and WC categories for all subjects in Figure 1. Compared with non-general obesity, the participants with general obesity had a significantly higher prevalence of different types of dyslipidemia (all $p < 0.001$). The prevalence of high TC, high TG, high LDL-C, low HDL-C, and dyslipidemia were 9.8%, 30.0%, 6.0%, 25.0%, and 52.7% among the participants with general obesity, respectively. The participants with central obesity had a similar higher prevalence of different types of dyslipidemia than those without central obesity (all $p < 0.001$).

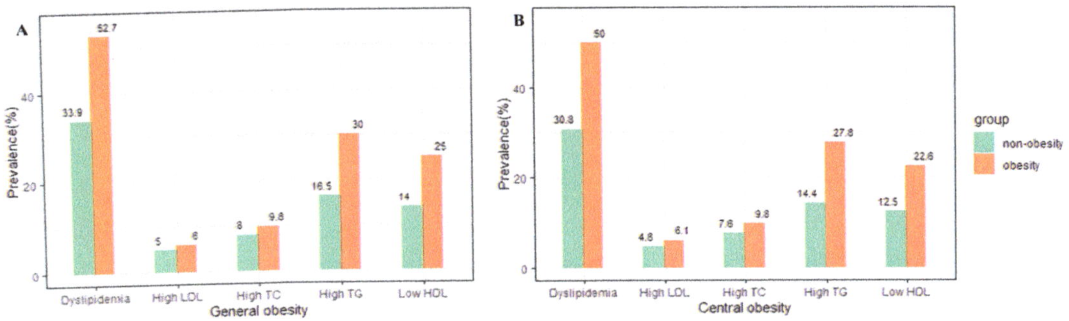

Figure 1. The prevalence of different forms of dyslipidemia categorized by general obesity (**A**) and central obesity (**B**) for all subjects.

3.3. *Association of Obesity with Different Forms of Dyslipidemia*

Figure 2 shows the associations of general and central obesity with different types of dyslipidemia. Subjects with obesity either by BMI (OR = 1.79, 95% CI: 1.69–1.91) or by WC (OR = 1.91, 95% CI: 1.83–2.00) had higher risk of dyslipidemia than those without obesity in the multivariable adjusted models. According to BMI categories, general obesity was associated with a 11%, 9%, 78%, and 79% increased risk of high TG, high TC, high LDL-C, and low HDL-C, respectively. According to WC categories, the adjusted OR for high TG, high TC, high LDL-C, and low HDL-C were 2.00 (95% CI: 1.90, 2.12), 1.13 (95% CI: 1.04, 1.22), 1.15 (95% CI: 1.05, 1.27), and 1.92 (95% CI: 1.81, 2.04), respectively, while the OR with high TG for central obesity was greatest.

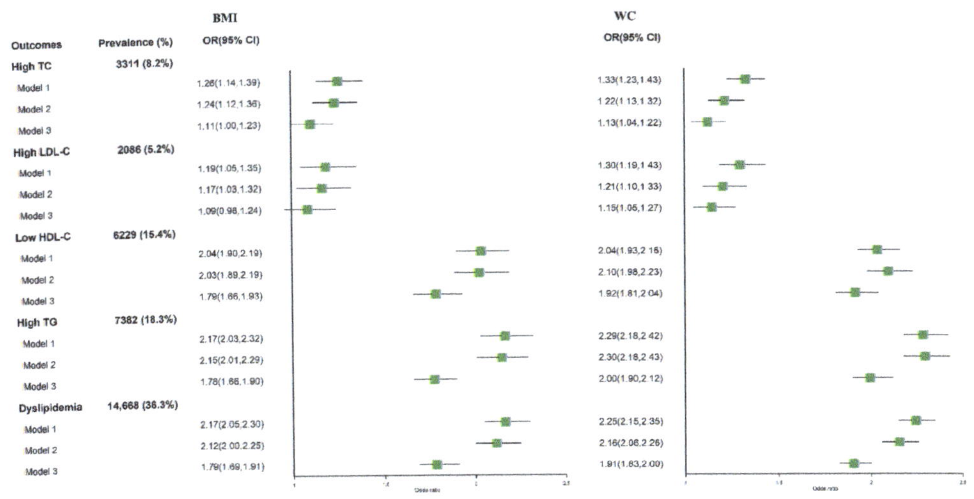

Figure 2. Odds ratios (OR) and 95% confidence intervals (CI) for BMI and WC categories with dyslipidemia. Model 1: unadjusted; model 2: adjusted for gender and age; model 3: additionally adjusted for education level, marriage, physical activity, smoking, alcohol drinking, CHDI, diabetes, and hypertension.

3.4. *Stratified Analysis*

Stratification by age, gender, hypertension and diabetes suggested that these factors might be potential modifiers on the association between obesity and different types of dyslipidemia (p for interaction < 0.05), with few exceptions (Tables 2 and 3). The associations between general or central obesity and different types of dyslipidemia were statistically significant among the individuals younger than 60 years old, with the OR ranging from 1.23 to 2.13. Compared with 60 years or older, general or central obesity among individuals younger than 60 years old may be associated with a greater risk of different types of dyslipidemia. Males, either with general obesity or central obesity, had significantly increased risk of high TG, TC, LDL-C, low HDL-C, and dyslipidemia, and the OR ranged from 1.28 to 2.25. However, the statistically significant associations were observed only for high TG, low HDL-C and dyslipidemia among females, and the OR is lower than these for males in all types of dyslipidemia. An interactive effect of hypertension and general or central obesity on low HDL-C, high TG and dyslipidemia was observed, but not in high TC and LDL-C (p for interaction > 0.05). Among participants without hypertension, the stronger associations were found between general obesity and low HDL-C, high TG, and dyslipidemia, with OR being 2.21, 2.16 and 2.10, respectively. Similar associations were also observed between central obesity and high TG, low HDL-C, and dyslipidemia. With subgroup analysis by diabetes, no effect modification was observed on the associations between obesity and high TG, high TC, and high LDL-C. Ann interaction of diabetes and general obesity on low HDL-C only was shown (p for interaction = 0.007), while the interaction of diabetes and central obesity on high TG, dyslipidemia was found (p for interaction < 0.05).

Table 2. Stratified analysis of general obesity with dyslipidemia.

Variables	High TC	High LDL-C	Low HDL-C	High TG	Dyslipidemia
Age (years)					
<60	1.26 (1.09, 1.45)	1.28 (1.07, 1.52)	1.87 (1.68, 2.07)	1.81 (1.65, 1.99)	1.88 (1.72, 2.05)
≥60	0.97 (0.84, 1.13)	0.89 (0.74, 1.07)	1.66 (1.49, 1.84)	1.65 (1.49, 1.82)	1.60 (1.47, 1.75)
p for Interaction	<0.001	<0.001	0.14	0.001	<0.001
Gender					
Male	1.53 (1.29, 1.81)	1.41 (1.14, 1.74)	1.94 (1.76, 2.14)	2.05 (1.85, 2.26)	2.24 (2.04, 2.46)
Female	0.90 (0.79, 1.02)	0.92 (0.78, 1.08)	1.48 (1.31, 1.67)	1.45 (1.31, 1.59)	1.40 (1.29, 1.53)
p for Interaction	<0.001	<0.001	0.01	<0.001	<0.001
Hypertension					
Yes	1.05 (0.94, 1.19)	1.05 (0.90, 1.21)	1.63 (1.49, 1.78)	1.63 (1.50, 1.76)	1.65 (1.53, 1.77)
No	1.25 (1.02, 1.53)	1.19 (0.93, 1.53)	2.21 (1.93, 2.54)	2.16 (1.90, 2.46)	2.10 (1.88, 2.34)
p for Interaction	0.15	0.40	<0.001	<0.001	0.001
Diabetes					
Yes	1.03 (0.85, 1.26)	1.07 (0.83, 1.37)	1.50 (1.30, 1.74)	1.62 (1.42, 1.85)	1.63 (1.43, 1.86)
No	1.13 (1.01, 1.28)	1.10 (0.94, 1.27)	1.89 (1.74, 2.06)	1.83 (1.69, 1.98)	1.83 (1.71, 1.96)
p for Interaction	0.46	0.84	0.007	0.06	0.17

Adjusted for gender, age, education level, marriage, physical activity, smoking, alcohol drinking, CHDI, diabetes, and hypertension, except for a stratifying variable.

Table 3. Stratified analysis of central obesity with dyslipidemia.

Variables	High TC	High LDL-C	Low HDL-C	High TG	Dyslipidemia
Age (years)					
<60	1.23 (1.10, 1.38)	1.37 (1.19, 1.58)	2.04 (1.87, 2.21)	2.13 (1.98, 2.30)	2.10 (1.96, 2.24)
≥60	1.02 (0.91, 1.13)	0.97 (0.85, 1.11)	1.68 (1.55, 1.83)	1.70 (1.57, 1.85)	1.64 (1.53, 1.75)
p for Interaction	<0.001	<0.001	0.008	<0.001	<0.001
Gender					
Male	1.31 (1.14, 1.51)	1.28 (1.08, 1.52)	2.03 (1.88, 2.19)	2.22 (2.05, 2.40)	2.25 (2.10, 2.42)
Female	0.99 (0.90, 1.09)	1.04 (0.93, 1.17)	1.60 (1.45, 1.76)	1.64 (1.51, 1.76)	1.52 (1.43, 1.62)
p for Interaction	0.003	0.09	0.15	0.04	<0.001
Hypertension					
Yes	1.11 (1.01, 1.22)	1.10 (0.98, 1.25)	1.75 (1.63, 1.89)	1.79 (1.67, 1.92)	1.75 (1.65, 1.85)
No	1.17 (1.02, 1.33)	1.24(1.06, 1.46)	2.20 (2.00, 2.42)	2.41 (2.20, 2.63)	2.18 (2.02, 2.35)
p for Interaction	0.11	0.09	<0.001	<0.001	<0.001
Diabetes					
Yes	1.08 (0.91, 1.28)	1.09 (0.88, 1.36)	1.76 (1.54, 2.00)	1.73 (1.54, 1.95)	1.74 (1.56, 1.94)
No	1.13 (1.04, 1.24)	1.16 (1.04, 1.30)	1.96 (1.83, 2.09)	2.07 (1.94, 2.20)	1.94 (1.85, 2.04)
p for Interaction	0.34	0.47	0.12	0.001	0.048

Adjusted for gender, age, education level, marriage, physical activity, smoking, alcohol drinking, CHDI, diabetes, and hypertension, except for a stratifying variable.

4. Discussion

This study aimed to comprehensively examine the positive associations of obesity and TC, TG and LDL-C levels, as well as the negative associations of obesity with HDL-C levels in a suburban area experiencing rapid urbanization of China. Though previous studies have been conducted on obesity or serum lipid, few studies have explored the association between obesity and various types of dyslipidemia concurrently in an area that has rapidly urbanized in China. Our results indicated that the prevalence of general obesity and central obesity were 13.5% and 28.9%, which was in line with previous large-scale surveys among Chinese [25,26]. Dyslipidemia was prevalent in most obese subjects and revealed nearly half of general obesity participants suffered dyslipidemia and higher values of LDL-C, TG, TC, and lower HDL-C than in normal-weight individuals [27]. Several previous regional epidemiological studies reported the prevalent dyslipidemia in obesity subjects differently [28,29]. The possible reasons for this discrepancy might be the socio-demographic characteristics of the subjects and the diagnostic criteria used [30]. Moreover, the differences in dietary pattern may also play a role in regional differences in the prevalence of dyslipidemia [31].

A recent study found that the incidence of deaths and disability-adjusted life years (DALYs) attributable to obesity has increased significantly [32]. Several studies have demonstrated that overweight and obesity are cardiometabolic risk factors [33–35]. Data from 97 prospective cohorts with 1.8 million participants indicate that obesity is associated with 31% coronary heart disease risk and 8% stroke mortality risk, due to elevated blood pressure and cholesterol together [36]. Therefore, effective control of lipid level can be expected to attenuate death from metabolic diseases. This study found that general or central obesity were associated with higher prevalence of dyslipidemia compared with non-obesity. The major types of dyslipidemia among obesity subjects are low HDL-C and high TG, which is consistent with other research and probably due to the elevated TG and Apo lipoprotein B from excess visceral fat in the abdomen and a low HDL-C production from inhibition of the liver [13,37]. A strong association has been identified between central obesity and metabolic risk factors, cardiovascular events and dyslipidemia [38]. As we know, higher levels of BMI and WC correlate with increased prevalence of abnormal lipids, depending on gender and age. It is clear that a higher BMI and WC contribute to the development of these metabolic diseases.

Our study suggested that subjects with obesity either by BMI or WC had higher risk of dyslipidemia than those without obesity, which were comparable to those of previous studies [39,40]. The adjusted multivariate logistic regression shows that the effect is the highest between general or central obesity and low HDL-C; in addition, the ORs between the WC and dyslipidemia are higher than that of BMI and dyslipidemia, indicating that WC has a greater influence on lipid level than BMI. Central obesity characterized by the accumulation of visceral fat in the abdomen is more closely associated with a global metabolic effect of insulin resistance than general obesity [41,42]. Insulin resistance might change the amount and composition of lipoprotein to cause abnormal blood lipid levels, which act on the metabolism of LDL, chylomicron, HDL, and very-low-density lipoprotein (VLDL) [43].

The modifying effects of gender, age, hypertension, and diabetes on the association between obesity and dyslipidemia were further observed. Compared with females, significantly higher risk of different forms of dyslipidemia was observed in obese males. A possible reason for this is the intensity of work pressure, unhealthy lifestyle and diets for males, which account for excessive fat accumulation [44]. The proportion of smoking and drinking among males is higher than that of females, which are risk factors of dyslipidemia [45]. Obese subjects younger than 60 years were at greater risk of different forms of dyslipidemia. Therefore, more targeted prevention should be formulated for residents, adapting to different genders and age groups. Hypertension exerted an effect modification on the association of obesity with low HDL-C, high TG and dyslipidemia. Non-hypertensive obese individuals may be more prone to low HDL-C, high TG and dyslipidemia. Similarly, diabetes exerted an effect modification on the association of general obesity with low HDL-C and the association of central obesity with high TG, dyslipidemia. However, the causality among obesity, hypertension, diabetes, and dyslipidemia remains association these may exert a comprehensive effect on each other. A previous study found that obesity can be caused by dyslipidemia and subjects with dyslipidemia are more likely to experience hypertension [46]. Other studies showed that weight gain, elevated blood pressure and blood glucose may be essential in incident dyslipidemia [47,48]. In short, we propose that effectively management of BMI or WC would be helpful in preventing and controlling different forms of dyslipidemia.

In this study, the main strengths included a large sample size and stratified analysis on the association between obesity and different types of dyslipidemia. We additionally considered the possible confounding effect of dietary quality on the associations and adjusted the CHDI in the final model. However, a few limitations should be taken into consideration. Firstly, the study is a cross-sectional study in Shanghai, which cannot provide causal relationships and may not be generalizable to different geographical regions. Thus, further prospective studies are necessary to verify the relationship of different indices of adiposity with dyslipidemia. Second, potential residual confounding factors (such as stress, sleep pattern, genetic factors) associated with dyslipidemia could not be considered, which seem to bias the results. In addition, the possible confounding factors of the lean and fat mass percentages or ratios that are directly connected to the LDL and TG levels were not considered because of limitations in budget and equipment. Nevertheless, BMI and WC are the most readily available indicators to estimate obesity. We did not account for the impact of these indicators (blood glucose, insulin levels, adiponectin, leptins) on dyslipidemia and may underestimate the prevalence of dyslipidemia. However, diabetes diagnosed based on FPG was adjusted in the models. Finally, health-related behaviors were reported by the participants themselves and we could not rule out subjects' error, which may lead to a reporting bias.

5. Conclusions

In conclusion, our study found that general or central obesity have a higher prevalence of different forms of dyslipidemia, and the major types were low HDL-C and high TG among adults in Shanghai. Central obesity may have a greater effect on lipid level than general obesity. Factors including age, gender, hypertension, and diabetes might be potential modifiers of the association. A significantly higher OR of various types of dyslipidemia was observed in obese younger than 60 years, males, without hypertension and diabetes. Effective management of obesity should be implemented to prevent and control the occurrence of dyslipidemia.

Author Contributions: J.Z. and Y.Z. designed the study, analyzed the data, and wrote the manuscript. Y.W., Y.X., X.T., Y.Y., Y.Q. and S.C. contributed to the data analysis and data acquisition. Q.Z., N.W. and Y.J. contributed to data acquisition. G.Z. and Y.J. supervised the study and revised the manuscript. All authors have read and agreed to the published version of the manuscript.

Funding: This work was supported by the National Key Research and Development Program of China (2017YFC0907000), the High Local Level Discipline Construction Project of Shanghai, the Shanghai New Three-year Action Plan for Public Health (Grant No. GWV-10.1-XK16).

Institutional Review Board Statement: The study protocol was approved by the Ethics Committee of the Fudan University, School of Public Health (IRB#2016-04-0586), and complied with the principles of the Declaration of Helsinki.

Informed Consent Statement: Informed written consents were obtained from all participants before-data collection.

Data Availability Statement: The dataset used and analyzed during the current study is available from the corresponding author upon reasonable request.

Acknowledgments: We thank all the members of the survey teams and the participants in the Shanghai Suburban Adult Cohort and Biobank (SSACB) study.

Conflicts of Interest: The authors declare that they have no conflict of interest.

Abbreviations

SSACB	The Shanghai Suburban Adult Cohort and Biobank study
OR	Odds ratios
CI	Confidence interval
TC	Total cholesterol
TG	Triglycerides
LDL-C	Low-density lipoprotein cholesterol
HDL-C	High-density lipoprotein cholesterol
WC	Waist circumference
FPG	Fasting plasma glucose
HbA1c	Glycated hemoglobin
BMI	Body mass index
SBP	Systolic blood pressure
DBP	Diastolic blood pressure
ADA	American Diabetes Association
CHDI	China Healthy Diet Index
MET	Metabolic equivalent of task
SD	Standard deviation
IQR	Interquartile range
DALYs	Disability-adjusted life years

References

1. Nagasawa, S.Y.; Okamura, T.; Iso, H.; Tamakoshi, A.; Yamada, M.; Watanabe, M.; Murakami, Y.; Miura, K.; Ueshima, H.; Evidence for Cardiovascular Prevention from Observational Cohorts in Japan Research Group. Relation between serum total cholesterol level and cardiovascular disease stratified by sex and age group: A pooled analysis of 65 594 individuals from 10 cohort studies in Japan. *J. Am. Heart Assoc.* **2012**, *1*, e001974. [CrossRef]
2. Di Angelantonio, E.; Gao, P.; Pennells, L.; Kaptoge, S.; Caslake, M.; Thompson, A.; Butterworth, A.S.; Sarwar, N.; Wormser, D.; Saleheen, D.; et al. Lipid-Related Markers and Cardiovascular Disease Prediction. *JAMA-J. Am. Med. Assoc.* **2012**, *307*, 2499–2506. [CrossRef]
3. Turgeon, R.D.; Anderson, T.J.; Gregoire, J.; Pearson, G.J. 2016 Guidelines for the management of dyslipidemia and the prevention of cardiovascular disease in adults by pharmacists. *Can. Pharm. J.* **2017**, *150*, 243–250. [CrossRef] [PubMed]
4. Katulanda, P.; Dissanayake, H.A.; De Silva, S.D.N.; Katulanda, G.W.; Liyanage, I.K.; Constantine, G.R.; Sheriff, R.; Matthews, D.R. Prevalence, patterns, and associations of dyslipidemia among Sri Lankan adults-Sri Lanka Diabetes and Cardiovascular Study in 2005–2006. *J. Clin. Lipidol.* **2018**, *12*, 447–454. [CrossRef]
5. Lee, J.S.; Chang, P.Y.; Zhang, Y.; Kizer, J.R.; Best, L.G.; Howard, B.V. Triglyceride and HDL-C Dyslipidemia and Risks of Coronary Heart Disease and Ischemic Stroke by Glycemic Dysregulation Status: The Strong Heart Study. *Diabetes Care* **2017**, *40*, 529–537. [CrossRef] [PubMed]
6. Bayram, F.; Kocer, D.; Gundogan, K.; Kaya, A.; Demir, O.; Coskun, R.; Sabuncu, T.; Karaman, A.; Cesur, M.; Rizzo, M.; et al. Prevalence of dyslipidemia and associated risk factors in Turkish adults. *J. Clin. Lipidol.* **2014**, *8*, 206–216. [CrossRef] [PubMed]
7. Latifi, S.M.; Moradi, L.; Shahbazian, H.; Aleali, A.M. A study of the prevalence of dyslipidemia among the adult population of Ahvaz, Iran. *Diabetes Metab. Synd.* **2016**, *10*, 190–193. [CrossRef]
8. Lu, Y.; Zhang, H.; Lu, J.; Ding, Q.; Li, X.; Wang, X.; Sun, D.; Tan, L.; Mu, L.; Liu, J.; et al. Prevalence of Dyslipidemia and Availability of Lipid-Lowering Medications Among Primary Health Care Settings in China. *JAMA Netw. Open* **2021**, *4*, e2127573. [CrossRef] [PubMed]
9. China Health and Nutrition Survey (CHNS) [DB/OL]. Available online: http://www.cpc.unc.edu/projects/china/data/datasets (accessed on 26 May 2022).
10. Opoku, S.; Gan, Y.; Fu, W.; Chen, D.; Addo-Yobo, E.; Trofimovitch, D.; Yue, W.; Yan, F.; Wang, Z.; Lu, Z. Prevalence and risk factors for dyslipidemia among adults in rural and urban China: Findings from the China National Stroke Screening and prevention project (CNSSPP). *BMC Public Health* **2019**, *19*, 1500. [CrossRef]
11. Huang, C.; Zhang, W.Q.; Tang, W.W.; Liu, Y.; Liu, J.X.; Xu, R.H.; Zhao, S.P.; Wang, T.D.; Huang, X.B. Prevalence and related factors of dyslipidemia among urban adults aged 35 to 79 years in Southwestern China. *Sci. Rep.* **2021**, *11*, 17579. [CrossRef]

12. Xi, Y.; Niu, L.; Cao, N.; Bao, H.; Xu, X.; Zhu, H.; Yan, T.; Zhang, N.; Qiao, L.; Han, K.; et al. Prevalence of dyslipidemia and associated risk factors among adults aged ≥35 years in northern China: A cross-sectional study. *BMC Public Health* **2020**, *20*, 1068. [CrossRef]
13. Vekic, J.; Zeljkovic, A.; Stefanovic, A.; Jelic-Ivanovic, Z.; Spasojevic-Kalimanovska, V. Obesity and dyslipidemia. *Metabolism* **2019**, *92*, 71–81. [CrossRef] [PubMed]
14. Sonestedt, E.; Hellstrand, S.; Drake, I.; Schulz, C.A.; Ericson, U.; Hlebowicz, J.; Persson, M.M.; Gullberg, B.; Hedblad, B.; Engstrom, G.; et al. Diet Quality and Change in Blood Lipids during 16 Years of Follow-up and Their Interaction with Genetic Risk for Dyslipidemia. *Nutrients* **2016**, *8*, 274. [CrossRef] [PubMed]
15. Wu, J.Y.; Duan, X.Y.; Li, L.; Dai, F.; Li, Y.Y.; Li, X.J.; Fan, J.G. Dyslipidemia in Shanghai, China. *Prev. Med.* **2010**, *51*, 412–415. [CrossRef] [PubMed]
16. Zheng, W.; Chen, Y.; Zhao, A.; Xue, Y.; Zheng, Y.; Mu, Z.; Wang, P.; Zhang, Y. Associations of sedentary behavior and physical activity with physical measurements and dyslipidemia in school-age children: A cross-sectional study. *BMC Public Health* **2016**, *16*, 1186. [CrossRef] [PubMed]
17. Zhao, Q.; Chen, B.; Wang, R.; Zhu, M.; Shao, Y.; Wang, N.; Liu, X.; Zhang, T.; Jiang, F.; Wang, W.; et al. Cohort profile: Protocol and baseline survey for the Shanghai Suburban Adult Cohort and Biobank (SSACB) study. *BMJ Open* **2020**, *10*, e035430. [CrossRef] [PubMed]
18. Chen, C.; Lu, F.C. The guidelines for prevention and control of overweight and obesity in Chinese adults. *Biomed. Environ. Sci.* **2004**, *17*, 1–36.
19. Li, J.C.; Lyu, J.; Gao, M.; Yu, C.Q.; Guo, Y.; Bian, Z.; Pei, P.; Du, H.D.; Chen, J.S.; Chen, Z.M.; et al. Association of body mass index and waist circumference with major chronic diseases in Chinese adults. *Zhonghua Liu Xing Bing Xue Za Zhi* **2019**, *40*, 1541–1547. [CrossRef]
20. Joint Committee for Developing Chinese guidelines on Prevention and Treatment of Dyslipidemia in Adults. Guidelines for the prevention and treatment of dyslipidemia in Chinese adults (2016 revision). *Chin. Circ. J.* **2016**, *31*, 937–953.
21. Chobanian, A.V.; Bakris, G.L.; Black, H.R.; Cushman, W.C.; Green, L.A.; Izzo, J.L., Jr.; Jones, D.W.; Materson, B.J.; Oparil, S.; Wright, J.T., Jr.; et al. The Seventh Report of the Joint National Committee on Prevention, Detection, Evaluation, and Treatment of High Blood Pressure: The JNC 7 report. *JAMA* **2003**, *289*, 2560–2572. [CrossRef]
22. American Diabetes Association. 2. Classification and Diagnosis of Diabetes: Standards of Medical Care in Diabetes—2019. *Diabetes Care* **2019**, *42*, S13–S28. [CrossRef] [PubMed]
23. Craig, C.L.; Marshall, A.L.; Sjöström, M.; Bauman, A.E.; Booth, M.L.; Ainsworth, B.E.; Pratt, M.; Ekelund, U.; Yngve, A.; Sallis, J.F.; et al. International physical activity questionnaire: 12-country reliability and validity. *Med. Sci. Sports Exerc.* **2003**, *35*, 1381–1395. [CrossRef] [PubMed]
24. Yu-na, H.; Yue-hui, F.; Xiao-guang, Y.; Gang-qiang, D. Establishment and application of china healthy diet index. *Acta Nutr. Sin.* **2017**, *39*, 436–441.
25. Chooi, Y.C.; Ding, C.; Magkos, F. The epidemiology of obesity. *Metabolism* **2019**, *92*, 6–10. [CrossRef]
26. Jaacks, L.M.; Vandevijvere, S.; Pan, A.; McGowan, C.J.; Wallace, C.; Imamura, F.; Mozaffarian, D.; Swinburn, B.; Ezzati, M. The obesity transition: Stages of the global epidemic. *Lancet Diabetes Endocrinol.* **2019**, *7*, 231–240. [CrossRef]
27. Zhang, M.; Deng, Q.; Wang, L.; Huang, Z.; Zhou, M.; Li, Y.; Zhao, Z.; Zhang, Y.; Wang, L. Prevalence of dyslipidemia and achievement of low-density lipoprotein cholesterol targets in Chinese adults: A nationally representative survey of 163,641 adults. *Int. J. Cardiol.* **2018**, *260*, 196–203. [CrossRef]
28. Kuwabara, M.; Kuwabara, R.; Niwa, K.; Hisatome, I.; Smits, G.; Roncal-Jimenez, C.A.; MacLean, P.S.; Yracheta, J.M.; Ohno, M.; Lanaspa, M.A.; et al. Different Risk for Hypertension, Diabetes, Dyslipidemia, and Hyperuricemia According to Level of Body Mass Index in Japanese and American Subjects. *Nutrients* **2018**, *10*, 1011. [CrossRef]
29. Pan, L.; Yang, Z.; Wu, Y.; Yin, R.X.; Liao, Y.; Wang, J.; Gao, B.; Zhang, L. The prevalence, awareness, treatment and control of dyslipidemia among adults in China. *Atherosclerosis* **2016**, *248*, 2–9. [CrossRef]
30. Camacho, P.A.; Otero, J.; Pérez, M.; Arcos, E.; García, H.; Narvaez, C.; Molina, D.I.; Sanchez, G.; Duran, M.; Cure, C.; et al. The spectrum of the dyslipidemia in Colombia: The PURE study. *Int. J. Cardiol.* **2019**, *284*, 111–117. [CrossRef]
31. Poggio, R.; Elorriaga, N.; Gutierrez, L.; Irazola, V.; Rubinstein, A.; Danaei, G. Associations between dietary patterns and serum lipids, apo and C-reactive protein in an adult population: Evidence from a multi-city cohort in South America. *Br. J. Nutr.* **2017**, *117*, 548–555. [CrossRef]
32. Dai, H.; Alsalhe, T.A.; Chalghaf, N.; Riccò, M.; Bragazzi, N.L.; Wu, J. The global burden of disease attributable to high body mass index in 195 countries and territories, 1990-2017: An analysis of the Global Burden of Disease Study. *PLoS Med.* **2020**, *17*, e1003198. [CrossRef]
33. Koliaki, C.; Liatis, S.; Kokkinos, A. Obesity and cardiovascular disease: Revisiting an old relationship. *Metabolism* **2019**, *92*, 98–107. [CrossRef] [PubMed]
34. Pagidipati, N.J.; Zheng, Y.; Green, J.B.; McGuire, D.K.; Mentz, R.J.; Shah, S.; Aschner, P.; Delibasi, T.; Rodbard, H.W.; Westerhout, C.M.; et al. Association of obesity with cardiovascular outcomes in patients with type 2 diabetes and cardiovascular disease: Insights from TECOS. *Am. Heart J.* **2020**, *219*, 47–57. [CrossRef] [PubMed]
35. Teixeira, F.D.C.; Pereira, F.E.F.; Pereira, A.F.; Ribeiro, B.G. Overweight or obesity and abdominal obesity and their association with cardiometabolic risk factors in Brazilian schoolchildren: A cross-sectional study. *Nutrition* **2020**, *78*, 110780. [CrossRef] [PubMed]

36. Global Burden of Metabolic Risk Factors for Chronic Diseases Collaboration (BMI Mediated Effects); Lu, Y.; Hajifathalian, K.; Ezzati, M.; Woodward, M.; Rimm, E.B.; Danaei, G. Metabolic Mediators of the Effects of Body-Mass Index, Overweight, and Obesity on Coronary Heart Disease and Stroke: A Pooled Analysis of 97 Prospective Cohorts with 1.8 Million Participants. *Lancet* **2014**, *383*, 970–983. [CrossRef] [PubMed]
37. Zhang, T.; Chen, J.; Tang, X.; Luo, Q.; Xu, D.; Yu, B. Interaction between adipocytes and high-density lipoprotein:new insights into the mechanism of obesity-induced dyslipidemia and atherosclerosis. *Lipids Health Dis.* **2019**, *18*, 223. [CrossRef] [PubMed]
38. Sangrós, F.J.; Torrecilla, J.; Giráldez-García, C.; Carrillo, L.; Mancera, J.; Mur, T.; Franch, J.; Díez, J.; Goday, A.; Serrano, R.; et al. Association of General and Abdominal Obesity With Hypertension, Dyslipidemia and Prediabetes in the PREDAPS Study. *Rev. Esp. Cardiol. (Engl. Ed.)* **2018**, *71*, 170–177. [CrossRef]
39. Shen, Z.; Munker, S.; Wang, C.Y.; Xu, L.; Ye, H.; Chen, H.T.; Xu, G.Y.; Zhang, H.; Chen, L.H.; Yu, C.H.; et al. Association between alcohol intake, overweight, and serum lipid Levels and the risk analysis associated with the development of dyslipidemia. *J. Clin. Lipidol.* **2014**, *8*, 273–278. [CrossRef]
40. Mirrakhimov, E.M.; Kerimkulova, A.S.; Lunegova, O.S.; Mirrakhimov, A.E.; Nabiev, M.P.; Neronova, K.V.; Bayramukova, A.A.; Alibaeva, N.T.; Satarov, N. The association of leptin with dyslipidemia, arterial hypertension and obesity in Kyrgyz (Central Asian nation) population. *BMC Res. Notes* **2014**, *7*, 411. [CrossRef]
41. Wu, H.; Ballantyne, C.M. Metabolic Inflammation and Insulin Resistance in Obesity. *Circ. Res.* **2020**, *126*, 1549–1564. [CrossRef]
42. Li, D.D.; Yang, Y.; Gao, Z.Y.; Zhao, L.H.; Yang, X.; Xu, F.; Yu, C.; Zhang, X.L.; Wang, X.Q.; Wang, L.H.; et al. Sedentary lifestyle and body composition in type 2 diabetes. *Diabetol. Metab. Syndr.* **2022**, *14*, 8. [CrossRef] [PubMed]
43. Muñoz Contreras, A.M.; Bedoya Berrío, G.; Velásquez, R.C. An approach to the etiology of metabolic syndrome. *Colomb. Med.* **2013**, *44*, 57–63. [CrossRef]
44. Lin, L.Y.; Hsu, C.Y.; Lee, H.A.; Tinkov, A.A.; Skalny, A.V.; Wang, W.H.; Chao, J.C. Gender difference in the association of dietary patterns and metabolic parameters with obesity in young and middle-aged adults with dyslipidemia and abnormal fasting plasma glucose in Taiwan. *Nutr. J.* **2019**, *18*, 75. [CrossRef] [PubMed]
45. Qiu, L.; Wang, W.; Sa, R.; Liu, F. Prevalence and Risk Factors of Hypertension, Diabetes, and Dyslipidemia among Adults in Northwest China. *Int. J. Hypertens.* **2021**, *2021*, 5528007. [CrossRef]
46. Jia, X.; Chen, Q.; Wu, P.; Liu, M.; Chen, X.; Xiao, J.; Chen, L.; Zhang, P.; Wang, S. Dynamic development of metabolic syndrome and its risk prediction in Chinese population: A longitudinal study using Markov model. *Diabetol. Metab. Syndr.* **2018**, *10*, 24. [CrossRef]
47. Erem, C.; Hacihasanoglu, A.; Deger, O.; Kocak, M.; Topbas, M. Prevalence of dyslipidemia and associated risk factors among Turkish adults: Trabzon lipid study. *Endocrine* **2008**, *34*, 36–51. [CrossRef]
48. Wang, S.; Xu, L.; Jonas, J.B.; You, Q.S.; Wang, Y.X.; Yang, H. Prevalence and associated factors of dyslipidemia in the adult Chinese population. *PLoS ONE* **2011**, *6*, e17326. [CrossRef]

nutrients

MDPI

Article

An Exploration of How Solar Radiation Affects the Seasonal Variation of Human Mortality Rates and the Seasonal Variation in Some Other Common Disorders

William B. Grant [1,*] and Barbara J. Boucher [2]

1 Sunlight, Nutrition, and Health Research Center, P.O. Box 641603, San Francisco, CA 94164-1603, USA
2 The Blizard Institute, Barts and The London School of Medicine and Dentistry, Queen Mary University of London, London E1 2AT, UK; bboucher@doctors.org.uk
* Correspondence: wbgrant@infionline.net; Tel.: +1-415-409-1980

Abstract: Many diseases have large seasonal variations in which winter overall mortality rates are about 25% higher than in summer in mid-latitude countries, with cardiovascular diseases and respiratory infections and conditions accounting for most of the variation. Cancers, by contrast, do not usually have pronounced seasonal variations in incidence or mortality rates. This narrative review examines the epidemiological evidence for seasonal variations in blood pressure, cardiovascular disease rates and respiratory viral infections in relation to atmospheric temperature and humidity, and solar UV exposure through vitamin D production and increased blood concentrations of nitric oxide. However, additional mechanisms most likely exist by which solar radiation reduces the risk of seasonally varying diseases. Some studies have been reported with respect to temperature without considering solar UV doses, although studies regarding solar UV doses, such as for respiratory infections, often consider whether temperature can affect the findings. More research is indicated to evaluate the relative effects of temperature and sun exposure on the seasonality of mortality rates for several diseases. Since solar ultraviolet-B (UVB) doses decrease to vanishingly small values at higher latitudes in winter, the use of safe UVB lamps for indoor use in winter may warrant consideration.

Keywords: blood pressure; COVID-19; humidity; influenza; nitric oxide; respiratory infection; temperature; UVA; UVB; viral infection; vitamin D

Citation: Grant, W.B.; Boucher, B.J. An Exploration of How Solar Radiation Affects the Seasonal Variation of Human Mortality Rates and the Seasonal Variation in Some Other Common Disorders. *Nutrients* **2022**, *14*, 2519. https://doi.org/10.3390/nu14122519

Academic Editor: Andrea Fabbri

Received: 10 May 2022
Accepted: 14 June 2022
Published: 17 June 2022

1. Introduction

Several diseases have seasonal variations with the highest rates in winter and the lowest rates in summer.

The seasonal variation in mortality rate is perhaps best illustrated by data assembled by Marti-Soler and colleagues [1]. Table 1 in [1] presents the seasonal variation (peak to nadir) in the ratio of observed to expected deaths for overall, cardiovascular disease (CVD), cancer and non-CVD/non-cancer for 19 countries ranging from Finland to New Zealand (60° N to 41° S). A plot of those data as a function of latitude is shown in Figure 1. CVD has an inverted U-shaped seasonal relation with respect to latitude, which is lower at both low and high latitudes than at mid-latitudes, whereas non-CVD/non-cancer death seasonal variations increase with respect to latitude, and cancer seasonal rates are higher inside than outside the tropics.

Several factors affect the risk of at least one of these disease categories and vary with latitude. Those factors include ambient atmospheric temperature [2,3], solar ultraviolet-B (UVB) production of vitamin D [4–6] and solar UVA induction of nitric oxide (NO) in the serum [7–9]. Figure 2 shows the seasonal variation in high minus low ambient atmospheric temperature for the approximate center of population for 18 countries from reference [1] (USA was omitted), and Figure 3 shows the seasonal variations in solar UVA, UVB and temperature for Geneva, NY, USA (https://uvb.nrel.colostate.edu/UVB/uvb-network.jsf, accessed on 1 June 2022).

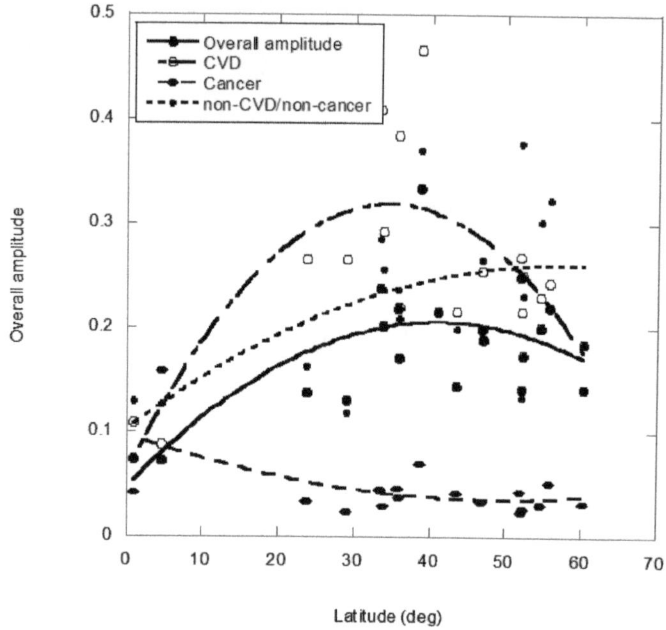

Figure 1. Plot of overall cardiovascular disease (CVD), cancer and non-CVD/non-cancer deaths showing seasonal variations versus 'absolute' latitude for 19 countries from Marti-Soler and colleagues [1]. The regression fits are second-order.

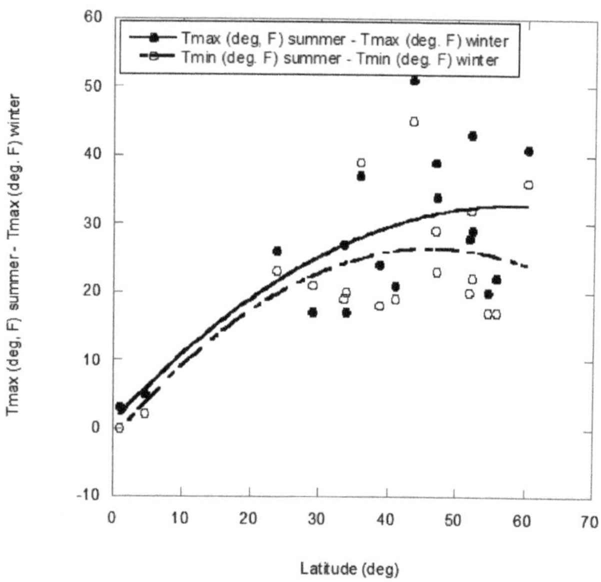

Figure 2. Seasonal variation in high minus low ambient atmospheric temperature for the approximate center of population for 18 countries from Marti-Soler and colleagues [1]. The regression fits are second-order.

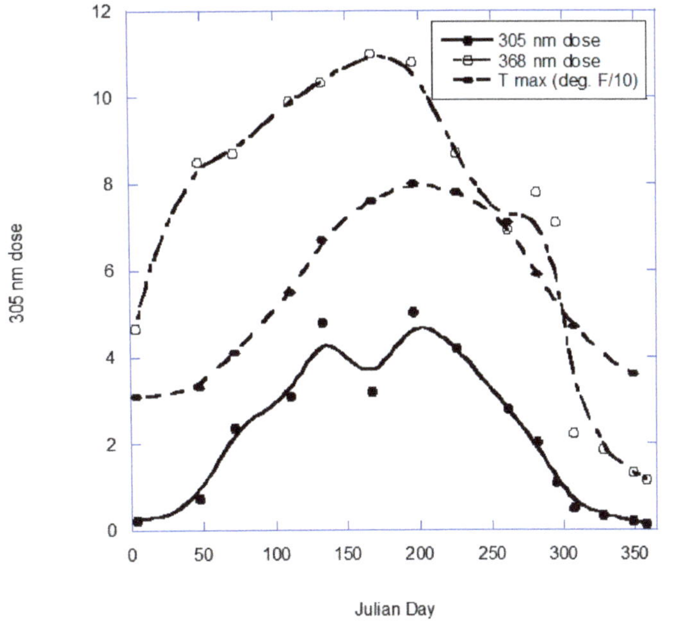

Figure 3. Annual variation for solar UVA and UVB in Geneva, NY, USA (42.9° N) for July 2006 to June 2007, as well as the temperature for 2021. UV data are from the UV-B Monitoring and Research Program operated by Colorado State University for the U.S. Department of Agriculture, https://uvb.nrel.colostate.edu/UVB/da_queryLampIrradiance.jsf (accessed on 1 June 2022). Temperature data are from https://www.usclimatedata.com/climate/geneva/new-york/united-states/usny0548 (accessed on 1 June 2022).

A book chapter [10] reported that, in Europe, from Greece (38° N) to Iceland (64° N), a U-shaped relationship exists between mean country serum 25-hydroxyvitamin D [25(OH)D] for adults in winter with higher concentrations at lower and higher latitudes than intermediate latitudes; however, mean concentrations in summer were near 68 nmol/L. The likely reasons for the inverted U-shape include that, at lower latitudes, solar UVB lasts longer than at higher latitudes [11], whereas at higher latitudes, national diets have more cold-water ocean fish and meat [12]. In addition, inhabitants in northern countries are more likely to take vitamin D supplements.

An article by Liu and colleagues [8] reported laboratory studies showing that UVA irradiation of the skin lowered BP, that UVA increases nitrite and reduces circulating nitrate, and that UVA increase forearm blood flow. NO is released by nitrite photolysis to nitrate. They predicted NO release from stores in the skin due to UV exposure as a function of latitude for summer and winter. The peak wavelength for the action spectrum was 330 nm, and it was estimated that the UVA spectrum (315–400 nm) is responsible for 80% of the release. The winter to summer difference in NO release rate peaks between 40° and 60° which correlates with the observed seasonal variation in BP and CVD rates and the summertime NO release rate decreases rapidly with increasing latitude in summer, which mirrors the relationship of blood pressure (BP) and hypertension prevalence.

The importance of temperature is underscored from analyses of the dependence of mortality rates when related to ambient temperature. Analyses based on 64.9 million deaths from 9 countries between 1 January 1980 and 31 December 2016, for 17 causes of death, were reported in 2021 [3]. Of particular interest for the present study was the finding that cold-related increased mortality rates are driven largely by CVD, chronic respiratory disease, metabolic disease and acute respiratory infections. Figure 4 shows the geographical variation in the percentage of deaths attributable to low temperatures for CVD, COPD, diabetes and kidney disease, respiratory infections and tuberculosis (TB) [3]. Only CVD as well as respiratory infections and TB have significantly higher percentages of deaths attributable to cold temperatures and higher latitudes.

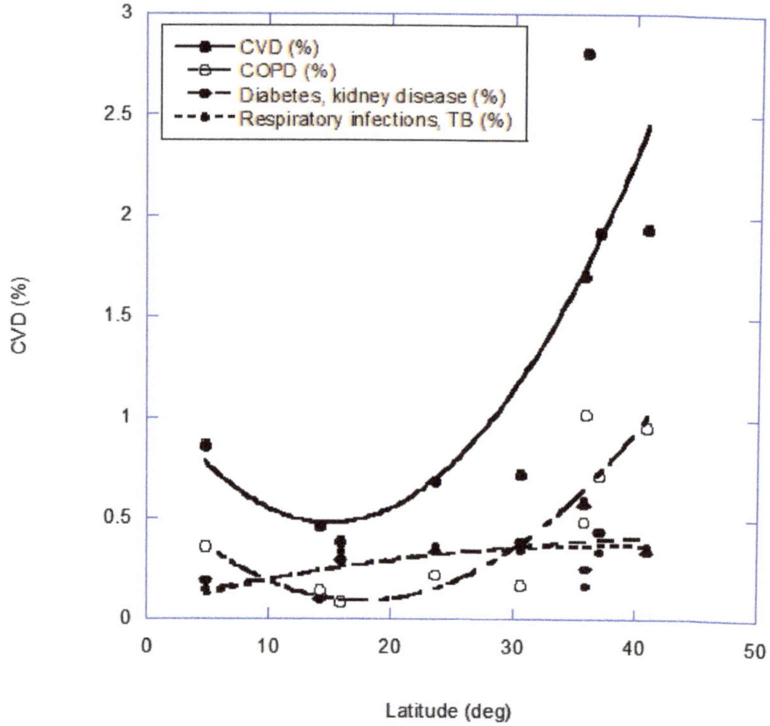

Figure 4. Percentage of deaths attributable to low temperatures for cardiovascular disease (CVD), chronic obstructive pulmonary disease (COPD), diabetes, kidney disease, respiratory infections and tuberculosis (TB) versus absolute latitude (doi:10.1016/S0140-6736(21)01700-1) [3]. The regression fits are second-order.

Solar UVB production of vitamin D has been linked to seasonal variations in several diseases. In 1981, Scragg proposed a mechanism to explain CVD's seasonal variations [13]. Rostand added blood BP in 1997 [14]. In 2002, Ponsonby and colleagues added three autoimmune diseases: multiple sclerosis, type one diabetes and rheumatoid arthritis [15]. In 2006, Cannell and colleagues proposed adding epidemic influenza [16]. In 2010, Lindqvist and colleagues added type two diabetes mellitus and metabolic control [17]. Studies in Sweden reported that avoiding sun exposure significantly increases mortality rates for women [18–20]. A review of mortality rates in the United States as a function of the day of the year suggested that seasonal variations in serum 25(OH)D concentrations are the primary driver of seasonal variations of mortality rates, although temperature variations may also play a role [6].

Notably, because the 25(OH)D action spectrum is in the solar ultraviolet-B (UVB) range (290–315 nm), vitamin D cannot be produced from solar radiation at high latitudes in winter. The classical study was reported by Webb and Holick in 1988 [21]. They showed that producing vitamin D from sun exposure is not possible for the darkest 6 months of the year in Boston (42.4° N) and Edmonton (53.5° N). Engelsen showed the time necessary to produce 400 IU of vitamin D for a young person with his or her face, neck and hands exposed as a function of latitude and time of the year to be 15 min [11]. Winter 25(OH)D concentrations are generally maintained at about 50–70% of summertime concentrations in Europe and the United States [22,23]. Vitamin D, as 25(OH)D, is recycled from the skeletal muscles by the action of parathyroid hormone (PTH) [24,25].

NO's role in human health was the subject of intense research starting in about 1977, with work by Murad and colleagues showing the stimulation of guanylyl cyclase by NO [26]. Stimulators of guanylate cyclase have been used to treat heart failure and pulmonary arterial hypertension. In 1979, Ignarro and colleagues [27] showed that NO was important in activating coronary arterial wall cell guanylyl cyclase. Furchgott and Zawadzki [28] showed that NO was involved in acetylcholine-induced relaxation and constriction of arterial walls through effects on the smooth muscle they contain. All three were awarded the Nobel Prize in 1998 [29]. A 20-year review of their work was published in 2019 [30].

The first review of NO produced in the skin through UV exposure was published in 1997 [31]. The proposed mechanism of production was the upregulation of the gene for inducible NO synthase (iNOS) enzyme. The benefits of NO mentioned included regulating blood flow, wound healing, preventing infections and improving eczema and psoriasis. In 2009, Oplander and colleagues reported that whole-body UVA exposure lowered BP by releasing NO from intracutaneous photolabile NO derivates [32]. Further early reviews of UV production of NO and its health benefits were published in 2011 by Juzeniene and colleagues [33] as well as by Juzeniene and Moan [34]. Later, Weller and colleagues explored the role of UV-induced NO production for various health outcomes, including lowering BP [8], improved physical performance [35], reducing liver inflammation [36] and reducing metabolic dysfunction experimentally [37].

The primary goal of the present report is to examine the roles of factors related to solar radiation that can affect the risk and seasonality of the three major diseases exhibiting pronounced seasonality of mortality: CVD, hypertension and respiratory tract infectious diseases. A secondary goal is to identify, and briefly discuss, other diseases and adverse health outcomes with seasonal variations, where the highest rates are seen in winter.

For this narrative review, literature searches were conducted with Google Scholar (scholar.google.com) and the National Library of Medicine's PubMed database (pubmed.gov). Searches were conducted for the seasonality of mortality and seasonality of other health outcomes. The primary factors investigated are the potential roles of solar UVB, vitamin D, solar UVA, NO and ambient temperature in explaining the observed seasonal variations. The search term "countries" was included in some of the searches in order to find multi-country studies of mortality rates and temperature effects on mortality rates. When appropriate multi-country studies were found, such as for seasonal mortality rates for the major diseases reviewed in this work, single-country studies were omitted. In addition, information on other factors that may affect seasonality was considered, including seasonal variations in gene expression and the effect of air pollution on disease incidence and mortality rates. Searches were also conducted regarding UVB lamps for indoor use with humans. Appropriate sources were found for data required regarding the seasonal variation of solar UVA and UVB doses, COVID-19 case rates, winter and summer temperatures of countries and latitudes of countries.

2. Results

2.1. Blood Pressure

Several environmentally related factors affect BP. This review considers temperature, vitamin D, sunlight and NO.

An analysis of the data of 38,589 participants in Harbin from the China Kadoorie Biobank during 2004–2008 reported an increase of 6.7 mmHg in systolic BP (SBP) and 2.1 mmHg in diastolic BP (DBP) for each 10 °C decrease in outdoor temperature when outdoor temperature was above 5 °C. An inverse association was evident between outdoor temperature and cardio-cerebrovascular event morbidity [38]. A study from Guangdong

Province, China, also reported reductions in BP with increasing ambient temperature up to 20 °C, as well as decreased rates of hypertension prevalence [39].

Rostand first reported latitudinal variations in BP and hypertension [14]. Using data from the INTERSALT study [40], he showed that a linear regression fit to SBP increased from 108 mmHg (95% confidence interval [CI], 104–113 mmHg) at the equator to 125 mmHg (95% CI, 121–129 mmHg) at 70° from the equator. Diastolic BP (DBP) increased from 67 mmHg (95% CI, 63–70 mmHg) at the equator to 78 mmHg (95% CI, 76–81 mmHg) at 70° from the equator. Using both INTERSALT and non-INTERSALT data, he found that hypertension rates increased from 8% (95% CI, 3–13%) at the equator to 25% (95% CI, 20–29%). He proposed that UVB, through effects on both circulating 25(OH)D and parathyroid hormone [PTH] concentrations as well as in part through affecting intracellular calcium, may explain the findings.

Rostand and colleagues [41] reported data on BP, sunlight and 25(OH)D concentrations among black and white participants in the Reasons for Geographic and Racial Differences in Stroke (REGARDS) Study conducted in the stroke belt states of the southeastern United States [41]. These researchers found inverse correlations between solar radiation and SBP for white people but not black people, and for women but not men. The group found no evidence that 25(OH)D concentrations are related to SBP. They then cited the possible role of UVA in increasing serum NO concentrations [8,32,42,43].

In 2020, Weller and colleagues published an article and correspondence regarding the role of UV exposure and changes in BP [44,45]. The article reported the results of a study of 342,459 patients (36% black, 64% white) at 2178 dialysis centers over 3 years. The researchers used incident UVA, UVB and temperature data from the National Oceanic and Atmospheric Administration database. They found linear inverse correlations between SBP and UVA, UVB and temperature. Although the BP findings for UVA and UVB were reduced when temperature was added to the analysis (by 57% for UVA and by 56% for UVB), the reductions with higher UVA and UVB were still significant. Predialysis SBP was about 4 mmHg higher for black patients. The correspondence reported that UVB's effect on BP was stronger than that of UVA. The authors argued that, because the evidence at that time for vitamin D in regulating BP was weak to nonexistent, the effect was due to increases in NO. However, that assumption has been overturned by a recent Mendelian randomization study that stratified the analysis by a genetically inferred 25(OH)D concentration [46]. In addition, supplementation by high-dose vitamin D_3 supplementation (~4000 IU/d) in an open label observational study in Canada that increased 25(OH)D concentrations above 100 nmol/L was found to reduce SBP by 14 to 18 mmHg and DBP by ~12 mmHg for hypertensives; however, no significant reduction in BP was found for pre-hypertensive participants. [47]. In addition, 71% of the 592 patients who were hypertensive at the beginning of te study were no longer hypertensive.

In summary, higher ambient temperatures (up to about 20 °C), higher NO concentrations, and higher 25(OH)D concentrations, i.e., higher vitamin D status, do appear to reduce BP. More research is indicated to assess each factor's relative contribution.

2.2. All-Cause Mortality Rate

Ambient temperature's effect on mortality rates has been known for many years. The historical review and analysis by Kutschenreuter [48] is a good source of information from the USA and serves as a good starting point. Figure 1 in that article uses monthly temperature data from 1921 to 1950 and monthly all-cause mortality rate data from 1949 to 1958 for three U.S. cities: Cincinnati, OH, USA; Los Angeles, CA, USA; and New York, NY, USA. The curves for all three cities show strong inverse correlations between ambient temperature and mortality rate. The amplitudes of mortality rate are +23% for a 46 °F change in temperature in Cincinnati, +25% for a 18 °F change in temperature in Los Angeles and +27% for a 44 °F change in temperature in New York. Figure 2 shows that the percent seasonal amplitude for all-cause mortality is in the low 30s for people older than 25 years but very low for those aged 1–24 years. Figure 4 shows that the seasonal variation in mortality rates is similar for white people and non-white people. All those results are consistent with the idea that changes in ambient temperature strongly influence mortality rates.

The article regarding the role of non-optimal temperature on daily mortality by Marti-Soler and colleagues [1] is very useful. Figure 1 in that article shows that, for lower respiratory infections, ischemic heart disease, stroke and chronic obstructive pulmonary disease, mortality increases for temperatures both below or above the theoretical minimum-risk exposure level (TMREL), which was calculated for temperatures from 6 °C to 28 °C. Evidently, the TMREL varies according to the mean temperature of the location. In a table, they estimated the attributable deaths and population attributable fractions (PAFs) for high and low temperatures for nine countries. Representative PAFs for low temperature in 2019 were 1.08% (95% CI, 0.92–1.31%) for Guatemala, 3.44% (95% CI, 3.03–3.75%) for the USA and 4.28% (3.88–4.66%) for China.

A study in the UK reported that winter mortality was explained almost entirely by a combination of monthly average temperatures over the previous 12 days, and weekly influenza A counts [49].

The role of UVB and vitamin D in reducing all-cause mortality rates is also well-known. For example, an observational study in Sweden involving 20,518 women aged 25–64 years recruited from 1990 to 1992 and followed up for 20 years reported that sun avoiders, in comparison with people with the highest sun exposure, had their life expectancy reduced by 0.6–2.1 years [19]. A pooled analysis of 32 observational studies showed that the hazard ratio for death decreased by 1.9 (95% CI, 1.6–2.2) for 25(OH)D < 10 ng/mL compared to 25(OH)D > 30 [50].

2.3. *Cardiovascular Disease*

Temperature has an important impact on CVD risk. In a study from Moscow, Russia, researchers found a relationship between average daily temperature and CVD mortality rates [51]. A plot of the basic mortality–temperature relationship indicated that this relationship was V-shaped with the minimum at ~18 °C. Each 1 °C increment of average daily temperature above 18 °C resulted in an increase in deaths from coronary heart disease (CHD) by 2.7%, from CVD by 4.7% and from respiratory diseases by 8.7%, with a lag of 0 to 1 day. Each 1 °C drop of average daily temperature from 18 °C to −10 °C resulted in an increase in deaths from CHD by 0.57%, from cerebrovascular diseases by 0.78%, and from respiratory diseases by 1.5%, with lags of maximum association varying from 3 days for non-accidental mortality to 6 days for cerebrovascular mortality.

Another article has reported exposure–response curves for ischemic heart disease and stroke [3], and it showed that the temperature effect depends on the daily mean temperature of residence, with people adapted for living with mean temperatures above 20 °C having reduced risks of mortality at up to 28 °C as compared to those of people living where mean temperatures were below 20 °C. The attributable percentage contribution of temperature to CVD mortality rate near 40° is 2.5%. A seasonal variation of 0.3 near 40° implies that the mean increase is near 15%. Thus, 17% (2.5/15) of the seasonal variability can be attributed to cold temperature. However, reducing the risk of CVD due to low temperatures is clearly possible by wearing warm clothing when outdoors in winter [52] or by staying in a heated home or work environment [53]. Similar results for CVD mortality rates with respect to temperature were also reported for residents of Spain [54]. The RR of CVD deaths was higher at 30 °C than at −5 °C, and the effect of low and high temperatures decreased from 1980–1994 to 2002–2016. The overall (all ages) RR at −5 °C compared with 20 °C was 1.5 in 2002–2016, whereas that for 30 °C was 1.8.

An important mechanism to explain why exposure to cold temperatures increases the risk of CVD mortality rates comes from Keatinge [55]. The body responds to cold by shifting up to a liter of blood from the skin to the internal organs to reduce heat loss. This shift leads to disposal of the extra fluid and salt, partly by the kidneys as urine and partly into the body's general intercellular spaces, thereby increasing blood viscosity and increasing the risks of thrombosis [55]. The effect of high temperature on risk of CVD mortality appears to relate to the effects of aging on thermoregulation during heat stress [56]. Compared with young adults during heat stress, older individuals typically respond with less individual sweat gland output, decreased skin blood flow, reduced cardiac output and smaller redistributions of blood flow from the splanchnic and renal circulations.

CHD rates vary with respect to the latitude of countries and season [57,58]. However, the geographical variation in data on a smaller scale, such as within a country, correlated most strongly with small particulate aerosol ($PM_{2.5}$) concentrations, as found in the United States [59].

Observational studies have shown significant inverse correlations between CVD incidence and 25(OH)D concentrations [46]. In addition, an observational study based on patients of the U.S. Veterans Health Administration system with a baseline serum 25(OH)D concentration of <20 ng/mL who were counseled to take vitamin D supplements to raised serum 25(OH)D to >30 ng/mL had a 35% (95% CI, 15–51%) reduced risk of myocardial infarction (MI) compared with those who remained at <20 ng/mL [60]. However, clinical trials have consistently failed to show that vitamin D supplementation reduces CVD incidence [61,62], casting doubt on findings from observational studies. By contrast, a randomized controlled trial (RCT) showed that high-dose vitamin D_3 supplementation reduces arterial stiffness in overweight African Americans [63].

In addition, two Mendelian randomization (MR) studies reported inverse correlations between genetic predictions of 25(OH)D concentrations by using data from the UK Biobank. Using nonlinear analysis, one study showed that CVD risk started to rise with lower genetic 25(OH)D concentrations in subjects with 25(OH)D concentrations below 20 ng/mL, progressing to a rapid increase in those with 25(OH)D values below 10 ng/mL [46]. The odds ratio (OR) for CVD for a 25(OH)D level of 10 ng/mL compared with 30 ng/mL was 1.11 (95% CI, 1.05–1.18). A similar relationship was also found for SBP. Thus, the authors estimate that approximately 6% of CVD risk in the UK can be prevented by raising serum 25(OH)D concentrations to above 40 ng/mL.

The second study used a stratified (nonlinear) analysis with respect to residual 25(OH)D concentrations [64] (calculated as the residual from the regression of seasonally adjusted 25(OH)D on the mean-centered generic risk score), which allowed for the comparison of individuals who would have had 25(OH)D concentrations in the same stratum if they had the same genotype. CVD mortality (6150 events) was significantly reduced in that MR analysis for 25(OH)D values of < 10 ng/mL by genotypes known to raise serum 25(OH)D (estimate = 0.69 [95% CI, 0.52–0.92]; $p = 0.01$). As noted in that article, MR uses genetic variants specifically related to a particular exposure to compare genetically defined population subgroups with different average levels of the exposure. The independent segregation of alleles at conception means that these genetically defined subgroups should not differ systematically with respect to confounding variables, creating a natural experiment analogous to that of a randomized trial. Thus, these two MR studies offer evidence that low 25(OH)D concentrations are significantly causal for CVD risk. Unfortunately, vitamin D RCTs have not yet been designed to be able to investigate this effect adequately [65,66].

Further investigation shows that high PTH values associated with low 25(OH)D concentrations, especially in the elderly, play an important role in CVD risk. A study in Utah reported that higher PTH at the baseline was noted in 26.1% of the study population. Highly significant differential CVD prevalence/incidence rates for most CVD risk factors, disease diagnoses and mortality were noted for PTH > 75 pg/mL (by 1.25- to 3-fold). PTH correlated only weakly with 25(OH)D and moderately with glomerular filtration rates. Risks related to PTH were attenuated slightly after adjusting for confounding factors, but they stayed significant [67].

An analysis of measurements of PTH and 25(OH)D in >300,000 U.S. patients confirmed that PTH increases with age and is inversely correlated with 25(OH)D concentrations. Patients most likely to have PTH > 75 pg/mL were older than 60 years and had 25(OH)D < 10 ng/mL [68]. PTH stimulates aldosterone secretion by increasing calcium concentrations in the cells of the adrenal zona glomerulosa as a result of binding to the PTH/PTH-rP receptor and indirectly by potentiating angiotensin-two-induced effects [69]. Meta-analyses of calcium supplementation studies reported significant increases in the risk of myocardial infarction compared with placebo and nonsignificant increases in the risk of stroke [70], which were not seen with comparable increases in dietary intakes [71].

A study of the seasonal variation of 25(OH)D concentrations in British adults aged 45 years indicated that 16% had values below 10 ng/mL in winter compared with 3% in summer [22]. Thus, this seasonal variation in vitamin D status appears large enough to contribute, in part, to the seasonal variation of CVD through a PTH-related mechanism.

A 2013 review by Lei and colleagues summarized the mechanisms by which NO reduces the risk of CVD. NO inhibits smooth muscle cell proliferation and migration; enhances the proliferation and migration of endothelial cells and inhibits their apoptosis; suppresses platelet aggregation; and prevents platelet, leukocyte and monocyte adhesion to endothelium. NO also inhibits the development of intimal hyperplasia in mechanically or immunologically injured vessels [72]. In 2016, Weller outlined the evidence that UVA-stimulated increase in levels of NO in the blood can play a role in reducing risk of CVD [73]. The hypothesis is supported by hypertension's importance as a risk factor for CVD [74].

A study in Scotland offered more support for UVA in reducing risk of MI [75]. A total of 56,370 MI patients were followed up from 2000 to 2011. Solar UVA and UVB doses were obtained from NASA satellite instruments. Monthly acute MI hospital admissions ranged from 6 to 11 out of a population of 100,000 during that period, whereas seasonal variations were about 1.5/100,000. The seasonal variations superimposed on the underlying trend had an amplitude of 0.31/100,000 (95% CI, 0.21–0.41/100,000). The log of UVA was significantly correlated with the amplitude after adjusting for UVB, with a correlation coefficient of –0.08 (95% CI, –0.13 to –0.02; $p = 0.008$). That amplitude was not significantly correlated with UVB nor temperature. Since the seasonal variations were small, UVA did not seem to have a large impact on seasonality.

A paper published in 2017 reviewed the environmental determinants of CVD [76]. The topics considered that could affect seasonality directly were discussed in this order: temperature, solar UVB and vitamin D, then solar UVA and NO. In addition, several other factors that affect CVD risk were discussed, of which air pollution and physical inactivity did impact the risk of CVD, with air pollution likely to be more important in the summer and physical inactivity more important in the winter.

In summary, temperature appears to be the most important factor regulating seasonal variations in CVD mortality rates. Temperature has the strongest support with respect to both short-term and seasonal variations in temperature [3]. However, the effect of 25(OH)D concentrations mediated by PTH and calcium is also important at older ages. UVA-stimulated release of NO is most likely important in reducing the risk of CVD death, but more research is required. More research is also indicated to better determine the relative strength of the effects of these factors on the risk of death from CVD.

2.4. Viral Infectious Diseases of the Respiratory Tract

Epidemic influenza has large seasonal variations, with peak rates 6 months apart in the northern and southern hemispheres [77]. Cannell proposed that solar UVB, through the production of vitamin D, may explain this epidemiology [16]. This hypothesis is supported by some clinical trials, such as one involving African American women in New York with a mean 25(OH)D baseline of 19 ± 8 ng/mL [78]. However, the hypothesis is not supported in another trial with African American women with a mean 25(OH)D baseline of 26 ± 12 ng/mL [79], where fewer cases of vitamin D deficiency must have been present. Support was also found for vitamin D supplementation in reducing the risk of influenza A but not influenza B in a study involving schoolchildren in Japan [80]. However, an analysis of seasonal influenza data from the Health Professionals Follow-up Study showed that absolute humidity and the school calendar better explained the seasonal patterns than did seasonal variations in 25(OH)D concentrations [81]. A later article reported that, where monthly average specific humidity or temperature decreased below thresholds of approximately 11–12 g/kg and 18 °C–21 °C during the year, influenza activity peaked during the cold–dry season (i.e., winter) when specific humidity and temperature are at minimal levels [82]. For tropical regions, seasonal influenza tends to peak in seasons of rainfall greater than 150 mm/month.

An observational study was conducted of the temporal variation in influenza and meteorological variables in several Northern European countries from 1 September 2017 to 31 August 2018 [83]. Temperature had the highest correlation with influenza, followed by the UV index, humidity, wind speed, precipitation and atmospheric temperature. The peak influenza rate occurred in calendar week 7 (i.e., mid-February), with only 10% of the peak rate seen in weeks 50 and 17. According to an analysis of solar UVB doses as a function of latitude and time of year, producing vitamin D from winter solar UVB is virtually impossible above 55° N from about week 53 to week 6 [11]. Similarly, a study of 45-year-old inhabitants of Great Britain from September 2002 to March 2004 reported that 25(OH)D concentrations were minimal in winter, with little change from January through April [22]. Thus, the UV effect on influenza rates observed was not due to UVB exposure or vitamin D production, suggesting that the UV effect was due to UVA rather than to UVB—something Cannell and colleagues did not evaluate [16].

COVID-19 rates also vary seasonally, with higher rates in winter but with occasional outbreaks in summer when a new variant of the SARS-CoV-2 virus arises. An analysis of the seasonality of viral infections noted that they spread faster in temperate regions when it is cold and dry, since the viruses remain viable longer ex vivo in such conditions [84,85]. However, in tropical regions, higher transmission is better supported by higher humidity, since the stability of the viruses ex vivo is maintained in droplets and on moister surfaces. Cold, dry conditions also reduce the body's immune response to virus infections by compromising nasal muco-ciliary clearance and local immune responses [86].

A recent article reported the correlations between meteorological and air quality variables and COVID-19 case rates in Morocco between 2 March and 31 December 2020 [87]. The factors that were associated with increased risk were relative humidity above 80%, wind speed above 20 m/s, ozone concentrations above 5 μg/m^3 and PM_{10} concentrations above 120 μg/m^3. Factors associated with lower risk were temperatures above 25 °C, precipitation above 25 mm and insolation above 10 h/day.

Grant and colleagues proposed in April 2020 that vitamin D reduces the risk of SARS-CoV-2 infection and COVID-19 [88]. Many studies have examined that hypothesis. The strongest support comes from two observational studies of SARS-CoV-2 infection or COVID-19 outcomes with respect to vitamin D supplementation. One prospective study conducted in Spain showed that patients prescribed with cholecalciferol or calcidiol and achieved 25(OH)D >30 ng/mL compared with untreated controls with 25(OH)D <20 ng/mL, had a significantly reduced risk of SARS-CoV-2 infection (multivariate analysis hazard ratio (maHR) = 0.66 [95% CI, 0.57–0.77; $p < 0.001$]) and COVID-19 mortality (maHR = 0.66 [95% CI, 0.46–0.93; $p < 0.001$]) and a significantly reduced risk of severe COVID-19 (maHR = 0.72 [95% CI, 0.52–1.00; $p = 0.05$]) [89]. Similar results were found for calcifediol prescriptions, with $p \leq 0.001$ for reductions in the risk of catching COVID-19 and in its severity and mortality.

The second observational study involved veteran patients receiving care at the U.S. Department of Veteran Affairs health care facilities between 20 February and 8 November 2020 [90]. Serum 25(OH)D data measured within 15–90 days before a positive SARS-CoV-2 test were available for 4599 patients. The adjusted probability of hospitalization fell from 0.25 ± 0.03 at 13 ng/mL to 0.18 ± 0.02 at 60 ng/mL. The adjusted probability of mortality fell from 0.11 ± 0.02 at 13 ng/mL to 0.06 ± 0.01 at 60 ng/mL. Since 60 ng/mL is higher than the likely mean 25(OH)D of ~35 ng/mL that the patients could have achieved from solar UVB exposure, the higher 25(OH)D concentrations most likely represent the result of vitamin D supplementation.

In an observational study in Spain, treating COVID-19 patients with calcifediol significantly reduced mortality rates [91]. The trial included 76 consecutive COVID-19 patients in a Cordoba hospital. A total of 50 patients were treated with high-dose calcifediol [25(OH)D_3], which has an advantage over vitamin D of rapidly increasing serum 25(OH)D concentrations. The multivariate risk estimate [OR] for admission to the intensive care unit for calcifediol treatment plus the conventional treatment versus the conventional treatment alone was 0.03 (95% CI, 0.003–0.25), and no patients treated with calcifediol died, whereas 2 not treated with calcifediol died.

A recent RCT involving health care workers in four Mexican hospitals reported that taking 4000 IU/d of vitamin D_3 for 1 month reduced SARS-CoV-2 infection rates by 75% [RR = 0.23 (95% CI, 0.09–0.55)] [92]. The effect was related to an increase in median 25(OH)D concentrations from 18 to 27 ng/mL

Chronic obstructive pulmonary disease (COPD) also has significant seasonal variations of exacerbations and mortality rates. The Towards a Revolution in COPD Health (TORCH) study involved 6112 COPD patients from 42 countries followed up for 3 years [93]. The patients were treated with standard drugs or placebo and visited clinics at 12-week intervals. In the northern hemisphere, exacerbations were 9.3% in winter and 5.3% in summer, whereas in the southern hemisphere, exacerbations were 12% in winter and 6% in summer. Findings of a review published in 2014 suggest that this seasonality was partly due to the increased prevalence of respiratory viral infections circulating in cold, damp conditions, along with increased airway inflammation or reduced 25(OH)D concentrations [94]. Vitamin D reduces inflammation through effects on cytokine production [95].

Evidence also exists that supports the role of UVA-induced increases in serum NO concentrations in reducing the risk of SARS-CoV-2 infection and COVID-19. An analysis of the geographical variations of COVID-19 mortality rates in England, Italy and the United States from January to April 2020 reported significant inverse correlations with respect to solar UVA doses in regions of each country when solar UVB doses were zero [9]. The adjusted mortality risk ratios per 100 kJ/m^2 increase in mean daily UVA were 0.49 (95% CI, 0.38–0.64) in England, 0.81 (95% CI, 0.71–0.93) in Italy and 0.71 (95% CI, 0.69–0.85) in the U.S. The pooled estimate was 0.68 (95% CI, 0.52–0.88).

Another study was conducted within Harvard's Nurses' Health Stufy II involving 39,315 participants, with 1768 testing positive for SARS-CoV-2 between May 2020 and March 2021 [96]. Higher predicted 25(OH)D concentrations, but not vitamin D intake, were associated with reduced SARS-CoV-2 infection rates. The highest quartiles of both UVB (annual) and UVA (winter) doses at residence locations were associated with a ~24% reduction in infection rates compared with the lowest doses.

A recent article reported that narrowband UVB (nUVB) treatment is very beneficial in treating hospitalized COVID-19 patients [97]. The study involved 30 COVID-19 patients in Louisiana in mid-2021, of whom 15 were randomized to be treated with nUVB and 15 with the same lamp but with the UVB blocked with UV-absorbing plexiglass. Patients were treated for up to 8 days. The end result was that two nUVB-treated patients died within 28 days compared to five untreated patients (p = 0.39). Interestingly, 25(OH)D concentrations on day 5 had decreased by −12 ng/mL (95% CI, −20 to −5 ng/mL) versus +1.7 ng/mL (95% CI, −12.1 to 7 ng/mL). The authors suggested that vitamin D was consumed by an nUVB-driven response to COVID-19. Although the primary endpoint was not significant, this study provided reasonable evidence that nUVB treatment effects were due to non-vitamin D effects. This concept is supported by Richard Weller, Centre for Inflammation Research, University of Edinburgh, Edinburgh, UK (private communication, June, 2022).

An article published in 2021 examined the surge dates for COVID-19 (based on SARS-CoV-2 seropositive case rates) in European countries during autumn of 2020 [98]. In Figure 3 of that article, the surge dates range from 10 September for Iceland to 22 October for Greece. The linear regression fit to the data has an r^2 value of 0.77. Figure 4 in that article shows that the surge date corresponds to solar UVB dose between 30% and 40% of the maximum for day 182 in July. A problem with attributing the effect to vitamin D is that serum 25(OH)D concentrations peak near 75 nmol/L in the UK (latitude, 52° N) in September and then decrease at a rate of 6–7 nmol/L to February or March [22]. Thus, the effect of vitamin D was too small and slow to explain the finding. A better explanation is that solar UVB induces various biological effects, both known and unknown, that reduce the risk of SARS-CoV-2 infection and COVID-19 [99], while UVA could also play a role.

In the winter in non-tropical regions, increases in solar UVA increases and in serum NO, atmospheric temperature and humidity appear to have the greatest impact on influenza and COVID-19 disease rates. In the summer, the non-vitamin D effects of UVB exposure and temperature/humidity seem to predominate, with some benefit from vitamin D.

2.5. Cancer

As reported by Marti-Soler and colleagues [1], cancer mortality rates do not have significant seasonal variations. Unlike CVD and respiratory tract infections, cancer mortality rates have significant inverse correlations with solar UVB in ecological studies [100]. Clinical trials have confirmed that vitamin D supplementation reduces the risk of cancer mortality [101]. Ambient temperature does not seem to affect cancer incidence or mortality rates. Two ecological studies, one in the entire U.S. [102] and one in Florida [37], showed an increase in cancer rates at colder temperatures. However, the studies did not consider solar UVB doses, which likely explain the findings. A recent review outlined the case for cold temperature increasing the risk of cancer [103]. Endothelial iNOS is a risk factor for several cancers, such as breast cancer [104], prostate cancer [105] and gastric cancer [106]. Since endothelial iNOS increases NO concentrations, these findings do not support UVA-induced increases in serum NO as reducing the risk of cancer.

Vitamin D deficiency is associated with increased colorectal cancer risks that appear likely to be causal [100]. Breast cancer survival is increased in people diagnosed in summer or autumn rather than in winter, although whether that is influenced by temperature, by vitamin D or by hormonal factors is unclear [107,108]. Vitamin D supplementation reduced cancer mortality though not cancer incidence in the VITAL study [109]. Data for the effects of NO, UVB or UVA are lacking, although mortality is reduced with greater sun exposure in Norway for several cancers, including breast, lung and prostate tumors [108]. Vitamin D supplementation has improved survival in breast cancer [110], but specific data for the effects of NO or UV are not available.

2.6. Other Health Outcomes

Further health outcomes are reported with seasonal variations in fatal health outcomes that are highest in winter (Table 1). Whether these risks are affected by the environmental factors identified above as those which have the greatest increases in mortality risks through CVD and respiratory illness (ambient temperature, atmospheric particulates, UVB production of vitamin D or other effects, or NO produced by UVA) is mostly unknown, though low 25(OH)D concentration has been considered a risk factor for each. A noteworthy feature of the diseases listed in this table is that only cancers have been found to have highly significant inverse correlations between incidence and mortality rates and solar UVB doses in mid-latitude countries [100], and it shows very little seasonal variation in mortality rates. Even though cancers develop slowly, serum 25(OH)D concentrations measured near the time of diagnosis are more strongly correlated with incidence than are concentrations measured more than a few months [111] or years [100] prior to diagnosis. Although RCTs have not found that supplementing with vitamin D reduces cancer incidence, it does reduce cancer mortality rates [101]. The failure of the many other RCTs in this area are likely attributable to poor design, conduct and analysis [100]. Thus, the lack of geographical ecological studies to support the role of vitamin D in reducing the risk of the other diseases already discussed supports the hypothesis that the non-vitamin-D-related effects of UVB exposure play important roles in reducing cancer and other health risks.

Table 1. Ratios of winter to summer mortality rate for common disorders in five countries.

Cause of Death	US 1951–1960 [112]	Australia 2015–2019 [113]	Japan 1970–1999 [114]	Netherlands 1979–1987 [115]	Scotland 1974–1988 [116]
All causes	1.17	1.10	1.04		1.33
Arteriosclerotic heart disease, CHD	1.28	1.15	1.14	1.34	1.28
Cancers	1.19	1.01	1.01	1.07	1.00
Cerebrovascular disease		1.11	1.09	1.25	1.30
Chronic respiratory disease		1.24		1.50	
Cirrhosis of liver	1.15				
Dementia		1.14			
Diabetes mellitus	1.10	1.12	1.20	1.28	
Digestive diseases			1.09		
Hypertensive heart disease	1.30				
Influenza				73	
Influenza, pneumonia (except newborn)	2.12	1.50	1.18		
Nephritis	1.22			1.30	
Nonrheumatic chronic endocarditis	1.24				
Pneumonia		1.33		1.88	
Respiratory					1.28
Rheumatic fever	1.21				
Septicemia				1.21	
Tuberculosis	1.17			1.59	
Vascular lesions, CNS	1.21				

CHD, coronary heart disease; CNS, central nervous system.

Both fever and high ambient temperatures are associated with exacerbations of multiple sclerosis and increased hospital admission rates [117,118]. Vitamin D deficiency is also associated with increased risks of MS, and the observational and epidemiological evidence for this association is increasingly supported by MR and mechanistic studies [119].

Type 1 diabetes results from autoimmune damage to islet beta cells that eventually destroys them. Variations occur in the risk of type 1 diabetes and of several other immune-mediated disorders with season of birth, which may be due to variations in either maternal or infant vitamin D status, in maternal or infant UVB exposure or both [120,121], but neither UVA exposure nor NO production has been investigated specifically as risk factors. Alhough vitamin D is necessary for normal insulin secretory responses to glucose and for maintaining healthy insulin sensitivity, it remains uncertain whether deficiency later in life increases type 1 diabetes risks [122].

Type 2 diabetes results from long periods of increased insulin resistance with high circulating insulin levels that lead to eventual islet beta cell failure. Since vitamin D is necessary for healthy insulin secretion and reduces abnormal insulin resistance through known mechanisms [123], vitamin D deficiency has long been thought likely to increase type 2 diabetes mellitus (T2DM) risks. That causal effect is supported by MR analyses [124] and by data from the Vitamin D and Type 2 Diabetes Study (D2d), showing up to 70% reductions in T2DM risk after 2.5 years of vitamin D supplementation of those subjects with prediabetes whose serum 25(OH)D values reached $\geq$100 nmol/l, a level only reached with intakes of 4000 IU/day [125].

Pregnancy-related disorders are more common in winter. Extremes of ambient temperature have adverse effects on birth outcomes, especially extremes of heat [126,127]. A recent literature review suggests that vitamin D deficiency has marked effects on pregnancy outcomes by increasing the risks of low birth weight, preterm birth and small-for-gestational-age births [128]. Deficiency is also associated with increased risks of gestational diabetes. Many trials of supplementation have failed to confirm causality in those disorders apart from gestational diabetes [129]. However, corrections of deficiency may require 4000 IU/day of vitamin D_3 rather than the much smaller doses usually given in trials, and further work with adequate replacements in deficiency should clarify the situation [130].

2.7. Gene Expression

A pair of studies published in 2015 reported significant seasonal variations in gene expression, including the vitamin D receptor [131,132]. Several studies reported by Holick and colleagues also showed that vitamin D supplementation significantly affects gene expression [133–135]. Although none of those studies examined the role of the many genes modified in disease risk, many of vitamin D's effects are due to controlling gene expression [136]. Since the risks of many diseases are lower in summer, when 25(OH)D concentrations are higher, it is reasonable to assume that genes with higher expression in summer are more likely to reduce the risk of disease. It is also likely that factors other than vitamin D affect seasonal variations in gene expression. For example, NO also regulates gene expression. By 2001, it was known that NO cannot only directly influence the activity of transcription factors, but it can also modulate upstream signaling cascades, mRNA stability and translation as well as the processing of the primary gene products [137]. A study published in 2003 reported that NO activates diverse signaling pathways to regulate gene expression experimentally [138].

2.8. Air Pollution

Air pollution is an important cause of mortality. Particulates smaller than 2.5 microns in diameter ($PM_{2.5}$) were estimated to cause 4.2 million (95% CI, 3.7–4.8 million) deaths globally in 2015 [139]. $PM_{2.5}$ is significantly correlated with chronic CHD mortality rates in the eastern U.S. during warm but not cold seasons [140]. Globally, 29% of the burden of stroke was attributed to air pollution in 2013 [141]. Air pollution is estimated to account for 790 thousand (95% CI, 645–934 thousand) deaths in Europe annually, with 40–80% from CVD [142]. A meta-analysis of the effects of air pollution on excess mortality rates reported in 2002 found that, for all common types of air pollutants, mortality rates were higher in warm seasons than in cold seasons [143]. It also found that excess mortality rates were about twice as high for respiratory diseases than for CVD for the pollutants, PM_{10}, carbon monoxide, nitrogen dioxide and sulfates, but not for ozone.

3. Discussion

This narrative review indicates that seasonal variations in atmospheric temperature strongly affect the risk and severity of high BP, CVD and respiratory viral diseases. However, both solar-UVB-induced changes in 25(OH)D and UVA-induced changes in NO appear likely to contribute to seasonal variations in the risk of these disorders, albeit to a more limited extent. However, ensuring vitamin D repletion remains worthwhile for reducing mortality rates, especially for the elderly in winter.

More research is indicated to evaluate the relative contributions of each of the factors discussed to the risks of these diseases and to death rates. Since the environmental factors that were discussed as affecting CVD, BP and the two viral illnesses considered—ambient temperature and exposure to UVB and UVA—cannot easily be changed at the population level, observational studies remain important for further studies. However, data on relevant environmental and biological risk factors near to the time of disease development or of death may prove especially useful. Furthermore, the designs of both observational studies and of the RCTs of potentially protective measures for disorders with seasonal variations in risk need to provide for collecting data on ambient temperature, humidity, exposure to UVA and UVB radiation and on changes in vitamin D status and in NO production concurrently with other potentially relevant variables throughout all such studies.

Other common disorders whose risks vary with season, such as those of multiple sclerosis and pregnancy disorders, have not yet been investigated in enough detail to identify the contributions of variations in atmospheric temperature, UVB, UVA or humidity to those risks. However, the existing data on the contributions of the environmental factors discussed here to seasonal variations in mortality, CVD and respiratory virus infections already suggest that effective public health measures to avoid temperature and humidity extremes and to optimize exposure to UVB and UVA can improve public health. In the future, better information on the contributions of these, or other, environmental factors to other common disorders with seasonal variations in their presentations could also prove to help in developing public health measures that are able to reduce CVD and respiratory system risks. Public health measures for reducing CVD and respiratory system risks from seasonal variations in temperature and humidity can already be stated, from the information reviewed, to be important for reducing the mortality rates seen from CVD and respiratory disease. Furthermore, this importance increases with global warming, adding to the urgency of reducing this global hazard.

Because mortality rates are inversely correlated with solar UVB doses, the lack of adequate solar UVB can be compensated for by the use of suitably regulated indoor UVB lamps. Studies conducted in England indicate that using a subliminal UVB source delivering the equivalent of 15 min of summer sunshine significantly increases the 25(OH)D concentrations of elderly residents of a residential of care homes [144,145]. The authors recommended additional studies before this approach could be used to recommend changes in health policy. UV fluorescent lamps have, however, already been used to increase vitamin D [146,147]. More recently, UVB sources using LEDs are being developed for stimulating vitamin D_3 production [148,149].

African Americans and other dark-skinned people living at higher latitudes than those of their ancestral homelands generally have poorer health outcomes than fair-skinned people in the same locations [150]. Although that article recommended vitamin D supplementation to help reduce ethnic health disparities, it now appears that UVB exposure should be considered as well.

Sunscreen can block UVB radiation from penetrating the skin. Since mounting evidence indicates that UVB has health benefits beyond vitamin D production, people should consider spending 15–30 min in the midday summer sun without sunscreen. Doing so daily can result in protective tanning in most, but not for all white subjects [151]. The protection from tanning is equivalent to a sun protection factor of 3–4 [152], meaning that one can stay in the sun 3–4 times longer without erythema than without a tan. In addition, raising serum 25(OH)D concentrations above 40 ng/mL can reduce the production of erythema from solar UVB exposure [153,154].

A recent article furnished data to help guide medical recommendations for sensible sun exposure to promote the cutaneous production of vitamin D_3 at different latitudes, seasonality, time of day and cloudiness status in Brazil [155] and these guidelines should be applicable elsewhere as well.

Hopefully, additional research can provide more information on the role of solar UVA and UVB in reducing the risks of seasonally variable diseases and mortality, leading to improved public health and clinical practice guidelines. Meanwhile, habitual increases in exposure to summer sunshine [with avoidance of sunburn] would also contribute to improving the public health [154].

Author Contributions: Writing, review and editing: W.B.G. and B.J.B. All authors have read and agreed to the published version of the manuscript.

Funding: This research received no external funding.

Acknowledgments: A presentation of this material was made at the Joint International Symposia Vitamin D in Prevention and Therapy and Biologic Effects of Light organized by J Reichrath, M.F. Holick, M. Friedrich and Th. Vogt, held in Homburg, Germany, on 4–6 May 2022 [156].

Conflicts of Interest: W.B.G.'s nonprofit organization, the Sunlight, Nutrition and Health Research Center, received funding from Bio-Tech Pharmacal, Inc. (Fayetteville, AR, USA). B.J.B. has no conflict of interest to declare.

References

1. Marti-Soler, H.; Gonseth, S.; Gubelmann, C.; Stringhini, S.; Bovet, P.; Chen, P.C.; Wojtyniak, B.; Paccaud, F.; Tsai, D.H.; Zdrojewski, T.; et al. Seasonal variation of overall and cardiovascular mortality: A study in 19 countries from different geographic locations. *PLoS ONE* **2014**, *9*, e113500. [CrossRef] [PubMed]
2. Bunker, A.; Wildenhain, J.; Vandenbergh, A.; Henschke, N.; Rocklov, J.; Hajat, S.; Sauerborn, R. Effects of Air Temperature on Climate-Sensitive Mortality and Morbidity Outcomes in the Elderly; a Systematic Review and Meta-analysis of Epidemiological Evidence. *EBioMedicine* **2016**, *6*, 258–268. [CrossRef] [PubMed]
3. Burkart, K.G.; Brauer, M.; Aravkin, A.Y.; Godwin, W.W.; Hay, S.I.; He, J.; Iannucci, V.C.; Larson, S.L.; Lim, S.S.; Liu, J.; et al. Estimating the cause-specific relative risks of non-optimal temperature on daily mortality: A two-part modelling approach applied to the Global Burden of Disease Study. *Lancet* **2021**, *398*, 685–697. [CrossRef]
4. Baggerly, C.A.; Cuomo, R.E.; French, C.B.; Garland, C.F.; Gorham, E.D.; Grant, W.B.; Heaney, R.P.; Holick, M.F.; Hollis, B.W.; McDonnell, S.L.; et al. Sunlight and Vitamin D: Necessary for Public Health. *J. Am. Coll. Nutr.* **2015**, *34*, 359–365. [CrossRef] [PubMed]
5. Holick, M.F. Biological Effects of Sunlight, Ultraviolet Radiation, Visible Light, Infrared Radiation and Vitamin D for Health. *Anticancer Res.* **2016**, *36*, 1345–1356.
6. Grant, W.B.; Bhattoa, H.P.; Boucher, B.J. Seasonal variations of U.S. mortality rates: Roles of solar ultraviolet-B doses, vitamin D, gene expression, and infections. *J. Steroid Biochem. Mol. Biol.* **2017**, *173*, 5–12. [CrossRef]
7. Feelisch, M.; Kolb-Bachofen, V.; Liu, D.; Lundberg, J.O.; Revelo, L.P.; Suschek, C.V.; Weller, R.B. Is sunlight good for our heart? *Eur. Heart J.* **2010**, *31*, 1041–1045. [CrossRef]
8. Liu, D.; Fernandez, B.O.; Hamilton, A.; Lang, N.N.; Gallagher, J.M.C.; Newby, D.E.; Feelisch, M.; Weller, R.B. UVA irradiation of human skin vasodilates arterial vasculature and lowers blood pressure independently of nitric oxide synthase. *J. Investig. Dermatol.* **2014**, *134*, 1839–1846. [CrossRef]
9. Cherrie, M.; Clemens, T.; Colandrea, C.; Feng, Z.; Webb, D.J.; Weller, R.B.; Dibben, C. Ultraviolet A radiation and COVID-19 deaths in the USA with replication studies in England and Italy. *Br. J. Dermatol.* **2021**, *185*, 363–370. [CrossRef]
10. Grant, W.B.; Bhattoa, H.P.; Pludowski, P. Determinants of Vitamin D Levels from Sun Exposure. In *Vitamin D, Ch. 56*, 5th ed.; Feldman, D., Pike, J.W., Eds.; Acadenic Press: London, UK, 2022.
11. Engelsen, O. The relationship between ultraviolet radiation exposure and vitamin D status. *Nutrients* **2010**, *2*, 482–495. [CrossRef]
12. Grant, W.B.; Fakhoury, H.M.A.; Karras, S.N.; Al Anouti, F.; Bhattoa, H.P. Variations in 25-Hydroxyvitamin D in Countries from the Middle East and Europe: The Roles of UVB Exposure and Diet. *Nutrients* **2019**, *11*, 2065. [CrossRef] [PubMed]
13. Scragg, R. Seasonality of cardiovascular disease mortality and the possible protective effect of ultra-violet radiation. *Int. J. Epidemiol.* **1981**, *10*, 337–341. [CrossRef] [PubMed]
14. Rostand, S.G. Ultraviolet light may contribute to geographic and racial blood pressure differences. *Hypertension* **1997**, *30*, 150–156. [CrossRef] [PubMed]
15. Ponsonby, A.L.; McMichael, A.; van der Mei, I. Ultraviolet radiation and autoimmune disease: Insights from epidemiological research. *Toxicology* **2002**, *181–182*, 71–78. [CrossRef]
16. Cannell, J.J.; Vieth, R.; Umhau, J.C.; Holick, M.F.; Grant, W.B.; Madronich, S.; Garland, C.F.; Giovannucci, E. Epidemic influenza and vitamin D. *Epidemiol. Infect.* **2006**, *134*, 1129–1140. [CrossRef]
17. Lindqvist, P.G.; Olsson, H.; Landin-Olsson, M. Are active sun exposure habits related to lowering risk of type 2 diabetes mellitus in women, a prospective cohort study? *Diabetes Res. Clin. Pract.* **2010**, *90*, 109–114. [CrossRef]
18. Lindqvist, P.G.; Epstein, E.; Olsson, H. Does an active sun exposure habit lower the risk of venous thrombotic events? A D-lightful hypothesis. *J. Thromb. Haemost.* **2009**, *7*, 605–610. [CrossRef]
19. Lindqvist, P.G.; Epstein, E.; Nielsen, K.; Landin-Olsson, M.; Ingvar, C.; Olsson, H. Avoidance of sun exposure as a risk factor for major causes of death: A competing risk analysis of the Melanoma in Southern Sweden cohort. *J. Intern. Med.* **2016**, *280*, 375–387. [CrossRef]
20. Lindqvist, P.G.; Epstein, E.; Landin-Olsson, M. Sun Exposure—Hazards and Benefits. *Anticancer Res.* **2022**, *42*, 1671–1677. [CrossRef]
21. Webb, A.R.; Kline, L.; Holick, M.F. Influence of season and latitude on the cutaneous synthesis of vitamin D3: Exposure to winter sunlight in Boston and Edmonton will not promote vitamin D3 synthesis in human skin. *J. Clin. Endocrinol. Metab.* **1988**, *67*, 373–378. [CrossRef]
22. Hypponen, E.; Power, C. Hypovitaminosis D in British adults at age 45 y: Nationwide cohort study of dietary and lifestyle predictors. *Am. J. Clin. Nutr.* **2007**, *85*, 860–868. [CrossRef] [PubMed]

23. Kroll, M.H.; Bi, C.; Garber, C.C.; Kaufman, H.W.; Liu, D.; Caston-Balderrama, A.; Zhang, K.; Clarke, N.; Xie, M.; Reitz, R.E.; et al. Temporal relationship between vitamin D status and parathyroid hormone in the United States. *PLoS ONE* **2015**, *10*, e0118108. [CrossRef] [PubMed]
24. Mason, R.S.; Rybchyn, M.S.; Abboud, M.; Brennan-Speranza, T.C.; Fraser, D.R. The Role of Skeletal Muscle in Maintaining Vitamin D Status in Winter. *Curr. Dev. Nutr.* **2019**, *3*, nzz087. [CrossRef] [PubMed]
25. Rybchyn, M.S.; Abboud, M.; Puglisi, D.A.; Gordon-Thomson, C.; Brennan-Speranza, T.C.; Mason, R.S.; Fraser, D.R. Skeletal Muscle and the Maintenance of Vitamin D Status. *Nutrients* **2020**, *12*, 3270. [CrossRef]
26. Katsuki, S.; Arnold, W.; Mittal, C.; Murad, F. Stimulation of guanylate cyclase by sodium nitroprusside, nitroglycerin and nitric oxide in various tissue preparations and comparison to the effects of sodium azide and hydroxylamine. *J. Cycl. Nucleotide Res.* **1977**, *3*, 23–35.
27. Gruetter, C.A.; Barry, B.K.; McNamara, D.B.; Gruetter, D.Y.; Kadowitz, P.J.; Ignarro, L. Relaxation of bovine coronary artery and activation of coronary arterial guanylate cyclase by nitric oxide, nitroprusside and a carcinogenic nitrosoamine. *J. Cycl. Nucleotide Res.* **1979**, *5*, 211–224.
28. Furchgott, R.F.; Zawadzki, J.V. The obligatory role of endothelial cells in the relaxation of arterial smooth muscle by acetylcholine. *Nature* **1980**, *288*, 373–376. [CrossRef]
29. SoRelle, R. Nobel prize awarded to scientists for nitric oxide discoveries. *Circulation* **1998**, *98*, 2365–2366. [CrossRef]
30. Ignarro, L.J. Nitric oxide is not just blowing in the wind. *Br. J. Pharmacol.* **2019**, *176*, 131–134. [CrossRef]
31. Weller, R. Nitric oxide—A newly discovered chemical transmitter in human skin. *Br. J. Dermatol.* **1997**, *137*, 665–672.
32. Oplander, C.; Volkmar, C.M.; Paunel-Gorgulu, A.; van Faassen, E.E.; Heiss, C.; Kelm, M.; Halmer, D.; Murtz, M.; Pallua, N.; Suschek, C.V. Whole body UVA irradiation lowers systemic blood pressure by release of nitric oxide from intracutaneous photolabile nitric oxide derivates. *Circ. Res.* **2009**, *105*, 1031–1040. [CrossRef] [PubMed]
33. Juzeniene, A.; Brekke, P.; Dahlback, A.; Andersson-Engels, S.; Reichrath, J.; Moan, K.; Holick, M.F.; Grant, W.B.; Moan, J. Solar radiation and human health. *Rep. Prog. Phys.* **2011**, *74*, 066701. [CrossRef]
34. Juzeniene, A.; Moan, J. Beneficial effects of UV radiation other than via vitamin D production. *Derm. Endocrinol.* **2012**, *4*, 109–117. [CrossRef] [PubMed]
35. Muggeridge, D.J.; Sculthorpe, N.; Grace, F.M.; Willis, G.; Thornhill, L.; Weller, R.B.; James, P.E.; Easton, C. Acute whole body UVA irradiation combined with nitrate ingestion enhances time trial performance in trained cyclists. *Nitric Oxide* **2015**, *48*, 3–9. [CrossRef] [PubMed]
36. Gorman, S.; Black, L.J.; Feelisch, M.; Hart, P.H.; Weller, R. Can skin exposure to sunlight prevent liver inflammation? *Nutrients* **2015**, *7*, 3219–3239. [CrossRef]
37. Fleury, N.; Feelisch, M.; Hart, P.H.; Weller, R.B.; Smoothy, J.; Matthews, V.B.; Gorman, S. Sub-erythemal ultraviolet radiation reduces metabolic dysfunction in already overweight mice. *J. Endocrinol.* **2017**, *233*, 81–92. [CrossRef]
38. Yu, B.; Jin, S.; Wang, C.; Yan, S.; Zhou, X.; Cui, X.; Tang, Z.; Luan, Q.; Guo, Y.; Bian, Z.; et al. The association of outdoor temperature with blood pressure, and its influence on future cardio-cerebrovascular disease risk in cold areas. *J. Hypertens.* **2020**, *38*, 1080–1089. [CrossRef]
39. Hu, J.; He, G.; Luo, J.; Xu, Y.; Xu, X.; Song, X.; Chen, S.; Ji, G.; Chen, Z.; Jiang, Q.; et al. Temperature-adjusted hypertension prevalence and control rate: A series of cross-sectional studies in Guangdong Province, China. *J. Hypertens.* **2021**, *39*, 911–918. [CrossRef]
40. Group, I.C.R. Intersalt: An international study of electrolyte excretion and blood pressure. Results for 24 hour urinary sodium and potassium excretion. Intersalt Cooperative Research Group. *BMJ* **1988**, *297*, 319–328.
41. Rostand, S.G.; McClure, L.A.; Kent, S.T.; Judd, S.E.; Gutierrez, O.M. Associations of blood pressure, sunlight, and vitamin D in community-dwelling adults. *J. Hypertens.* **2016**, *34*, 1704–1710. [CrossRef]
42. Warren, J.B. Nitric oxide and human skin blood flow responses to acetylcholine and ultraviolet light. *FASEB J.* **1994**, *8*, 247–251. [CrossRef] [PubMed]
43. Andrukhova, O.; Slavic, S.; Zeitz, U.; Riesen, S.C.; Heppelmann, M.S.; Ambrisko, T.D.; Markovic, M.; Kuebler, W.M.; Erben, R.G. Vitamin D is a regulator of endothelial nitric oxide synthase and arterial stiffness in mice. *Mol. Endocrinol.* **2014**, *28*, 53–64. [CrossRef] [PubMed]
44. Weller, R.B.; Wang, Y.; He, J.; Maddux, F.W.; Usvyat, L.; Zhang, H.; Feelisch, M.; Kotanko, P. Does Incident Solar Ultraviolet Radiation Lower Blood Pressure? *J. Am. Heart Assoc.* **2020**, *9*, e013837. [CrossRef] [PubMed]
45. Weller, R.B.; Feelisch, M.; Kotanko, P. Correspondence on 'Seasonal variation in blood pressure: Evidence, consensus and recommendations for clinical practice. Consensus statement by the ESH Working Group on Blood Pressure Monitoring and Cardiovascular Variability'. *J. Hypertens.* **2020**, *38*, 2077–2079. [CrossRef]
46. Zhou, A.; Selvanayagam, J.B.; Hypponen, E. Non-linear Mendelian randomization analyses support a role for vitamin D deficiency in cardiovascular disease risk. *Eur. Heart J.* **2021**, *43*, 1731–1739. [CrossRef]
47. Mirhosseini, N.; Vatanparast, H.; Kimball, S.M. The Association between Serum 25(OH)D Status and Blood Pressure in Participants of a Community-Based Program Taking Vitamin D Supplements. *Nutrients* **2017**, *9*, 1244. [CrossRef]
48. Kutschenreuter, P.H. A study of the effect of weather on mortality. *Trans. N. Y. Acad. Sci.* **1959**, 126–138. [CrossRef]
49. Wilkinson, P.; Pattenden, S.; Armstrong, B.; Fletcher, A.; Kovats, R.S.; Mangtani, P.; McMichael, A.J. Vulnerability to winter mortality in elderly people in Britain: Population based study. *BMJ* **2004**, *329*, 647. [CrossRef]

50. Garland, C.F.; Kim, J.J.; Mohr, S.B.; Gorham, E.D.; Grant, W.B.; Giovannucci, E.L.; Baggerly, L.; Hofflich, H.; Ramsdell, J.W.; Zeng, K.; et al. Meta-analysis of all-cause mortality according to serum 25-hydroxyvitamin D. *Am. J. Public Health* **2014**, *104*, e43–e50. [CrossRef]
51. Revich, B.; Shaposhnikov, D. Temperature-induced excess mortality in Moscow, Russia. *Int. J. Biometeorol.* **2008**, *52*, 367–374. [CrossRef]
52. Donaldson, G.C.; Ermakov, S.P.; Komarov, Y.M.; McDonald, C.P.; Keatinge, W.R. Cold related mortalities and protection against cold in Yakutsk, eastern Siberia: Observation and interview study. *BMJ* **1998**, *317*, 978–982. [CrossRef] [PubMed]
53. Keatinge, W.R.; Coleshaw, S.R.; Holmes, J. Changes in seasonal mortalities with improvement in home heating in England and Wales from 1964 to 1984. *Int. J. Biometeorol.* **1989**, *33*, 71–76. [CrossRef] [PubMed]
54. Achebak, H.; Devolder, D.; Ballester, J. Trends in temperature-related age-specific and sex-specific mortality from cardiovascular diseases in Spain: A national time-series analysis. *Lancet Planet. Health* **2019**, *3*, e297–e306. [CrossRef]
55. Keatinge, W.R. Winter mortality and its causes. *Int. J. Circumpolar Health* **2002**, *61*, 292–299. [CrossRef]
56. Kenney, W.L.; Munce, T.A. Invited review: Aging and human temperature regulation. *J. Appl. Physiol.* **2003**, *95*, 2598–2603. [CrossRef]
57. Grimes, D.S.; Hindle, E.; Dyer, T. Sunlight, cholesterol and coronary heart disease. *QJM* **1996**, *89*, 579–589. [CrossRef]
58. Zittermann, A.; Schleithoff, S.S.; Koerfer, R. Putting cardiovascular disease and vitamin D insufficiency into perspective. *Br. J. Nutr.* **2005**, *94*, 483–492. [CrossRef]
59. Thurston, G.D.; Burnett, R.T.; Turner, M.C.; Shi, Y.; Krewski, D.; Lall, R.; Ito, K.; Jerrett, M.; Gapstur, S.M.; Diver, W.R.; et al. Ischemic Heart Disease Mortality and Long-Term Exposure to Source-Related Components of U.S. Fine Particle Air Pollution. *Environ. Health Perspect.* **2016**, *124*, 785–794. [CrossRef]
60. Acharya, P.; Dalia, T.; Ranka, S.; Sethi, P.; Oni, O.A.; Safarova, M.S.; Parashara, D.; Gupta, K.; Barua, R.S. The Effects of Vitamin D Supplementation and 25-Hydroxyvitamin D Levels on the Risk of Myocardial Infarction and Mortality. *J. Endocr. Soc.* **2021**, *5*, bvab124. [CrossRef]
61. Manson, J.E.; Cook, N.R.; Lee, I.M.; Christen, W.; Bassuk, S.S.; Mora, S.; Gibson, H.; Gordon, D.; Copeland, T.; D'Agostino, D.; et al. Vitamin D Supplements and Prevention of Cancer and Cardiovascular Disease. *N. Engl. J. Med.* **2019**, *380*, 33–44. [CrossRef]
62. Barbarawi, M.; Kheiri, B.; Zayed, Y.; Barbarawi, O.; Dhillon, H.; Swaid, B.; Yelangi, A.; Sundus, S.; Bachuwa, G.; Alkotob, M.L.; et al. Vitamin D Supplementation and Cardiovascular Disease Risks in More Than 83000 Individuals in 21 Randomized Clinical Trials: A Meta-analysis. *JAMA Cardiol.* **2019**, *4*, 765–776. [CrossRef] [PubMed]
63. Raed, A.; Bhagatwala, J.; Zhu, H.; Pollock, N.K.; Parikh, S.J.; Huang, Y.; Havens, R.; Kotak, I.; Guo, D.H.; Dong, Y. Dose responses of vitamin D3 supplementation on arterial stiffness in overweight African Americans with vitamin D deficiency: A placebo controlled randomized trial. *PLoS ONE* **2017**, *12*, e0188424. [CrossRef] [PubMed]
64. Emerging Risk Factors Collaboration/EPIC-CVD/Vitamin D Studies Collaboration. Estimating dose-response relationships for vitamin D with coronary heart disease, stroke, and all-cause mortality: Observational and Mendelian randomisation analyses. *Lancet Diabetes Endocrinol.* **2021**, *9*, 837–846. [CrossRef]
65. Heaney, R.P. Design and analysis of clinical trials of nutrients: Author reply. *Nutr. Rev.* **2014**, *72*, 354. [CrossRef]
66. Grant, W.B.; Boucher, B.J.; Bhattoa, H.P.; Lahore, H. Why vitamin D clinical trials should be based on 25-hydroxyvitamin D concentrations. *J. Steroid Biochem. Mol. Biol.* **2018**, *177*, 266–269. [CrossRef]
67. Anderson, J.L.; Vanwoerkom, R.C.; Horne, B.D.; Bair, T.L.; May, H.T.; Lappe, D.L.; Muhlestein, J.B. Parathyroid hormone, vitamin D, renal dysfunction, and cardiovascular disease: Dependent or independent risk factors? *Am. Heart J.* **2011**, *162*, 331–339.e2. [CrossRef]
68. Valcour, A.; Blocki, F.; Hawkins, D.M.; Rao, S.D. Effects of age and serum 25-OH-vitamin D on serum parathyroid hormone levels. *J. Clin. Endocrinol. Metab.* **2012**, *97*, 3989–3995. [CrossRef]
69. Tomaschitz, A.; Ritz, E.; Pieske, B.; Rus-Machan, J.; Kienreich, K.; Verheyen, N.; Gaksch, M.; Grubler, M.; Fahrleitner-Pammer, A.; Mrak, P.; et al. Aldosterone and parathyroid hormone interactions as mediators of metabolic and cardiovascular disease. *Metabolism* **2014**, *63*, 20–31. [CrossRef]
70. Bolland, M.J.; Avenell, A.; Baron, J.A.; Grey, A.; MacLennan, G.S.; Gamble, G.D.; Reid, I.R. Effect of calcium supplements on risk of myocardial infarction and cardiovascular events: Meta-analysis. *BMJ* **2010**, *341*, c3691. [CrossRef]
71. Boucher, B.J. Calcium supplements may increase the risk of cardiovascular events in postmenopausal women. *Evid. Based Med.* **2012**, *17*, 16–17. [CrossRef]
72. Lei, J.; Vodovotz, Y.; Tzeng, E.; Billiar, T.R. Nitric oxide, a protective molecule in the cardiovascular system. *Nitric Oxide* **2013**, *35*, 175–185. [CrossRef] [PubMed]
73. Weller, R.B. Sunlight Has Cardiovascular Benefits Independently of Vitamin D. *Blood Purif.* **2016**, *41*, 130–134. [CrossRef] [PubMed]
74. Fuchs, F.D.; Whelton, P.K. High Blood Pressure and Cardiovascular Disease. *Hypertension* **2020**, *75*, 285–292. [CrossRef] [PubMed]
75. Mackay, D.F.; Clemens, T.L.; Hastie, C.E.; Cherrie, M.P.C.; Dibben, C.; Pell, J.P. UVA and Seasonal Patterning of 56 370 Myocardial Infarctions Across Scotland, 2000–2011. *J. Am. Heart Assoc.* **2019**, *8*, e012551. [CrossRef]
76. Bhatnagar, A. Environmental Determinants of Cardiovascular Disease. *Circ. Res.* **2017**, *121*, 162–180. [CrossRef]
77. Hope-Simpson, R.E. The role of season in the epidemiology of influenza. *J. Hyg.* **1981**, *86*, 35–47. [CrossRef]
78. Aloia, J.F.; Li-Ng, M. Re: Epidemic influenza and vitamin D. *Epidemiol. Infect.* **2007**, *135*, 1095–1096. [CrossRef]
79. Li-Ng, M.; Aloia, J.F.; Pollack, S.; Cunha, B.A.; Mikhail, M.; Yeh, J.; Berbari, N. A randomized controlled trial of vitamin D3 supplementation for the prevention of symptomatic upper respiratory tract infections. *Epidemiol. Infect.* **2009**, *137*, 1396–1404. [CrossRef]

80. Urashima, M.; Segawa, T.; Okazaki, M.; Kurihara, M.; Wada, Y.; Ida, H. Randomized trial of vitamin D supplementation to prevent seasonal influenza A in schoolchildren. *Am. J. Clin. Nutr.* **2010**, *91*, 1255–1260. [CrossRef]
81. Shaman, J.; Jeon, C.Y.; Giovannucci, E.; Lipsitch, M. Shortcomings of vitamin D-based model simulations of seasonal influenza. *PLoS ONE* **2011**, *6*, e20743. [CrossRef]
82. Tamerius, J.D.; Shaman, J.; Alonso, W.J.; Bloom-Feshbach, K.; Uejio, C.K.; Comrie, A.; Viboud, C. Environmental predictors of seasonal influenza epidemics across temperate and tropical climates. *PLoS Pathog.* **2013**, *9*, e1003194. [CrossRef]
83. Ianevski, A.; Zusinaite, E.; Shtaida, N.; Kallio-Kokko, H.; Valkonen, M.; Kantele, A.; Telling, K.; Lutsar, I.; Letjuka, P.; Metelitsa, N.; et al. Low Temperature and Low UV Indexes Correlated with Peaks of Influenza Virus Activity in Northern Europe during 2010–2018. *Viruses* **2019**, *11*, 207. [CrossRef]
84. Harper, G.J. Airborne micro-organisms: Survival tests with four viruses. *J. Hyg.* **1961**, *59*, 479–486. [CrossRef] [PubMed]
85. Audi, A.; AlIbrahim, M.; Kaddoura, M.; Hijazi, G.; Yassine, H.M.; Zaraket, H. Seasonality of Respiratory Viral Infections: Will COVID-19 Follow Suit? *Front. Public Health* **2020**, *8*, 567184. [CrossRef] [PubMed]
86. Eccles, R. An explanation for the seasonality of acute upper respiratory tract viral infections. *Acta Otolaryngol.* **2002**, *122*, 183–191. [CrossRef] [PubMed]
87. Khalis, M.; Toure, A.B.; El Badisy, I.; Khomsi, K.; Najmi, H.; Bouaddi, O.; Marfak, A.; Al-Delaimy, W.K.; Berraho, M.; Nejjari, C. Relationship between Meteorological and Air Quality Parameters and COVID-19 in Casablanca Region, Morocco. *Int. J. Environ. Res. Public Health* **2022**, *19*, 4989. [CrossRef]
88. Grant, W.B.; Lahore, H.; McDonnell, S.L.; Baggerly, C.A.; French, C.B.; Aliano, J.L.; Bhattoa, H.P. Evidence that Vitamin D Supplementation Could Reduce Risk of Influenza and COVID-19 Infections and Deaths. *Nutrients* **2020**, *12*, 988. [CrossRef]
89. Oristrell, J.; Oliva, J.C.; Casado, E.; Subirana, I.; Dominguez, D.; Toloba, A.; Balado, A.; Grau, M. Vitamin D supplementation and COVID-19 risk: A population-based, cohort study. *J. Endocrinol. Investig.* **2022**, *45*, 167–179. [CrossRef]
90. Seal, K.H.; Bertenthal, D.; Carey, E.; Grunfeld, C.; Bikle, D.D.; Lu, C.M. Association of Vitamin D Status and COVID-19-Related Hospitalization and Mortality. *J. Gen. Intern. Med.* **2022**, *37*, 853–861. [CrossRef]
91. Entrenas Castillo, M.; Entrenas Costa, L.M.; Vaquero Barrios, J.M.; Alcala Diaz, J.F.; Lopez Miranda, J.; Bouillon, R.; Quesada Gomez, J.M. Effect of calcifediol treatment and best available therapy versus best available therapy on intensive care unit admission and mortality among patients hospitalized for COVID-19: A pilot randomized clinical study. *J. Steroid Biochem. Mol. Biol.* **2020**, *203*, 105751. [CrossRef]
92. Villasis-Keever, M.A.; Lopez-Alarcon, M.G.; Miranda-Vovales, G.; Zurita-Cruz, J.N.; Barrada-Vazquez, A.Z. Efficacy and Safety of Vitamin D Supplementation to Prevent COVID-19 in Frontline Healthcare Workers. A Randomized Clinical Trial. *Arch. Med. Res.* **2022**, *53*, 423–430. [CrossRef] [PubMed]
93. Jenkins, C.R.; Celli, B.; Anderson, J.A.; Ferguson, G.T.; Jones, P.W.; Vestbo, J.; Yates, J.C.; Calverley, P.M. Seasonality and determinants of moderate and severe COPD exacerbations in the TORCH study. *Eur. Respir. J.* **2012**, *39*, 38–45. [CrossRef] [PubMed]
94. Donaldson, G.C.; Wedzicha, J.A. The causes and consequences of seasonal variation in COPD exacerbations. *Int. J. Chron. Obstruct. Pulmon. Dis.* **2014**, *9*, 1101–1110. [CrossRef] [PubMed]
95. Guillot, X.; Semerano, L.; Saidenberg-Kermanac'h, N.; Falgarone, G.; Boissier, M.C. Vitamin D and inflammation. *Jt. Bone Spine* **2010**, *77*, 552–557. [CrossRef]
96. Ma, W.; Nguyen, L.H.; Yue, Y.; Ding, M.; Drew, D.A.; Wang, K.; Merino, J.; Rich-Edwards, J.W.; Sun, Q.; Camargo, C.A.; et al. Associations between predicted vitamin D status, vitamin D intake, and risk of severe acute respiratory syndrome coronavirus 2 (SARS-CoV-2) infection and coronavirus disease 2019 (COVID-19) severity. *Am. J. Clin. Nutr.* **2022**, *115*, 1123–1133. [CrossRef]
97. Lau, F.H.; Powell, C.E.; Adonecchi, G.; Danos, D.M.; DiNardo, A.R.; Chugden, R.J.; Wolr, P.; Castilla, C.F. Pilot Phase Results of a Prospective, Randomized Controlled Trial of Narrowband Ultraviolet B Phototherapy in Hospitalized COVID-19 Patients. *Exp. Dermatol.* **2022**. [CrossRef]
98. Walrand, S. Autumn COVID-19 surge dates in Europe correlated to latitudes, not to temperature-humidity, pointing to vitamin D as contributing factor. *Sci. Rep.* **2021**, *11*, 1981. [CrossRef]
99. Gorman, S.; Weller, R.B. Investigating the Potential for Ultraviolet Light to Modulate Morbidity and Mortality From COVID-19: A Narrative Review and Update. *Front. Cardiovasc. Med.* **2020**, *7*, 616527. [CrossRef]
100. Muñoz, A.; Grant, W.B. Vitamin D and Cancer: An Historical Overview of the Epidemiology and Mechanisms. *Nutrients* **2022**, *14*, 1448. [CrossRef]
101. Keum, N.; Lee, D.H.; Greenwood, D.C.; Manson, J.E.; Giovannucci, E. Vitamin D supplementation and total cancer incidence and mortality: A meta-analysis of randomized controlled trials. *Ann. Oncol.* **2019**, *30*, 733–743. [CrossRef]
102. Sharma, A.; Sharma, T.; Panwar, M.S.; Sharma, D.; Bundel, R.; Hamilton, R.T.; Radosevich, J.A.; Mandal, C.C. Colder environments are associated with a greater cancer incidence in the female population of the United States. *Tumour Biol.* **2017**, *39*, 1010428317724784. [CrossRef] [PubMed]
103. Bandyopadhayaya, S.; Ford, B.; Mandal, C.C. Cold-hearted: A case for cold stress in cancer risk. *J. Therm. Biol.* **2020**, *91*, 102608. [CrossRef] [PubMed]
104. Lu, J.; Wei, Q.; Bondy, M.L.; Yu, T.K.; Li, D.; Brewster, A.; Shete, S.; Sahin, A.; Meric-Bernstam, F.; Wang, L.E. Promoter polymorphism (-786t>C) in the endothelial nitric oxide synthase gene is associated with risk of sporadic breast cancer in non-Hispanic white women age younger than 55 years. *Cancer* **2006**, *107*, 2245–2253. [CrossRef] [PubMed]

105. Lee, K.M.; Kang, D.; Park, S.K.; Berndt, S.I.; Reding, D.; Chatterjee, N.; Chanock, S.; Huang, W.Y.; Hayes, R.B. Nitric oxide synthase gene polymorphisms and prostate cancer risk. *Carcinogenesis* **2009**, *30*, 621–625. [CrossRef] [PubMed]
106. Goto, T.; Haruma, K.; Kitadai, Y.; Ito, M.; Yoshihara, M.; Sumii, K.; Hayakawa, N.; Kajiyama, G. Enhanced expression of inducible nitric oxide synthase and nitrotyrosine in gastric mucosa of gastric cancer patients. *Clin. Cancer Res.* **1999**, *5*, 1411–1415.
107. Mason, B.H.; Holdaway, I.M.; Stewart, A.W.; Neave, L.M.; Kay, R.G. Season of tumour detection influences factors predicting survival of patients with breast cancer. *Breast Cancer Res. Treat.* **1990**, *15*, 27–37. [CrossRef]
108. Porojnicu, A.C.; Dahlback, A.; Moan, J. Sun exposure and cancer survival in Norway: Changes in the risk of death with season of diagnosis and latitude. *Adv. Exp. Med. Biol.* **2008**, *624*, 43–54.
109. Chandler, P.D.; Chen, W.Y.; Ajala, O.N.; Hazra, A.; Cook, N.; Bubes, V.; Lee, I.M.; Giovannucci, E.L.; Willett, W.; Buring, J.E.; et al. Effect of Vitamin D3 Supplements on Development of Advanced Cancer: A Secondary Analysis of the VITAL Randomized Clinical Trial. *JAMA Netw. Open* **2020**, *3*, e2025850. [CrossRef]
110. Madden, J.M.; Murphy, L.; Zgaga, L.; Bennett, K. De novo vitamin D supplement use post-diagnosis is associated with breast cancer survival. *Breast Cancer Res. Treat.* **2018**, *172*, 179–190. [CrossRef]
111. Mohr, S.B.; Gorham, E.D.; Alcaraz, J.E.; Kane, C.I.; Macera, C.A.; Parsons, J.K.; Wingard, D.L.; Horst, R.; Garland, C.F. Serum 25-hydroxyvitamin D and breast cancer in the military: A case-control study utilizing pre-diagnostic serum. *Cancer Causes Control.* **2013**, *24*, 495–504. [CrossRef]
112. Rosenwaike, I. Seasonal variation of deaths in the United States, 1951–1960. *J. Am. Stat. Assoc.* **1966**, *61*, 706–719.
113. Gregory, G.; Zhu, L.; Hayen, A.; Bell, K.J.L. Learning from the pandemic: Mortality trends and seasonality of deaths in Australia in 2020. *Int. J. Epidemiol.* **2022**, *51*, 718–726. [CrossRef] [PubMed]
114. Nakaji, S.; Parodi, S.; Fontana, V.; Umeda, T.; Suzuki, K.; Sakamoto, J.; Fukuda, S.; Wada, S.; Sugawara, K. Seasonal changes in mortality rates from main causes of death in Japan (1970–1999). *Eur. J. Epidemiol.* **2004**, *19*, 905–913. [CrossRef] [PubMed]
115. Mackenbach, J.P.; Kunst, A.E.; Looman, C.W. Seasonal variation in mortality in The Netherlands. *J. Epidemiol. Community Health* **1992**, *46*, 261–265. [CrossRef]
116. Douglas, A.S.; Allan, T.M.; Rawles, J.M. Composition of seasonality of disease. *Scott. Med. J.* **1991**, *36*, 76–82. [CrossRef]
117. Ebi, K.L.; Capon, A.; Berry, P.; Broderick, C.; de Dear, R.; Havenith, G.; Honda, Y.; Kovats, R.S.; Ma, W.; Malik, A.; et al. Hot weather and heat extremes: Health risks. *Lancet* **2021**, *398*, 698–708. [CrossRef]
118. Vandebergh, M.; Degryse, N.; Dubois, B.; Goris, A. Environmental risk factors in multiple sclerosis: Bridging Mendelian randomization and observational studies. *J. Neurol.* **2022**. [CrossRef]
119. Wang, R. Mendelian randomization study updates the effect of 25-hydroxyvitamin D levels on the risk of multiple sclerosis. *J. Transl. Med.* **2022**, *20*, 3. [CrossRef]
120. Sloka, S.; Grant, M.; Newhook, L.A. The geospatial relation between UV solar radiation and type 1 diabetes in Newfoundland. *Acta Diabetol.* **2010**, *47*, 73–78. [CrossRef]
121. Disanto, G.; Chaplin, G.; Morahan, J.M.; Giovannoni, G.; Hypponen, E.; Ebers, G.C.; Ramagopalan, S.V. Month of birth, vitamin D and risk of immune-mediated disease: A case control study. *BMC Med.* **2012**, *10*, 69. [CrossRef]
122. Boucher, B.J.; Mannan, N.; Noonan, K.; Hales, C.N.; Evans, S.J. Glucose intolerance and impairment of insulin secretion in relation to vitamin D deficiency in east London Asians. *Diabetologia* **1995**, *38*, 1239–1245. [CrossRef] [PubMed]
123. Niroomand, M.; Fotouhi, A.; Irannejad, N.; Hosseinpanah, F. Does high-dose vitamin D supplementation impact insulin resistance and risk of development of diabetes in patients with pre-diabetes? A double-blind randomized controlled trial. *Diabetes Res. Clin. Pract.* **2019**, *148*, 1–9. [CrossRef] [PubMed]
124. Bejar, C.A.; Goyal, S.; Afzal, S.; Mangino, M.; Zhou, A.; van der Most, P.J.; Bao, Y.; Gupta, V.; Smart, M.C.; Walia, G.K.; et al. A Bidirectional Mendelian Randomization Study to evaluate the causal role of reduced blood vitamin D levels with type 2 diabetes risk in South Asians and Europeans. *Nutr. J.* **2021**, *20*, 71. [CrossRef]
125. Dawson-Hughes, B.; Staten, M.A.; Knowler, W.C.; Nelson, J.; Vickery, E.M.; LeBlanc, E.S.; Neff, L.M.; Park, J.; Pittas, A.G.; Group, D.d.R. Intratrial Exposure to Vitamin D and New-Onset Diabetes Among Adults With Prediabetes: A Secondary Analysis From the Vitamin D and Type 2 Diabetes (D2d) Study. *Diabetes Care* **2020**, *43*, 2916–2922. [CrossRef]
126. Zhang, Y.; Yu, C.; Wang, L. Temperature exposure during pregnancy and birth outcomes: An updated systematic review of epidemiological evidence. *Environ. Pollut.* **2017**, *225*, 700–712. [CrossRef] [PubMed]
127. Samuels, L.; Nakstad, B.; Roos, N.; Bonell, A.; Chersich, M.; Havenith, G.; Luchters, S.; Day, L.T.; Hirst, J.E.; Singh, T.; et al. Physiological mechanisms of the impact of heat during pregnancy and the clinical implications: Review of the evidence from an expert group meeting. *Int. J. Biometeorol.* **2022**. [CrossRef] [PubMed]
128. Zhao, R.; Zhou, L.; Wang, S.; Yin, H.; Yang, X.; Hao, L. Effect of maternal vitamin D status on risk of adverse birth outcomes: A systematic review and dose-response meta-analysis of observational studies. *Eur. J. Nutr.* **2022**. [CrossRef]
129. Nausheen, S.; Habib, A.; Bhura, M.; Rizvi, A.; Shaheen, F.; Begum, K.; Iqbal, J.; Ariff, S.; Shaikh, L.; Raza, S.S.; et al. Impact evaluation of the efficacy of different doses of vitamin D supplementation during pregnancy on pregnancy and birth outcomes: A randomised, controlled, dose comparison trial in Pakistan. *BMJ Nutr. Prev. Health* **2021**, *4*, 425–434. [CrossRef]
130. Wagner, C.L.; Hollis, B.W. The Implications of Vitamin D Status During Pregnancy on Mother and her Developing Child. *Front. Endocrinol.* **2018**, *9*, 500. [CrossRef]

131. Dopico, X.C.; Evangelou, M.; Ferreira, R.C.; Guo, H.; Pekalski, M.L.; Smyth, D.J.; Cooper, N.; Burren, O.S.; Fulford, A.J.; Hennig, B.J.; et al. Widespread seasonal gene expression reveals annual differences in human immunity and physiology. *Nat. Commun.* **2015**, *6*, 7000. [CrossRef]
132. Goldinger, A.; Shakhbazov, K.; Henders, A.K.; McRae, A.F.; Montgomery, G.W.; Powell, J.E. Seasonal effects on gene expression. *PLoS ONE* **2015**, *10*, e0126995. [CrossRef] [PubMed]
133. Hossein-nezhad, A.; Spira, A.; Holick, M.F. Influence of vitamin D status and vitamin D3 supplementation on genome wide expression of white blood cells: A randomized double-blind clinical trial. *PLoS ONE* **2013**, *8*, e58725. [CrossRef] [PubMed]
134. Shirvani, A.; Kalajian, T.A.; Song, A.; Holick, M.F. Disassociation of Vitamin D's Calcemic Activity and Non-calcemic Genomic Activity and Individual Responsiveness: A Randomized Controlled Double-Blind Clinical Trial. *Sci. Rep.* **2019**, *9*, 17685. [CrossRef] [PubMed]
135. Shirvani, A.; Kalajian, T.A.; Song, A.; Allen, R.; Charoenngam, N.; Lewanczuk, R.; Holick, M.F. Variable Genomic and Metabolomic Responses to Varying Doses of Vitamin D Supplementation. *Anticancer Res.* **2020**, *40*, 535–543. [CrossRef]
136. Wimalawansa, S.J. Vitamin D Deficiency: Effects on Oxidative Stress, Epigenetics, Gene Regulation, and Aging. *Biology* **2019**, *8*, 30. [CrossRef]
137. Bogdan, C. Nitric oxide and the regulation of gene expression. *Trends Cell Biol.* **2001**, *11*, 66–75. [CrossRef]
138. Hemish, J.; Nakaya, N.; Mittal, V.; Enikolopov, G. Nitric oxide activates diverse signaling pathways to regulate gene expression. *J. Biol. Chem.* **2003**, *278*, 42321–42329. [CrossRef]
139. Cohen, A.J.; Brauer, M.; Burnett, R.; Anderson, H.R.; Frostad, J.; Estep, K.; Balakrishnan, K.; Brunekreef, B.; Dandona, L.; Dandona, R.; et al. Estimates and 25-year trends of the global burden of disease attributable to ambient air pollution: An analysis of data from the Global Burden of Diseases Study 2015. *Lancet* **2017**, *389*, 1907–1918. [CrossRef]
140. Hu, Z. Spatial analysis of MODIS aerosol optical depth, PM2.5, and chronic coronary heart disease. *Int. J. Health Geogr.* **2009**, *8*, 27. [CrossRef]
141. Feigin, V.L.; Roth, G.A.; Naghavi, M.; Parmar, P.; Krishnamurthi, R.; Chugh, S.; Mensah, G.A.; Norrving, B.; Shiue, I.; Ng, M.; et al. Global burden of stroke and risk factors in 188 countries, during 1990–2013: A systematic analysis for the Global Burden of Disease Study 2013. *Lancet Neurol.* **2016**, *15*, 913–924. [CrossRef]
142. Lelieveld, J.; Klingmuller, K.; Pozzer, A.; Poschl, U.; Fnais, M.; Daiber, A.; Munzel, T. Cardiovascular disease burden from ambient air pollution in Europe reassessed using novel hazard ratio functions. *Eur. Heart J.* **2019**, *40*, 1590–1596. [CrossRef] [PubMed]
143. Stieb, D.M.; Judek, S.; Burnett, R.T. Meta-analysis of time-series studies of air pollution and mortality: Effects of gases and particles and the influence of cause of death, age, and season. *J. Air Waste Manag. Assoc.* **2002**, *52*, 470–484. [CrossRef] [PubMed]
144. Corless, D.; Gupta, S.P.; Switala, S.; Barragry, J.M.; Boucher, B.J.; Cohen, R.D.; Diffey, B.L. Response of plasma-25-hydroxyvitamin D to ultraviolet irradiation in long-stay geriatric patients. *Lancet* **1978**, *2*, 649–651. [CrossRef]
145. Chuck, A.; Todd, J.; Diffey, B. Subliminal ultraviolet-B irradiation for the prevention of vitamin D deficiency in the elderly: A feasibility study. *Photodermatol. Photoimmunol. Photomed.* **2001**, *17*, 168–171. [CrossRef]
146. Chandra, P.; Wolfenden, L.L.; Ziegler, T.R.; Tian, J.; Luo, M.; Stecenko, A.A.; Chen, T.C.; Holick, M.F.; Tangpricha, V. Treatment of vitamin D deficiency with UV light in patients with malabsorption syndromes: A case series. *Photodermatol. Photoimmunol. Photomed.* **2007**, *23*, 179–185. [CrossRef]
147. Dabai, N.S.; Pramyothin, P.; Holick, M.F. The effect of ultraviolet radiation from a novel portable fluorescent lamp on serum 25-hydroxyvitamin D3 levels in healthy adults with Fitzpatrick skin types II and III. *Photodermatol. Photoimmunol. Photomed.* **2012**, *28*, 307–311. [CrossRef]
148. Park, D.-H.; Oh, S.-T.; Lim, J.-H. Development of UVB LED Lighting System Based on UV Dose Calculation Algorithm to Meet Individual Daily UV Dose. *Appl. Sci.* **2019**, *9*, 2479. [CrossRef]
149. Oh, S.Y.; Lim, J.-H. Development and Effect Analysis of UVB-LED General Lighting to Support Vitamin D Synthesis. *Appl. Sci.* **2020**, *10*, 889. [CrossRef]
150. Ames, B.N.; Grant, W.B.; Willett, W.C. Does the High Prevalence of Vitamin D Deficiency in African Americans Contribute to Health Disparities? *Nutrients* **2021**, *13*, 499. [CrossRef]
151. Jablonski, N.G.; Chaplin, G. Colloquium paper: Human skin pigmentation as an adaptation to UV radiation. *Proc. Natl. Acad. Sci. USA* **2010**, *107* (Suppl. 2), 8962–8968. [CrossRef]
152. Mohania, D.; Chandel, S.; Kumar, P.; Verma, V.; Digvijay, K.; Tripathi, D.; Choudhury, K.; Mitten, S.K.; Shah, D. Ultraviolet Radiations: Skin Defense-Damage Mechanism. *Adv. Exp. Med. Biol.* **2017**, *996*, 71–87. [PubMed]
153. Scott, J.F.; Das, L.M.; Ahsanuddin, S.; Qiu, Y.; Binko, A.M.; Traylor, Z.P.; Debanne, S.M.; Cooper, K.D.; Boxer, R.; Lu, K.Q. Oral Vitamin D Rapidly Attenuates Inflammation from Sunburn: An Interventional Study. *J. Investig. Dermatol.* **2017**, *137*, 2078–2086. [CrossRef] [PubMed]
154. Scott, J.F.; Lu, K.Q. Vitamin D as a Therapeutic Option for Sunburn: Clinical and Biologic Implications. *DNA Cell Biol.* **2017**, *36*, 879–882. [CrossRef] [PubMed]
155. Leal, A.; Correa, M.P.; Holick, M.F.; Melo, E.V.; Lazaretti-Castro, M. Sun-induced production of vitamin D3 throughout 1 year in tropical and subtropical regions: Relationship with latitude, cloudiness, UV-B exposure and solar zenith angle. *Photochem. Photobiol. Sci.* **2021**, *20*, 265–274. [CrossRef] [PubMed]
156. Reichrath, J.; Vogt, T.; Holick, M.F.; Friedrich, M. Abstracts of the Joint International Symposia "Vitamin D in Prevention and Therapy" and "Biologic Effects of Light". *Anticancer Res.* **2022**, *42*, 2193–2222. [CrossRef]

nutrients

MDPI

Article

Temporal Association of Reduced Serum Vitamin D with COVID-19 Infection: Two Single-Institution Case–Control Studies

Diviya Gupta †, Sahit Menon †, Michael H. Criqui and Bryan K. Sun *

School of Medicine, University of California San Diego, La Jolla, CA 92093, USA; digupta@health.ucsd.edu (D.G.); s1menon@health.ucsd.edu (S.M.); mcriqui@health.ucsd.edu (M.H.C.)
* Correspondence: bryansun@health.ucsd.edu; Tel.: +1-(858)-534-9418
† These authors contributed equally to this work.

Abstract: (1) Background: Vitamin D supplementation has been proposed for the prevention and treatment of COVID-19, but it is not clear if reduced serum vitamin D predisposes individuals to COVID-19 and/or is a secondary consequence of infection. This study assessed the temporal association between serum vitamin D and COVID-19 with two single-institution case–control studies through the University of California San Diego (UCSD) Health System. (2) Methods: This study included patients who tested positive for COVID-19 from 1 January to 30 September 2020 with serum 25-hydroxy-vitamin D (25(OH)D) measured within 180 days of diagnosis. Patients were separated based on whether 25(OH)D was measured before (n = 107 cases, 214 controls) or after (n = 203 cases, 406 controls) COVID-19 diagnosis. COVID-19 infection status was the outcome variable in the pre-diagnosis study, whereas serum 25(OH)D level was the outcome variable in the post-diagnosis study. (3) Results: Serum 25(OH)D levels were not associated with the odds of subsequent COVID-19 infection (OR 1.0, 95% CI: 1.0 to 1.0, p = 0.98). However, COVID-19-positive individuals had serum 25(OH)D measurements that were 2.7 ng/mL lower than the controls (95% CI: −5.2 to −0.2, p = 0.03). (4) Conclusions: In our study population, serum 25(OH)D levels were not associated with the risk of acquiring COVID-19 infection but were reduced in subjects after COVID-19 infection. These results support the possibility that reduced serum 25(OH)D is a consequence and not a cause of COVID-19 infection.

Keywords: COVID-19; vitamin D; 25(OH)D; case–control study

Citation: Gupta, D.; Menon, S.; Criqui, M.H.; Sun, B.K. Temporal Association of Reduced Serum Vitamin D with COVID-19 Infection: Two Single-Institution Case–Control Studies. *Nutrients* **2022**, *14*, 2757. https://doi.org/10.3390/nu14132757

Academic Editors: William B. Grant and Ronan Lordan

Received: 19 June 2022
Accepted: 28 June 2022
Published: 2 July 2022

1. Introduction

SARS-CoV-2, the coronavirus that causes COVID-19, has claimed over 6 million lives globally. In addition to vaccination efforts, there has been widespread public interest in complementary measures to mitigate the risk of viral infection and mortality. Vitamin D, available as an inexpensive supplement, gained extensive attention for its potential role in the prevention and treatment of COVID-19. However, the relationship between vitamin D and COVID-19 is unclear.

Mechanistically, vitamin D may enhance the immune response to SARS-CoV-2 in several ways. Vitamin D boosts innate immunity by augmenting the expression of human cathelicidin antimicrobial peptide (CAMP) [1] in lung and skin epithelial cells [2], which has been shown to attenuate the infectivity and viability of viruses [3]. In addition, vitamin D prevents excessive adaptive immune responses [4]. Because systemic inflammatory responses have been associated with respiratory distress and mortality in patients with severe COVID-19 [5], it has been proposed that vitamin D supplementation could mitigate these harmful inflammatory reactions.

Some medical experts and public figures have recommended vitamin D supplementation for COVID-19 [6], using serum 25-hydroxy-vitamin D (25(OH)D), a standard laboratory serum marker of vitamin D stores, as a primary biomarker for vitamin sufficiency. Recommendations for supplementation have particularly been endorsed for populations with the most elevated COVID-19 risk and who also have higher rates of vitamin D deficiency, including the elderly, individuals with chronic diseases, darker-complected individuals such as African Americans, and those who live at high latitudes [7].

The studies that have evaluated the correlation between serum vitamin D and COVID-19 have shown mixed results. Some reports have concluded that there is an association between vitamin D deficiency and increased susceptibility to COVID-19 infection [8,9], but others have not [10,11]. Other studies, focusing on different endpoints, have found that 25(OH)D deficiency is correlated with a higher risk of intensive care unit admission, ventilation dependency, and a lower survival rate [12–16]. Beneficial outcomes from vitamin D supplementation trials have been reported [17], although these therapeutic studies have not always been conducted in randomized groups [18] and have had modest sample sizes. It has been postulated that any beneficial effect of vitamin D on severe COVID-19 could be masked by the effect of other adjunctive treatments such as dexamethasone [19]. Highly powered, randomized controlled trials will be needed to definitively test for causality.

The differences in conclusions among the published results may be caused, at least in part, by significant methodological differences. Some reports examined serum 25(OH)D based on country-wide averages [20,21] or inferred vitamin D status based on geographic latitude [22] and did not directly assess serum vitamin D in individual subjects. Other studies used vitamin D results that were measured years prior to COVID-19 testing [23], which may have resulted in inaccurate representations because 25(OH)D levels can change significantly with time [24] and season [25]. Finally, some reports assessed 25(OH)D levels drawn prior to COVID-19 diagnosis [8], while others relied on measurements taken after diagnosis [12,26,27], and others included both [28]. Because severe illness itself can cause the rapid reduction of serum 25(OH)D [29], the timing of laboratory measurement is critical: vitamin D deficiency may predispose individuals to COVID-19, but it is also possible that COVID-19 infection can reduce 25(OH)D.

In this report, we sought to address these methodological differences by examining the temporal correlation between serum 25(OH)D and COVID-19 infection. We performed two complementary single-center studies examining patients testing positive for COVID-19 between 1 January 2020 to 30 September 2020 in the University of California San Diego (UCSD) Health system who had a serum 25(OH)D assessment within 180 days of diagnosis. These dates capture most COVID-19 cases in the UCSD Health system from the onset of the COVID-19 pandemic until the initiation of vaccinations.

In the first study, we used a case–control design to compare the 25(OH)D levels drawn prior to COVID-19-positive diagnosis with COVID-19-negative controls matched by age, sex, body mass index (BMI), diabetes, hypertension, time from vitamin D draw, and the season that the 25(OH)D test was performed. In the second study, we applied a case–control design to assess serum 25(OH)D levels drawn after COVID-19 diagnosis, applying the same matching criteria. Finally, we performed a subgroup analysis of the second study to specifically examine COVID-19 patients whose disease was severe enough to require hospitalization, comparing them against a matched hospitalized cohort that was COVID-19-negative.

Our overarching approach was based on the reasoning that if vitamin D deficiency increased susceptibility to COVID-19 infection, then serum 25(OH)D levels drawn prior to diagnosis would be significantly lower in cases vs. controls. However, if 25(OH)D was lower in COVID-19 cases only if measured after diagnosis, then COVID-19 infection itself may have led to a reduction in vitamin D levels. The null hypotheses were that there was no correlation between 25(OH)D levels and COVID-19 in either study. Finally, by focusing on the subgroup of patients requiring hospitalization for COVID-19 and comparing it to a matched hospitalized cohort, we sought to examine if any correlation between 25(OH)D and COVID-19 was amplified among patients affected by severe forms of COVID-19 illness.

2. Materials and Methods

The UCSD Institutional Review Board approved this study and approved a waiver for informed consent based on the requirements outlined in the Code of Federal Regulations on the Protection of Human Subjects (45 CFR 46). All data collection and analysis were performed in accordance with relevant guidelines and regulations. Data were collected for all subjects who tested for COVID-19 through the UCSD Health system from 1 January 2020 to 30 September 2020 (n = 6050), capturing an interval from the onset of local infections to the time prior to the initiation of COVID-19 vaccinations.

For primary analysis (Figure 1), cases were identified as patients who had serum 25(OH)D drawn within 180 days of COVID-19 diagnosis (n = 346). Seventeen patients had serum 25(OH)D drawn within 14 days of COVID-19 diagnosis and were excluded to account for the SARS-CoV-2 incubation period and to minimize confounding by possible early manifestations of COVID-19 [30]. Additionally, 18 patients in the pre-diagnosis study received vitamin D supplementation after serum 25(OH)D testing, but before COVID-19 diagnosis; these subjects were excluded due to probable changes in serum 25(OH)D levels from supplementation. One patient was removed from the pre-diagnosis study due to incomplete data on matching covariates.

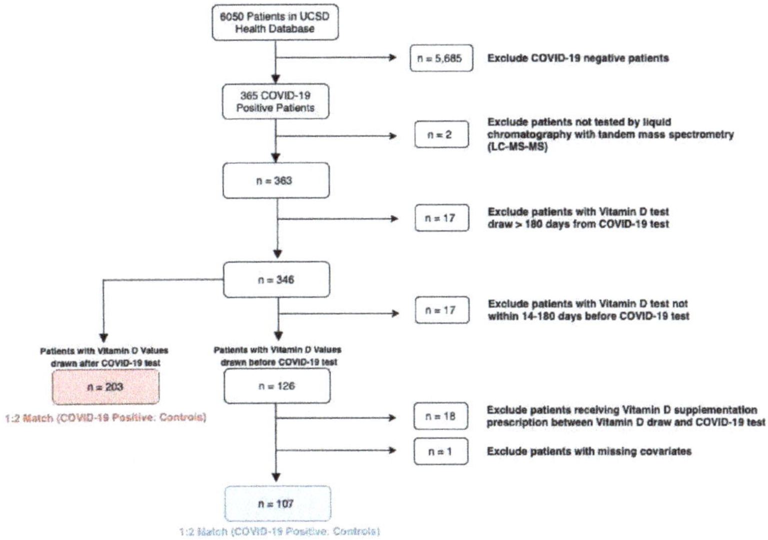

Figure 1. Flow diagram showing the selection of study subjects and controls.

After exclusions, there were 107 cases in the pre-diagnosis study and 203 cases in the post-diagnosis study. Cases in both studies were matched 1:2 to COVID-19-negative controls by age, sex, BMI, diagnosis of diabetes (ICD-10 codes E11.0–E11.9), diagnosis of hypertension (ICD-10 code I10), the number of days between vitamin D draw and COVID-19 diagnosis, and the meteorological season of 25(OH)D laboratory draw. Patient ethnicity and skin phototype data were not available. Matching was performed using nearest-neighbor matching based on Mahalanobis distance.

For the pre-diagnosis study, a case–control analysis was performed with 25(OH)D as the independent variable and COVID-19 infection status as the dependent variable. A conditional logistic regression model was performed to estimate the odds ratio (OR) and 95% confidence interval (95% CI) for serum 25(OH)D as a continuous variable. Results are reported as the change in odds of COVID-19 positivity for every 1 ng/mL increase in 25(OH)D.

For the post-diagnosis study, a case–control analysis was performed in which COVID-19 infection status served as the binary independent variable to assess serum 25(OH)D levels as the outcome. An ordinary least squares (OLS) regression model was performed to estimate the change in serum 25(OH)D per ng/mL associated with COVID-19 positivity.

In a subset analysis of the second study, serum 25(OH)D levels in hospitalized COVID-19 patients were compared to COVID-19-negative hospitalized patients. Here, cases were defined as patients who had serum 25(OH)D drawn up to 180 days after inpatient admission due to COVID-19 ($n = 120$). Matching criteria were identical to other analyses except for two changes. First, subjects were matched by the number of days between vitamin D draw and COVID-19 hospitalization, not COVID-19 diagnosis, and second, length of hospital stay was included as an additional matching covariate to promote the comparison of hospitalized patients with similar disease severities [31]. An OLS regression model was performed to estimate the change in serum 25(OH)D per ng/mL associated with COVID-19 illness.

Analyses were performed using RStudio software, version 1.3.959. Continuous covariates are reported as mean ± standard deviation (SD) and compared using unpaired *t*-tests; categorical variables are reported as numbers and percentages and compared using Chi-square tests. The significance level for all analyses was set to a two-sided p-value < 0.05.

Subjects or the public were not involved in the design, conduct, reporting, or dissemination plans of this study.

3. Results

3.1. Serum 25(OH)D Was Lower in COVID-19 Subjects Tested after, but Not before Diagnosis

Baseline characteristics of the study cohorts are shown in Tables 1 and 2. Covariates consisted of established risk factors for COVID-19 susceptibility [32] that included age, sex, obesity, and medical comorbidities. In addition, due to seasonal variation in serum vitamin D levels, control subjects were matched by the meteorological season that the serum vitamin D was measured, as well as length of the delay between serum vitamin D assessment and COVID-19 testing. Matched study populations were balanced between groups.

Table 1. Characteristics of the COVID-19 study population and controls: serum 25(OH)D measured before COVID-19 diagnosis.

Characteristic	COVID-19-Positive ($n = 107$)	COVID-19-Negative ($n = 214$)	p-Value
Age (years) ± SD	51.2 ± 16.0	52.6 ± 15.3	0.49
Sex—no. (%)			
Male	53 (49.5)	106 (49.5)	1.00
Female	54 (50.5)	108 (50.5)	
Body Mass Index ± SD	27.8 ± 5.8	27.5 ± 5.4	0.67
Diabetes—no. (%)			
Yes	22 (20.6)	44 (20.6)	1.00
No	85 (79.4)	170 (79.4)	
Hypertension—no. (%)			
Yes	50 (46.7)	100 (46.7)	1.00
No	57 (53.3)	114 (53.3)	
Days between vitamin D and COVID-19 test ± SD	77.4 ± 39.9	77.7 ± 39.6	0.95
Season—no. (%)			
Winter (1 January–29 February)	36 (33.6)	72 (33.6)	
Spring (1 March–31 May)	33 (30.8)	68 (31.8)	0.98
Summer (1 June–31 August)	38 (35.5)	74 (34.6)	
Fall (1 September–30 September)	-	-	

Table 2. Characteristics of the COVID-19 study population and controls: serum 25(OH)D measured after COVID-19 diagnosis.

Characteristic	COVID-19-Positive (n = 203)	COVID-19-Negative (n = 406)	p-Value
Age (years) ± SD	52.7 ± 15.7	53.4 ± 15.2	0.60
Sex—no. (%)			
Male	123 (60.6)	246 (60.6)	1.00
Female	80 (39.4)	160 (39.4)	
Body Mass Index ± SD	28.0 ± 6.5	27.6 ± 5.9	0.44
Diabetes—no. (%)			
Yes	70 (34.5)	140 (34.5)	1.00
No	133 (65.5)	266 (65.5)	
Hypertension—no. (%)			
Yes	109 (53.7)	218 (53.7)	1.00
No	94 (46.3)	188 (46.3)	
Days between vitamin D and COVID-19 test ± SD	32.3 ± 40.4	32.3 ± 38.9	0.99
Season—no. (%)			
Winter (1 January–29 February)	-	-	
Spring (1 March–31 May)	42 (20.7)	82 (20.2)	
Summer (1 June–31 August)	124 (61.1)	251 (61.8)	0.98
Fall (1 September–30 September)	37 (18.2)	73 (18.0)	

For subjects in which 25(OH)D serum levels were assessed prior to COVID-19 testing, the mean serum 25(OH)D was 35.5 ng/mL (SD 13.7) for cases and 35.4 ng/mL (SD 13.8) for controls. A one-unit increase in serum 25(OH)D did not affect the odds of contracting COVID-19 (OR 1.0, 95% CI 1.0 to 1.0, $p = 0.98$). These data revealed no significant association between serum 25(OH)D and the odds of subsequent COVID-19 positivity.

In contrast, in subjects for whom 25(OH)D was measured after diagnosis, subsequent assessment of 25(OH)D showed a mean serum 25(OH)D of 30.5 ng/mL (SD 15.5) for cases and 33.2 ng/mL (SD 15.7) for controls. COVID-19 positivity was associated with serum 25(OH)D levels that were lower by 2.7 ng/mL on average (95% CI −5.2 to −0.2, $p = 0.03$) (Tables 3 and 4). These data indicated that COVID-19-positive subjects showed a significant reduction in 25(OH)D compared to matched COVID-19-negative subjects.

Table 3. Pre-diagnosis—conditional logistic regression.

Predictor	Cases (n = 107) Serum 25(OH)D (ng/mL) ± SD	Controls (n = 214) Serum 25(OH)D (ng/mL) ± SD	Odds Ratio	95% CI	p-Value
Vitamin D (ng/mL)	35.5 ± 13.7	35.4 ± 13.8	1.0	1.0 to 1.0	0.98

Table 4. Post-diagnosis—ordinary least squares regression.

Predictor	Cases (n = 203) Serum 25(OH)D (ng/mL) ± SD	Controls (n = 406) Serum 25(OH)D (ng/mL) ± SD	Beta Estimate	95% CI	p-Value
COVID-19 Infection	30.5 ± 15.5	33.2 ± 15.7	−2.7	−5.2 to −0.2	0.03

Continuous covariates are reported as mean ± standard deviation (SD) and compared using unpaired t-tests. Categorical variables are reported as numbers (no.) and percentages (%) and compared using Chi-square tests. A p-value < 0.05 was considered a significant difference for covariates. Season dates were defined by the meteorological seasons within the study interval.

Serum 25(OH)D for cases and controls are reported as mean levels in ng/mL ± standard deviation (SD). In the pre-diagnosis study, a conditional logistic regression was performed in which the independent variable is continuous (i.e., serum 25(OH)D level) and the dependent variable is binary (i.e., COVID-19 infection status).

In the post-diagnosis study, an ordinary least squares regression was performed, in which the independent variable is binary (i.e., COVID-19 infection status) and the dependent variable is continuous (i.e., serum 25(OH)D level). Abbreviations: CI = confidence interval; 25(OH)D = 25-hydroxy vitamin D.

3.2. *Reduced 25(OH)D in COVID-19-Positive Hospitalized Patients Compared to a COVID-19-Negative Hospitalized Cohort*

It has been reported that vitamin D deficiency may correlate to severe outcomes from COVID-19 infection. To assess for potential correlation between serum 25(OH)D and severe COVID-19 infection, a subgroup analysis was performed to compare COVID-19 subjects requiring hospitalization against a COVID-negative hospitalized control group. To promote matching patients with comparable disease severity, control patients were also matched by the length of hospital stay, an indicator of disease severity [31]. Baseline characteristics of the matched analysis for the hospitalized cohort are shown in Table 5. No significant differences were observed for matching characteristics between cases and controls.

Table 5. Characteristics of hospitalized COVID-19 study population and controls.

Characteristic	COVID-19-Positive (n = 120)	COVID-19-Negative (n = 240)	p-Value
Age (years) ± SD	55.9 ± 14.6	57.9 ± 14.0	0.22
Sex—no. (%)			
Male	82 (68.3)	165 (68.8)	0.94
Female	38 (31.7)	75 (31.3)	
Body Mass Index ± SD	28.1 ± 6.8	26.8 ± 6.3	0.07
Diabetes—no. (%)			
Yes	59 (49.2)	115 (47.9)	0.82
No	61 (50.8)	125 (52.1)	
Hypertension—no. (%)			
Yes	83 (69.2)	165 (68.8)	0.94
No	37 (30.8)	75 (31.4)	
Days between vitamin D test and COVID-19 hospitalization ± SD	21.0 ± 34.4	19.9 ± 30.0	0.76
Season—no. (%)			
Winter (1 January–29 February)	7 (5.8)	16 (6.7)	
Spring (1 March–31 May)	37 (30.8)	77 (32.1)	
Summer (1 June–31 August)	67 (55.8)	128 (53.3)	0.97
Fall (1 September–30 September)	9 (7.5)	19 (7.9)	
Length of hospitalization (days) ± SD	21.4 ± 21.4	17.2 ± 17.8	0.07

Continuous covariates are reported as mean ± standard deviation (SD) and compared using unpaired t-tests. Categorical variables are reported as numbers (no.) and percentages (%) and compared using Chi-square tests. A p-value less than 0.05 was considered a significant difference for covariates. Season dates were defined by the meteorological seasons within the study interval.

Within a 180-day window following inpatient admission, Table 6 shows that COVID-19-positive hospitalized cases had a mean 25(OH)D of 23.9 ng/mL (SD 13.5), while COVID-19-negative hospitalized controls had a mean 25(OH)D level of 27.3 ng/mL (SD 15.4). Thus, patient hospitalization due to COVID-19 infection was associated with 3.3 ng/mL (95% CI −6.3 to −0.4, p = 0.03) lower serum 25(OH)D levels compared to hospitalized COVID-19-negative patients (Table 6). These data indicated that COVID-19-positive hospitalized subjects showed a significant reduction in 25(OH)D compared to COVID-19-negative hospitalized subjects.

Table 6. Association of serum 25-hydroxy vitamin D with severe COVID-19 infection.

Predictor	Cases (n = 120) Serum 25(OH)D (ng/mL) ± SD	Controls (n = 240) Serum 25 (OH)D (ng/mL) ± SD	Beta Estimate	95% CI	p-Value
COVID-19 Hospitalization	23.9 ± 13.5	27.3 ± 15.4	−3.3	−6.3 to −0.4	0.03

Serum 25(OH)D for cases and controls is reported as mean levels in ng/mL ± standard deviation (SD). An ordinary least squares regression was performed in which the independent variable is binary (i.e., hospitalization due to COVID-19 or hospitalization due to another cause) and the dependent variable is continuous (i.e., serum 25(OH)D level). Abbreviations: CI = confidence interval; 25(OH)D = 25-hydroxy vitamin D.

4. Discussion

Defining the temporal association between serum vitamin D and COVID-19 infection provides a basis for evaluating the potential use of vitamin D supplementation to prevent COVID-19 infection and/or mitigate disease severity. Our single-center study found that COVID-19 infection and hospitalization were associated with lower serum vitamin D levels drawn after diagnosis or hospital admission. However, serum 25(OH)D levels did not affect the odds of initially testing positive for COVID-19, indicating that lower vitamin D was not a risk factor for COVID-19 infection in our study population. Among the published studies, some reports have proposed that vitamin D insufficiency, defined as serum concentrations 20–30 ng/mL, or vitamin D deficiency, defined as concentrations <20 ng/mL, may be risk factors for COVID-19 infection [33]. To explore this possibility, we also stratified our pre-diagnosis subjects into these categories. We found no statistically significant association with increased odds of COVID-19 with either vitamin D insufficiency or deficiency compared to the sufficient (>30 ng/mL) reference group. Viewed together, our results are consistent with other reports that identified lower 25(OH)D in association with COVID-19 [8,12,16,34], but suggest that lower 25(OH)D levels may be an outcome of COVID-19 infection rather than a cause of it.

The relationship between vitamin D levels and medical conditions is complex. In certain contexts, including a study that analyzed the relationship between 25(OH)D concentration and non-severe community-acquired pneumonia, there was no difference in serum (OH)D levels during the acute phase and up to 90 days after recovery [35]. Similarly, a study of 14 patients in the acute phase response to malaria demonstrated that vitamin D levels remained unaffected over the course of hospital stay and 2–6 weeks after discharge [36]. In contrast, in studies on pancreatitis, vitamin D has been shown to have statistically significant decreases due to inflammatory processes [37,38]. A systematic review cautioned against the notion that inflammatory conditions generate rapid decreases in vitamin D, partly because its findings illustrated the possibility of heterogeneity; however, many studies still showed a reduction in vitamin D following inflammatory insult [39]. Given the plausible relationship between inflammatory conditions and changes in vitamin D levels, reverse causality may explain some of the changes in vitamin D levels, and that preexisting vitamin D status alone is not solely responsible [40].

Our findings are consistent with prior reports that examined the relationship between vitamin D and respiratory illnesses. While vitamin D deficiency has been associated with an increased risk of respiratory infections [41], acute illness itself can also reduce serum 25(OH)D through fluid shifts, the depletion of serum binding proteins, and renal wasting [29]. Acute inflammation following surgery has been associated with a reduction in serum 25(OH)D within 48 h [42]. Inflammatory mediators lead to an increase in the activity of CYP24A1 and CYP27B1, enzymes that metabolize vitamin D pathway compounds [43]. Vitamin D metabolism is dysregulated in patients with asthma and chronic obstructive pulmonary diseases, leading some investigators to suggest that the relationship between airway inflammation and vitamin D deficiency is bidirectional [44]. These findings support the biological plausibility that elevated rates of vitamin D deficiency observed in subjects infected by COVID-19 could be, at least in part, secondary to the respiratory infection itself [40].

Our study has several strengths. First, our data from a single institution directly assessed 25(OH)D and COVID-19 laboratory results using uniform, standardized assays, minimizing the significant variations that can occur between different testing methodologies [45]. Second, our restriction of 25(OH)D values to a 180-day window, matching controls by season, and matching by the time interval between 25(OH)D and COVID-19 testing addresses important time-dependent effects that affect 25(OH)D levels [23,25]. Third, our stratification of 25(OH)D data from the same institutional population into subjects who were tested before vs. after COVID-19 diagnosis allowed us to assess the temporal relationship between vitamin D and COVID-19.

Our study also has several limitations. The retrospective design does not allow for the determination of causality, and our sample size is not powered to detect smaller, but potentially significant, correlations. Additionally, although we matched baseline characteristics between cases and controls, other unaccounted confounders could have affected the results. Data on the racial identity of our study subjects was incomplete, which did not allow us to include this factor in cohort matching. Racial disparities in COVID-19 illness have been observed, with one retrospective study finding that a positive COVID-19 test was associated with lower vitamin D levels in Black but not White individuals [8]. Our health system's catchment area (San Diego County) has a lower Black population (~5.5%) than the U.S. national average (~13.4%) and has a relatively high level of sun and UV exposure. UV exposure may provide a protective effect against COVID-19 both through vitamin D-dependent and -independent mechanisms [46]. Therefore, the results from our analysis may differ from other study populations and do not argue against the potential utility of vitamin D supplementation for specific populations.

Ultimately, intervention trials could provide the most conclusive insight to the therapeutic value of vitamin D supplementation both prior to and after COVID-19 infection. Early reports have shown mixed results: A small ($n = 76$) randomized trial indicated that oral calcifediol supplementation reduced the need for intensive care unit admission in COVID-19-infected subjects [17], though the trial was not placebo-controlled and did not measure baseline or post-treatment serum vitamin D levels. By contrast, a randomized, double-blind, placebo-controlled trial on hospitalized COVID-19 patients found no benefit to a single 200,000 IU dose of vitamin D3 on the length of hospital stay [47]. Viewed together with the results from our study and taken in context with other published studies to date, we recommend caution in the therapeutic expectations for vitamin D supplementation in the prevention of COVID-19.

5. Conclusions

We performed two complementary case–control studies at UCSD Health to evaluate the temporal association of serum vitamin D and COVID-19 infection, examining 25(OH)D levels drawn before or after COVID-19 diagnosis. Our main finding is that serum 25(OH)D drawn before COVID-19 diagnosis did not differ between cases and controls, but serum 25(OH)D drawn in subjects who tested positive for COVID-19 was significantly lower than matched controls. This result was even more pronounced in subjects with severe COVID-19 infection requiring hospitalization. These results support the possibility that lower serum vitamin D levels may not predispose individuals to COVID-19 infection, but may be a consequence of it.

Author Contributions: Study conception and design, D.G., S.M., M.H.C. and B.K.S.; acquisition of the data, D.G. and B.K.S.; data analysis and interpretation, D.G., S.M., M.H.C. and B.K.S.; writing—original draft preparation, D.G. and S.M.; writing—review and editing, M.H.C. and B.K.S.; created visualizations, D.G. and S.M.; project administration, B.K.S. All authors have read and agreed to the published version of the manuscript.

Funding: This work was supported by the UC San Diego Office of Research Affairs. Award/grant number is not applicable.

Institutional Review Board Statement: The UCSD Institutional Review Board approved this study and approved a waiver for informed consent based on the requirements outlined in the Code of Federal Regulations on the Protection of Human Subjects (45 CFR 46).

Informed Consent Statement: Not applicable.

Data Availability Statement: The datasets for this study are not publicly published due to the presence of potentially patient-identifiable information, but will be made available from the corresponding author on reasonable request.

Acknowledgments: We would like to thank Amy Sitapati for her expertise in informatics and for offering her advice in developing this publication.

Conflicts of Interest: The authors declare no conflict of interest.

Abbreviations

25(OH)D	25-hydroxy-vitamin D
BMI	body mass index
CAMP	cathelicidin antimicrobial peptide
CI	confidence interval
OLS	ordinary least squares
SD	standard deviation
UCSD	University of California San Diego (UCSD)

References

1. Wei, R.; Christakos, S. Mechanisms Underlying the Regulation of Innate and Adaptive Immunity by Vitamin D. *Nutrients* **2015**, *7*, 8251–8260. [CrossRef] [PubMed]
2. Ginde, A.A.; Mansbach, J.M.; Camargo, C.A. Vitamin D, Respiratory Infections, and Asthma. *Curr. Allergy Asthma Rep.* **2009**, *9*, 81–87. [CrossRef] [PubMed]
3. Barlow, P.G.; Svoboda, P.; Mackellar, A.; Nash, A.A.; York, I.A.; Pohl, J.; Davidson, D.J.; Donis, R.O. Antiviral Activity and Increased Host Defense against Influenza Infection Elicited by the Human Cathelicidin LL-37. *PLoS ONE* **2011**, *6*, e25333. [CrossRef]
4. Aranow, C. Vitamin D and the Immune System. *J. Investig. Med.* **2012**, *59*, 881–886. [CrossRef]
5. Sinha, P.; Matthay, M.A.; Calfee, C.S. Is a "Cytokine Storm" Relevant to COVID-19? *JAMA Intern. Med.* **2020**, *180*, 1152. [CrossRef] [PubMed]
6. Grant, W.B.; Lahore, H.; McDonnell, S.L.; Baggerly, C.A.; French, C.B.; Aliano, J.L.; Bhattoa, H.P. Evidence That Vitamin D Supplementation Could Reduce Risk of Influenza and COVID-19 Infections and Deaths. *Nutrients* **2020**, *12*, 988. [CrossRef] [PubMed]
7. Kimlin, M.G. Geographic Location and Vitamin D Synthesis. *Mol. Aspects Med.* **2008**, *29*, 453–461. [CrossRef] [PubMed]
8. Meltzer, D.O.; Best, T.J.; Zhang, H.; Vokes, T.; Arora, V.; Solway, J. Association of Vitamin D Status and Other Clinical Characteristics With COVID-19 Test Results. *JAMA Netw. Open* **2020**, *3*, e2019722. [CrossRef]
9. Merzon, E.; Tworowski, D.; Gorohovski, A.; Vinker, S.; Golan Cohen, A.; Green, I.; Frenkel-Morgenstern, M. Low Plasma 25(OH) Vitamin D Level Is Associated with Increased Risk of COVID-19 Infection: An Israeli Population-based Study. *FEBS J.* **2020**, *287*, 3693–3702. [CrossRef]
10. Hastie, C.E.; Mackay, D.F.; Ho, F.; Celis-Morales, C.A.; Katikireddi, S.V.; Niedzwiedz, C.L.; Jani, B.D.; Welsh, P.; Mair, F.S.; Gray, S.R.; et al. Vitamin D Concentrations and COVID-19 Infection in UK Biobank. *Diabet. Metab. Syndr. Clin. Res. Rev.* **2020**, *14*, 561–565. [CrossRef]
11. Ferrari, D.; Locatelli, M. No Significant Association between Vitamin D and COVID-19. A Retrospective Study from a Northern Italian Hospital. *Int. J. Vitamin Nutr. Res.* **2020**, *91*, 200–203. [CrossRef]
12. Radujkovic, A.; Hippchen, T.; Tiwari-Heckler, S.; Dreher, S.; Boxberger, M.; Merle, U. Vitamin D Deficiency and Outcome of COVID-19 Patients. *Nutrients* **2020**, *12*, 2757. [CrossRef] [PubMed]
13. Maghbooli, Z.; Sahraian, M.A.; Ebrahimi, M.; Pazoki, M.; Kafan, S.; Tabriz, H.M.; Hadadi, A.; Montazeri, M.; Nasiri, M.; Shirvani, A.; et al. Vitamin D Sufficiency, a Serum 25-Hydroxyvitamin D at Least 30 Ng/ML Reduced Risk for Adverse Clinical Outcomes in Patients with COVID-19 Infection. *PLoS ONE* **2020**, *15*, e0239799. [CrossRef] [PubMed]
14. Macaya, F.; Espejo Paeres, C.; Valls, A.; Fernández-Ortiz, A.; González Del Castillo, J.; Martín-Sánchez, F.J.; Runkle, I.; Rubio Herrera, M.Á. Interaction between Age and Vitamin D Deficiency in Severe COVID-19 Infection. *Nutr. Hosp.* **2020**, *37*, 1039–1042. [CrossRef]

15. Ye, K.; Tang, F.; Liao, X.; Shaw, B.A.; Deng, M.; Huang, G.; Qin, Z.; Peng, X.; Xiao, H.; Chen, C.; et al. Does Serum Vitamin D Level Affect COVID-19 Infection and Its Severity?-A Case-Control Study. *J. Am. Coll. Nutr.* **2020**, *40*, 724–731. [CrossRef]
16. Panagiotou, G.; Tee, S.A.; Ihsan, Y.; Athar, W.; Marchitelli, G.; Kelly, D.; Boot, C.S.; Stock, N.; Macfarlane, J.; Martineau, A.R.; et al. Low Serum 25-hydroxyvitamin D (25[OH]D) Levels in Patients Hospitalized with COVID-19 Are Associated with Greater Disease Severity. *Clin. Endocrinol.* **2020**, *93*, 508–511. [CrossRef]
17. Entrenas Castillo, M.; Entrenas Costa, L.M.; Vaquero Barrios, J.M.; Alcalá Díaz, J.F.; López Miranda, J.; Bouillon, R.; Quesada Gomez, J.M. Effect of Calcifediol Treatment and Best Available Therapy versus Best Available Therapy on Intensive Care Unit Admission and Mortality among Patients Hospitalized for COVID-19: A Pilot Randomized Clinical Study. *J. Steroid Biochem. Mol. Biol.* **2020**, *203*, 105751. [CrossRef] [PubMed]
18. Annweiler, C.; Hanotte, B.; Grandin de l'Eprevier, C.; Sabatier, J.-M.; Lafaie, L.; Célarier, T. Vitamin D and Survival in COVID-19 Patients: A Quasi-Experimental Study. *J. Steroid Biochem. Mol. Biol.* **2020**, *204*, 105771. [CrossRef]
19. Martineau, A.R.; Forouhi, N.G. Vitamin D for COVID-19: A Case to Answer? *Lancet Diabet. Endocrinol.* **2020**, *8*, 735–736. [CrossRef]
20. Ali, N. Role of Vitamin D in Preventing of COVID-19 Infection, Progression and Severity. *J. Infection Public Health* **2020**, *13*, 1373–1380. [CrossRef]
21. Ilie, P.C.; Stefanescu, S.; Smith, L. The Role of Vitamin D in the Prevention of Coronavirus Disease 2019 Infection and Mortality. *Aging Clin. Exp. Res.* **2020**, *32*, 1195–1198. [CrossRef] [PubMed]
22. Rhodes, J.; Dunstan, F.; Laird, E.; Subramanian, S.; Kenny, R.A. COVID-19 Mortality Increases with Northerly Latitude after Adjustment for Age Suggesting a Link with Ultraviolet and Vitamin D. *BMJ Nutr. Prev. Health* **2020**, *3*, 118–120. [CrossRef] [PubMed]
23. Kasahara, A.K.; Singh, R.J.; Noymer, A. Vitamin D (25OHD) Serum Seasonality in the United States. *PLoS ONE* **2013**, *8*, e65785. [CrossRef] [PubMed]
24. Meng, J.E.; Hovey, K.M.; Wactawski-Wende, J.; Andrews, C.A.; LaMonte, M.J.; Horst, R.L.; Genco, R.J.; Millen, A.E. Intraindividual Variation in Plasma 25-Hydroxyvitamin D Measures 5 Years Apart among Postmenopausal Women. *Cancer Epidemiol. Biomark. Prev.* **2012**, *21*, 916–924. [CrossRef] [PubMed]
25. Klingberg, E.; Oleröd, G.; Konar, J.; Petzold, M.; Hammarsten, O. Seasonal Variations in Serum 25-Hydroxy Vitamin D Levels in a Swedish Cohort. *Endocrine* **2015**, *49*, 800–808. [CrossRef] [PubMed]
26. Campi, I.; Gennari, L.; Merlotti, D.; Mingiano, C.; Frosali, A.; Giovanelli, L.; Torlasco, C.; Pengo, M.F.; Heilbron, F.; Soranna, D.; et al. Vitamin D and COVID-19 Severity and Related Mortality: A Prospective Study in Italy. *BMC Infect. Dis.* **2021**, *21*, 566. [CrossRef] [PubMed]
27. Davoudi, A.; Najafi, N.; Aarabi, M.; Tayebi, A.; Nikaeen, R.; Izadyar, H.; Salar, Z.; Delavarian, L.; Vaseghi, N.; Daftarian, Z.; et al. Lack of Association between Vitamin D Insufficiency and Clinical Outcomes of Patients with COVID-19 Infection. *BMC Infect. Dis.* **2021**, *21*, 450. [CrossRef]
28. D'Avolio, A.; Avataneo, V.; Manca, A.; Cusato, J.; De Nicolò, A.; Lucchini, R.; Keller, F.; Cantù, M. 25-Hydroxyvitamin D Concentrations Are Lower in Patients with Positive PCR for SARS-CoV-2. *Nutrients* **2020**, *12*, 1359. [CrossRef]
29. Quraishi, S.A.; Camargo, C.A. Vitamin D in Acute Stress and Critical Illness. *Curr. Opin. Clin. Nutr. Metab. Care* **2012**, *15*, 625–634. [CrossRef]
30. CDC COVID-19 Response Team; Bialek, S.; Boundy, E.; Bowen, V.; Chow, N.; Cohn, A.; Dowling, N.; Ellington, S.; Gierke, R. Severe Outcomes Among Patients with Coronavirus Disease 2019 (COVID-19) — United States, February 12–March 16, 2020. *MMWR Morb. Mortal. Wkly. Rep.* **2020**, *69*, 343–346. [CrossRef]
31. Hicks, N.; Kammerling, R. The Relationship between a Severity of Illness Indicator and Mortality and Length-of-Stay. *Health Trends* **1993**, *25*, 65–68. [PubMed]
32. Jordan, R.E.; Adab, P.; Cheng, K.K. Covid-19: Risk Factors for Severe Disease and Death. *BMJ* **2020**, *368*, m1198. [CrossRef] [PubMed]
33. Rubin, R. Sorting Out Whether Vitamin D Deficiency Raises COVID-19 Risk. *JAMA* **2021**, *325*, 329–330. [CrossRef] [PubMed]
34. Alpcan, A.; Tursun, S.; Kandur, Y. Vitamin D Levels in Children with COVID-19: A Report from Turkey. *Epidemiol. Infect.* **2021**, *149*, e180. [CrossRef] [PubMed]
35. Haugen, J.; Chandyo, R.K.; Ulak, M.; Mathisen, M.; Basnet, S.; Brokstad, K.A.; Valentiner-Branth, P.; Shrestha, P.S.; Strand, T.A. 25-Hydroxy-Vitamin D Concentration Is Not Affected by Severe or Non-Severe Pneumonia, or Inflammation, in Young Children. *Nutrients* **2017**, *9*, 52. [CrossRef]
36. Newens, K.; Filteau, S.; Tomkins, A. Plasma 25-Hydroxyvitamin D Does Not Vary over the Course of a Malarial Infection. *Transact. Royal Soc. Trop. Med. Hygiene* **2006**, *100*, 41–44. [CrossRef]
37. Bang, U.C.; Novovic, S.; Andersen, A.M.; Fenger, M.; Hansen, M.B.; Jensen, J.-E.B. Variations in Serum 25-Hydroxyvitamin D during Acute Pancreatitis: An Exploratory Longitudinal Study. *Endocrine Res.* **2011**, *36*, 135–141. [CrossRef]
38. Kruit, A.; Zanen, P. The Association between Vitamin D and C-Reactive Protein Levels in Patients with Inflammatory and Non-Inflammatory Diseases. *Clin. Biochem.* **2016**, *49*, 534–537. [CrossRef]
39. Silva, M.C.; Furlanetto, T.W. Does Serum 25-Hydroxyvitamin D Decrease during Acute-Phase Response? A Systematic Review. *Nutr. Res.* **2015**, *35*, 91–96. [CrossRef]

40. Smolders, J.; van den Ouweland, J.; Geven, C.; Pickkers, P.; Kox, M. Letter to the Editor: Vitamin D Deficiency in COVID-19: Mixing up Cause and Consequence. *Metabolism* **2021**, *115*, 154434. [CrossRef]
41. Hughes, D.A.; Norton, R. Vitamin D and Respiratory Health. *Clin. Exp. Immunol.* **2009**, *158*, 20–25. [CrossRef] [PubMed]
42. Waldron, J.L.; Ashby, H.L.; Cornes, M.P.; Bechervaise, J.; Razavi, C.; Thomas, O.L.; Chugh, S.; Deshpande, S.; Ford, C.; Gama, R. Vitamin D: A Negative Acute Phase Reactant. *J. Clin. Pathol.* **2013**, *66*, 620–622. [CrossRef] [PubMed]
43. Wang, J.; Liu, X.; Wang, H.; Li, Y.; Lan, N.; Yuan, X.; Wu, M.; Liu, Z.; Li, G. Allergen Specific Immunotherapy Enhanced Defense against Bacteria via TGF-β1-Induced CYP27B1 in Asthma. *Oncotarget* **2017**, *8*, 68681–68695. [CrossRef]
44. Jolliffe, D.A.; Stefanidis, C.; Wang, Z.; Kermani, N.Z.; Dimitrov, V.; White, J.H.; McDonough, J.E.; Janssens, W.; Pfeffer, P.; Griffiths, C.J.; et al. Vitamin D Metabolism Is Dysregulated in Asthma and Chronic Obstructive Pulmonary Disease. *Am. J. Respir. Crit. Care Med.* **2020**, *202*, 371–382. [CrossRef] [PubMed]
45. Lai, J.K.C.; Lucas, R.M.; Banks, E.; Ponsonby, A.-L. Ausimmune Investigator Group Variability in Vitamin D Assays Impairs Clinical Assessment of Vitamin D Status: Variability in Vitamin D Assays. *Intern. Med. J.* **2012**, *42*, 43–50. [CrossRef] [PubMed]
46. Lau, F.H.; Powell, C.E.; Adonecchi, G.; Danos, D.M.; DiNardo, A.R.; Chugden, R.J.; Wolf, P.; Castilla, C.F. Pilot Phase Results of a Prospective, Randomized Controlled Trial of Narrowband Ultraviolet B Phototherapy in Hospitalized COVID-19 Patients. *Exp. Dermatol.* **2022**. [CrossRef]
47. Murai, I.H.; Fernandes, A.L.; Sales, L.P.; Pinto, A.J.; Goessler, K.F.; Duran, C.S.C.; Silva, C.B.R.; Franco, A.S.; Macedo, M.B.; Dalmolin, H.H.H.; et al. Effect of a Single High Dose of Vitamin D3 on Hospital Length of Stay in Patients With Moderate to Severe COVID-19: A Randomized Clinical Trial. *JAMA* **2021**, *325*, 1053–1060. [CrossRef]

nutrients

MDPI

Systematic Review

Vitamin D-Related Risk Factors for Maternal Morbidity during Pregnancy: A Systematic Review

Maria M. Morales-Suárez-Varela [1,2], Nazlı Uçar [1], Isabel Peraita-Costa [1,2], María Flores Huertas [1], Jose Miguel Soriano [3], Agustin Llopis-Morales [1] and William B. Grant [4,*]

1 Area of Preventive Medicine and Public Health, Department of Preventive Medicine and Public Health, Food Sciences, Toxicology and Legal Medicine, School of Pharmacy, University de Valencia, Avenida Vicent Andres Estelles s/n, 46100 Valencia, Spain; maria.m.morales@uv.es (M.M.M.-S.-V.); unaz@alumni.uv.es (N.U.); ipecos@alumni.uv.es (I.P.-C.); marcia.f.huertas@uv.es (M.F.H.); allomo@alumni.uv.es (A.L.-M.)
2 Biomedical Research Center Network on Epidemiology and Public Health (CIBERESP), Institute of Health Carlos III, Avenida Monforte de Lemos, 3-5, Pabellón 11, Planta 0, 28029 Madrid, Spain
3 Unit of Nutrition and Bromatology, Department of Preventive Medicine and Public Health, Food Sciences, Toxicology and Forensic Medicine, Universitat de València, Avda. Vicente Andrés Estellés s/n, Burjassot, 46100 Valencia, Spain; jose.soriano@uv.es
4 Sunlight, Nutrition, and Health Research Center, P.O. Box 641603, San Francisco, CA 94164-1603, USA
* Correspondence: wbgrant@infionline.net

Abstract: Vitamin D has well-defined classical functions related to metabolism and bone health but also has non-classical effects that may influence pregnancy. Maternal morbidity remains a significant health care concern worldwide, despite efforts to improve maternal health. Nutritional deficiencies of vitamin D during pregnancy are related to adverse pregnancy outcomes, but the evidence base is difficult to navigate. The primary purpose of this review is to map the evidence on the effects of deficiencies of vitamin D on pregnancy outcome and the dosage used in such studies. A systematic search was performed for studies on vitamin D status during pregnancy and maternal outcomes. A total of 50 studies came from PubMed, 15 studies came from Cochrane, and 150 studies came from Embase, for a total of 215 articles. After screening, 34 were identified as candidate studies for inclusion. Finally, 28 articles met the inclusion criteria, which originated from 15 countries. The studies included 14 original research studies and 13 review studies conducted between 2012 and 2021. This review was finally limited to the 14 original studies. This systematic review was conducted according to the Preferred Reporting Items for Systematic Review and Meta-Analysis (PRISMA) guidelines, and the quality and strength of the evidence was evaluated using the Navigation Guide Systematic Review Methodology (SING). We found evidence that supports the idea that supplementary vitamin D for pregnant women is important for reducing the risk of gestational diabetes, hypertension, preeclampsia, early labor, and other complications. The data retrieved from this review are consistent with the hypothesis that adequate vitamin D levels might contribute to a healthy pregnancy.

Keywords: gestational diabetes; hypertension; maternal morbidity; preeclampsia; pregnancy; supplementation; vitamin D; 25-hydroxyvitamin D

Citation: Morales-Suárez-Varela, M.M.; Uçar, N.; Peraita-Costa, I.; Huertas, M.F.; Soriano, J.M.; Llopis-Morales, A.; Grant, W.B. Vitamin D-Related Risk Factors for Maternal Morbidity during Pregnancy: A Systematic Review. *Nutrients* **2022**, *14*, 3166. https://doi.org/10.3390/nu14153166

Academic Editor: Marloes Dekker Nitert

Received: 22 June 2022
Accepted: 28 July 2022
Published: 31 July 2022

1. Introduction

There is evidence of early interest in the relationship between vitamin D status and maternal health outcomes [1].

Vitamin D (D_2 or ergocalciferol, D_3 or cholecalciferol, or both) is a fat-soluble lipophilic prohormone proven to have many metabolic and biological functions. This vitamin is mainly synthetized in the skin as cholecalciferol through the action of ultraviolet light (vitamin D3), but it is also obtained from diet sources and food supplements such as ergocalciferol (vitamin D2) [2] and food materials such as fish oil, fish flesh, dietary supplements, eggs, butter, fortified foods, liver, and mushrooms. Vitamin D deficiency (serum

25-hydroxyvitamin D [25(OH)D] < 20 ng/mL) [3–5] is a major public health concern that is widespread among the general population and highly prevalent in pregnant women; it is found in 60% of them [6–9]. Maintaining serum concentrations between 30 and 50 ng/mL is recommended to achieve the health benefits of vitamin D [10–13].

Globally, it has been estimated that a billion people may be affected by vitamin D deficiency or insufficiency [14]. Studies in Ethiopia and India have also found that more than 80% and 60% of pregnant women suffered from vitamin D deficiency, using a cutoff of <50 nmol/L vitamin D, indicating the need for more research on the potential outcome and benefits of supplementation in developing countries [15,16].

Severe maternal morbidity during pregnancy is identified and reported worldwide. Its rising rates remain a large healthcare concern [17]. In 2005, worldwide, there were around 535,900 maternal deaths reported, which translates to a mortality ratio of about 402 maternal deaths per 100,000 live births [18]. The majority of these maternal deaths occurred in sub-Saharan Africa, with 270,500 deaths, and Asia, with 240,600 deaths [18]. Just five countries—India (117,100), Nigeria (58,800), the Democratic Republic of Congo (32,300), Afghanistan (26,000), and Ethiopia (22,200)—accounted for almost half (48%) of all maternal deaths [18].

Maternal morbidity is an unintended outcome of labor and delivery that results in significant short- or long-term consequences to woman's health [19]. Severe maternal morbidity (SMM) affects around an estimated 50,000 women per year in the United States—0.5–1.3% of pregnancies [19,20]. However, determining the true rates of SMM in the United States and worldwide is difficult because of the lack of standard definitions of such cases as well as the difficulty in identifying cases [21].

During pregnancy, there are significant alterations in phosphate and calcium metabolism owing to calcium accumulating in the fetal skeleton, and the fetus relies exclusively on the maternal supply of vitamin D, which it receives across the placenta, as it is not capable of synthesizing vitamin D on its own for adequate bone mineral formation [22,23]. A low level of vitamin D during the pregnancy and special attention during the early stage of pregnancy produce less bone mineral content in the fetal skeleton. Calcitriol cord blood concentrations tend to be lower than those found in maternal serum [2–13] due to the fact that calcitriol cannot easily cross the placental barrier [24,25], and parathyroid hormone concentrations are low in the fetus [26]. The high levels of phosphorus and calcium concentrations found in serum also contribute to lower fetal calcitriol concentrations because these factors suppress the expression of renal 25OHD-1-α-hydroxylase (CYP27B1) in the fetus [27].

The recommended daily allowance (RDA) of vitamin D for women in the United States aged 19–50 years, including during pregnancy, is established at 600 IU per day [27]. This recommendation was based on the amount of intake necessary to sustain blood levels of vitamin D above 50 nmol/L for a population with minimal sunlight exposure and was developed solely based on outcomes related to bone health [27]. According to the US Institute of Medicine, it is considered that 1000–1600 IU (25–40 g/day) of supplemental vitamin D is necessary during pregnancy to obtain the highest level of vitamin D3 during this period [28]. This recommendation was contentious, as many researchers have argued that insufficiency should be defined at thresholds of 75 nmol/L or even higher, which would require a much higher intake to reach [29,30]. Nevertheless, some studies [31–33] established that the safe and maximal production of vitamin D (at least 32 ng/mL) is achieved with a supplementation of 4000 IU/day until delivery.

Vitamin D can also be referred to as 25-hidroxyvitamin D or calcidiol, and it is transformed into its active form 1,25-dihydroxyvitamin D by CYP27B1 [33]. This enzyme is mainly located in the kidney but is also significantly expressed in the placenta. Pregnancy represents a special physiological situation due to the important role played by the placenta in the metabolism of this vitamin [34]. The placenta is thought to be the major site of vitamin D metabolism in pregnancy. The 1a-hydroxylase, the 24-hydroxylase, the 25-hydroxylase (CYP2R1), the vitamin D binding protein (VDB), and the vitamin D receptor

(VDR) have all been detected either in trophoblast cultures or in freshly obtained placental tissue [35–38]. Undoubtedly, the placenta can metabolize vitamin D, providing active 1,25-(OH)2 vitamin D in vitro. However, it is unclear to what extent placental vitamin D metabolism contributes to maternal vitamin D status in pregnancy.

Numerous functions have been attributed to vitamin D due to the pleiotropic properties of the vitamin D receptor (VDR) [39]. Increasing scientific evidence points to the role of vitamin D in maternal mortality and morbidity, in addition to its implication in several pathologies. Allergic and autoimmune diseases and even cancer implications have also been postulated [40]. The vitamin D deficiency during pregnancy cause maternal and fetal side effects [41], such as increases the risk of preeclampsia, glucose intolerance, gestational diabetes, preterm birth and hypocalcemia crisis in the mother. As poor skeletal development, dysfunction in both the mother and newborn and increase the risk birth of a small child for gestational age (SGA) [42]. Also in the fetus it is related to an inadequate immune system, wheezing and eczema, and respiratory infections in infants [43,44].

An area of study that has garnered significant attention is the role of vitamin D and its effect on pregnancy. There is a lack of evidence from systematic reviews and meta- analyses to evaluate the association between vitamin D during pregnancy and maternal morbidity. Given the high prevalence of low vitamin D level status during pregnancy and the public health importance of clarifying the role of vitamin D during pregnancy in offspring health, a better understanding of the nonclassical functions of vitamin D in preventing adverse health outcomes in high-risk populations is needed. The aim of the present review is to summarize the primary outcome in order to identify a cut-off value for a serum vitamin D concentration that increases the risk of maternal morbidity during pregnancy and to determine the possibility of supplementation to avoid it.

2. Materials and Methods

This systematic review was conducted according to the Preferred Reporting Items for Systematic Reviews and Meta-Analyses (PRISMA) guidelines [45,46]. The quality and strength of the evidence was evaluated using the Navigation Guide Systematic Review Methodology (SING) [47–49]. Systematic review registration PROSPERO (CDR42022343174).

2.1. Question PECO

The PECO question (P: population; E: exposure; C: comparison; O: outcome) of the study was "Is there more morbidity in pregnant women with low levels of vitamin D compared to those with adequate levels of vitamin D?", in which P is pregnancy women; E is a low intake/level of vitamin D; C is an adequate intake/level of vitamin D; and O is pregnancy morbidity.

2.2. Literature Search

The goal of the search strategy was to identify studies that reported the associations between serum vitamin D concentrations or the intake of vitamin D from supplementation or diet during pregnancy and its maternal morbidity affects. First, we performed a literature search to identify publications eligible for inclusion in the PubMed and Embase databases. The keywords included "pregnancy" OR "gestation" AND "vitamin D" AND "morbidity." The search was limited to human subjects and English and Spanish language articles published between 2010 and May 2022. A total of 50 studies were recovered from PubMed, 15 were recovered from Cochrane, and 150 were recovered from Embase, for a total of 215. In the first phase, duplicates were removed, and the reference lists of relevant publications were searched for fresh research that fulfilled the inclusion requirements. Following the first literature search, the reviewers examined the titles and abstracts to locate those that fulfilled the selection criteria. These articles were assessed for eligibility, with the first screening of the articles based on the information available in the abstract and results sections of each study. The initial screening identified 34 candidate studies, of which 28 met the inclusion and exclusion criteria. The PRISMA flowchart (Figure 1) shows the number of articles at each stage of the screening process.

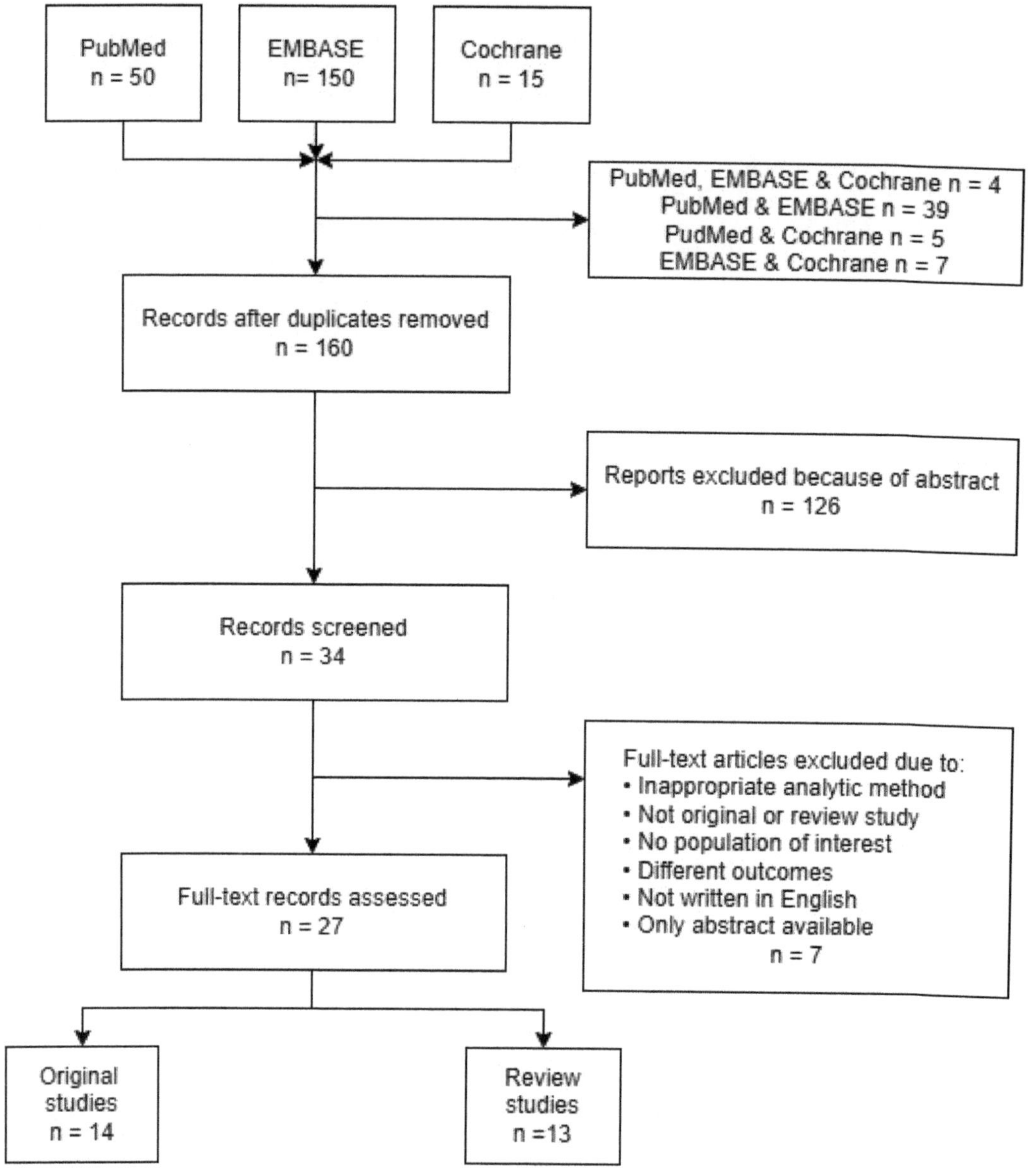

Figure 1. Search strategy: PRISMA flowchart.

2.3. Study Inclusion/Exclusion Criteria and Data Extraction

The types of studies included in this review meet the following criteria: controlled trials, both randomized and nonrandomized; prospective cohorts; case-control studies; and systematic reviews looking at the effects of vitamin D on maternal morbidity. All studies were longitudinal in nature and focused on how vitamin D levels in pregnancy were related to maternal morbidity. Specific inclusion/exclusion criteria were developed for the selection of studies to be included in this work, and only published works that met all the criteria were included for review. The selection criteria were the following:

1. Original research article or review (abstracts, case reports, ecological studies, and comments were excluded)
2. Available in English and Spanish
3. Published between 2010 and May 2022
4. Study carried out on humans
5. Exposure of interest is vitamin D status or supplementation during pregnancy
6. Data on vitamin D or metabolite concentration in maternal blood during pregnancy available
7. Main outcomes of interest are the incidence of maternal morbidity.

After a thorough assessment by all the authors of the candidate studies, 26 were included in this review.

2.4. Data Extraction

The data for the present review were retrieved from the previous research articles published earlier. The following data were extracted for the present study: (i) Study characteristics: authors, location and year, type of study, and source of data collection; (ii) sample size; (iii) primary outcome; (iv) findings (maternal morbidity & vitamin D level) (Table 1). The relevant data of the reviews were also summarized in a second table, including: (i) factors analyzed; (ii) gestational week when sample was collected; (iii) vitamin D cutoff (blood sample nmol/L); (iv) vitamin D collected (serum or supplementation); (v) average maternal age (Table 2).

Table 1. Original studies that show vitamin D-related risk factors for maternal morbidity during pregnancy.

Author	Location, Year(s)	Study Type	Data Source	Sample Size	Primary Outcome	Findings	SING&		NOS
							LE	GR	
Rezende et al., 2012 [50]	Brazil	Case-control; observational	IRB at the Faculty of Medicine of Ribeirao Preto, University of São Paulo	n = 529: n = 154 (GH) n = 162 (PE) n = 213 (healthy)	PE and GH	Similar genotype distributions were found for the 3 VDR polymorphisms in both the PE and GH groups compared with the HP group (all $p > 0.05$). VDR haplotype frequency distribution was similar in both the PE and GH groups compared with the HP group (all $p > 0.05$).	2^{++}	B	8
Lechtermann et al., 2014 [51]	Northern Hemisphere, 2005–2008	Cohort; observational	Department of Gynecology and Obstetrics, UK-Essen, University of Duisburg-Essen, Germany	n = 63: n = 20 (PE) n = 43 (healthy)	PE	In patients with PE, vitamin D levels were lower but differed significantly from the controls only in the summer (18.21 ± 17.1 vs. 49.2 ± 29.2 ng/mL; $p < 0.001$), whereas 1,25-$(OH)_2$ vitamin D levels were significantly lower only in the winter (291 ± 217 vs. 612.3 ± 455 pmol/mL; $p < 0.05$). A two-factorial ANOVA produced a statistically significant model ($p < 0.0001$) with an effect of season ($p < 0.01$) and PE ($p = 0.01$) on maternal vitamin D levels, as well as a significant interaction between the two variables ($p = 0.02$).	2^{++}	B	8
Achkar et al., 2015 [52]	Canada, 2014	Nested case-control	Canadian cohort studies of pregnant women, Quebec City, Nova Scotia, and Halifax, 2002–2010	n = 169 (PE) n = 1975 (control)	PE	Women who developed PE had a significantly lower vitamin D concentration (47.2 ± 17.7 vs. 52.3 ± 17.2 nmol/L; $p < 0.0001$). Women with vitamin D <30 nmol/L, compared with those with at least 50 nmol/L, had a greater risk of developing PE (adjusted OR = 2.23; 95% CI, 1.29–3.83) after adjustment for pre-pregnancy BMI, maternal age, smoking, parity, season and year of blood collection, gestational week at blood collection, and cohort site. An exploratory analysis with cubic splines showed a dose–response relationship between maternal vitamin D and the risk of PE, up to levels ~50 nmol/L, where the association appears to plateau.	2^{++}	B	8

Table 1. *Cont.*

Author	Location, Year(s)	Study Type	Data Source	Sample Size	Primary Outcome	Findings	SING&		NOS
							LE	GR	
Lawal et al., 2016 [53]	Nigeria, 2014	Case-control; observational	Department of Chemical Pathology of the tertiary health care facility	$n = 100$ (GDM) $n = 100$ (control)	GDM	Overall mean values of plasma 25-hydroxycholecalciferol were 28.77 ± 12.42 ng/mL. Overall, 58% of subjects had plasma 25-hydroxycholecalciferol levels < 30 ng/mL. The proportion of cases with vitamin D insufficiency was 62% (54% for controls). The OR for GDM was 1.39 (95% CI, 0.79–2.44) and $p = 0.3159$.	2^{++}	B	8
Mirzakhani et al., 2016 [54]	USA, 2009–2011	Randomized, double-blind, placebo-controlled clinical trial; experimental	Boston University Medical Center; Washington University in St. Louis, Missouri; and Kaiser Permanente Southern California Region in San Diego	$n = 440$ (4400 IU) $n = 436$ (placebo 400 IU)	PE	No significant difference was found between the treatment or control groups in terms of incidence of PE (8.08% vs. 8.33%, respectively; relative risk: 0.97; 95% CI, 0.61–1.53). In a cohort analysis and after adjustment for confounders, a significant effect of sufficient vitamin D status (≥30 ng/mL was observed in both early and late pregnancy compared with insufficient levels (adjusted OR, 0.28; 95% CI, 0.10–0.96). The differential expression of 348 vitamin D-associated genes (158 upregulated) was found in the peripheral blood of women who developed PE (FDR <0.05 in the Vitamin D Antenatal Asthma Reduction Trial [VDAART]; $p < 0.05$ in a replication cohort).	2^{++}	B	8
Brodowski et al., 2017 [55]	Germany	Cohort; observational	Hannover Medical Center (Germany)	$n = 12$ (PE) $n = 13$ (NC)	PE	Vitamin D_3 improved HUVEC function in neither group. No effect of vitamin D_3 on VEGF expression was found.	2^{++}	B	8
Accortt et al., 2017 [56]	USA, 2004–2016	Nested cohort; observational	Community Child Health Network	$n = 164$ (cohort)	PE and GDM	Serum vitamin D was significantly inversely correlated with the AL index (Spearman's $r = -0.247$; $p = 0.002$).	2^{+}	B	8

Table 1. *Cont.*

Author	Location, Year(s)	Study Type	Data Source	Sample Size	Primary Outcome	Findings	SING&		NOS
							LE	GR	
Singla et al., 2019 [57]	India, 2017–2018	Prospective comparative; observational	Department of Obstetrics and Gynaecology, Adesh Institute of Medical Sciences and Research, Bathinda, Punjab	$n = 60$: $n = 30$ (PE) $n = 30$ (NC)	PE	Vitamin D deficiency was found in all participants, but the mean vitamin D level was significantly lower in the PE group (8.7 ± 5.32 vs. 14.2 ± 7.88 ng/mL, $p < 0.05$).	2^{++}	B	8
Nandi et al., 2020 [58]	India	Cross-sectional; observational	Department of Obstetrics and Gynecology, Bharati Medical College and Hospital, Pune	$n = 50$ (PE) $n = 69$ (NC)	PE	Vitamin D levels were lower ($p < 0.01$ for both) in women with PE. PUFA levels were lower ($p < 0.05$), whereas SFA and total MUFA were higher ($p < 0.05$ for both) in women with PE. Cord erythrocyte PUFA levels were higher ($p < 0.01$) in PE women. Vitamin D levels were negatively associated with maternal systolic and diastolic blood pressure ($p < 0.01$ for both). Vitamin D levels were positively associated with PUFA ($p < 0.01$) and negatively associated with SFA ($p < 0.05$), MUFA ($p < 0.01$).	2^{++}	B	8
Rohr Thomsen et al., 2020 [59]	Denmark, 1989–2010	Cohort; observational	Aarhus Birth Cohort at the Department of Gynecology and Obstetrics, Aarhus University Hospital	$n = 50{,}665$ (cohort)	GH and PE	Seasonal variation was found for GH ($p = 0.01$), PE ($p = 0.001$), and early-onset PE ($p = 0.014$). Increased risk was observed when conceiving during spring and early summer, peaking in midsummer, and decreasing steadily during late summer and fall to reach the nadir by winter.	2^{++}	B	8

Table 1. *Cont.*

Author	Location, Year(s)	Study Type	Data Source	Sample Size	Primary Outcome	Findings	SING& LE	SING& GR	NOS
Osman et al., 2020 [60]	Egypt, 2019	Case-control; observational	—	n = 200 (PE) n = 100 (eclampsia) n = 200 (NC)	Eclampsia and PE	Mean vitamin D level was lower in the PE group (14.8 ± 5.4 ng/mL) and the eclampsia group (10.5 ± 1.6 ng/mL) than in the pregnant controls (19.5 ± 6.5 ng/mL) (p = 0.002). The difference was significant only between the eclampsia group and the pregnant controls (p = 0.02). All eclampsia cases had vitamin D insufficiency, compared with 17.5% of the PE group and 39.5% of the controls. Deficiency of vitamin D (<12 ng/mL) was 47.5% in the PE group, 80% in the eclampsia group, and 10.5% in the control group (p = 0.04).	2^{++}	B	8
Nandi et al., 2020 [61]	India	Cross-sectional	Department of Obstetrics and Gynecology, Bharati Medical College and Hospital	n = 50 (PE) n = 69 (NC)	PE	Vitamin D deficiency increases oxidative stress through alterations in one-carbon metabolism, which can result in an imbalance in LCPUFA metabolites and contribute to placental inflammation and endothelial dysfunction in PE.	2^{+}	C	8
Schoenmakers et al., 2020 [62]	Sweden, 2013–2014	Nested case-control; retrospective	Antenatal care units and medical records	n = 1827 (cohort) n = 30 (normo-calcemic)	Hypercalcemia crisis	Hypercalcemic women had a relatively high serum 1,25(OH)2D concentration despite appropriately suppressed PTH, which is suggestive of abnormal gestational adaptations. The prevalence of gestational hypercalcemia was 1.7% in the third trimester. Primary hyperparathyroidism and vitamin D toxicity were not found as main causes of hypercalcemia.	2^{+}	C	8

Table 1. *Cont.*

Author	Location, Year(s)	Study Type	Data Source	Sample Size	Primary Outcome	Findings	SING&		NOS
							LE	GR	
Olmos-Ortiz et al., 2021 [63]	Mexico	Cross-sectional	Department of Reproductive Biology, Instituto Nacional de Ciencias Médicas y Nutrición Salvador Zubirán	$n = 48$ (UTI) $n = 44$ (normal pregnancy)	UTIs and GH	Vitamin D deficiency might predispose women to maternal cardiovascular risk and perinatal infections, especially in male-carrying pregnancies, probably owing to lower placental CYP27B1 and cathelicidin expression. Strong negative correlations were found between calcitriol and maternal systolic and diastolic blood pressure in the UTI cohort ($p < 0.002$). Cathelicidin gene expression was positively correlated with gestational age in the UTI cohort and with newborn anthropometric parameters.	2^{+}	C	8

Abbreviations: 25(OH)D, 25-hydroxyvitamin D; 95% CI, 95% confidence interval; AL, allostatic load; ANOVA, analysis of variance; BMI, body mass index; CM, explant conditioned media; FDR, Food & Drug Administration; GDM, gestational diabetes mellitus; GH, gestational hypertension; HP, healthy pregnant; HUVEC, human umbilical vein endothelial cells; IRB, institutional review board; IUGR, intrauterine growth retardation; MUFA, monounsaturated fatty acids; NC, normotensive control; NOS, Newcastle–Ottawa Scale; OR, odds ratio; PE, preeclampsia or preeclamptic; PTH, parathyroid hormone; PUFA, polyunsaturated fatty acids; RFLP, restriction fragment length polymorphism; SFA, saturated fatty acids; UTI, urinary tract infection; UV, ultraviolet; VDR, vitamin D receptor; VEGF, vascular endothelial growth factor.

Table 2. Vitamin D-related information in original studies that show vitamin D-related risk factors for maternal morbidity during pregnancy.

Author	Factor	Vitamin D Analysis Time	Assay Method	Cutoff Values, nmol/L in Blood Sample	25(OH)D Measured or Vitamin D Supplementation Studied	Maternal Age
Rezende et al., 2012 [61]	VDR polymorphisms with PE or GH	—	Genotypes for FokI, ApaI, and BsmI determined by RFLP	—	Serum sample	27–28
Lechtermann et al., 2014 [59]	Season on maternal vitamin D status and placental vitamin D metabolism	—	ELISA; 25(OH)D ELISA (Immunodiagnostik, Bensheim, Germany)	50	Serum sample	31–32
Achkar et al., 2015 [60]	PE and vitamin D status	20 weeks	Automated chemiluminescence immunoassay (DiaSorin Liaison, Stillwater, MN, USA)	75	Serum sample	25–>35
Lawal et al., 2016 [58]	Vitamin D status and GDM	—	Cobas e411 (Roche Diagnostics, GmbH) analyzer	75	Serum sample	31.73
Mirzakhani et al., 2016 [55]	PE and vitamin D supplementation	Initiated between 10–18 weeks	Supplementation vitamin D study (4400 vs. 400 IU/day)	75	Supplementation comparison	18–39
Brodowski et al., 2017 [57]	Vitamin D status and its relationship with postpartum AL	Either 6 or 12 months postpartum	Highly selective liquid chromatography–tandem mass spectrometry using Zrt laboratory methods	50	Serum sample	27.8
Accortt et al., 2017 [56]	PE and 1,25(OH)$_2$ vitamin D$_3$	Delivery	LIAISON 25(OH) Vitamin D$_3$ TOTAL Assay (DiaSorin, USA)	50	Maternal and cord serum sample	32.2
Singla et al., 2019 [52]	PE	—	Immune fluorescence assay test using a vitamin D kit on a Tosho AIA 360 fully automatic hormone analyzer	50	Serum sample	20–40
Nandi et al., 2020 [51]	Maternal and cord serum vitamin D levels in women with PE	Delivery	EIA method using an AC-57SF1, 25-Hydroxy Vitamin D EIA kit (AC-57SF1, IDS, Boldon, UK)	75	Maternal and cord serum sample	18–35
Rohr Thomsen et al., 2020 [54]	hypertensive disorders and PE	—	No direct measurements	—	Serum sample	<20–>35
Osman et al., 2020 [62]	Hypertensive disorders of pregnancy	—	25(OH)D$_3$/D$_2$ Orgentec Diagnostika ELISA Kit GmbH	50	Serum sample	20–35

Table 2. *Cont.*

Author	Factor	Vitamin D Analysis Time	Assay Method	Cutoff Values, nmol/L in Blood Sample	25(OH)D Measured or Vitamin D Supplementation Studied	Maternal Age
Nandi et al., 2020 [53]	Maternal and cord serum vitamin D levels in women with PE	Delivery	ELISA Serum TXB2 levels (Cayman Chemicals, item No. 501020; Ann Arbor, MI, USA)	—	Maternal and cord serum sample	18–35
Schoenmakers et al., 2020 [63]	Gestational hypercalcemia	Pregnant women in trimester 1 (before gestational week 16) and in trimester 3 (after gestational week 31).	ELISA Free vitamin D (DIASource Immunoassays, Louvain-la Neuve, Belgium)	30–50	Serum sample	33.2
Olmos-Ortiz et al., 2021 [64]	Vitamin D_3 (calcitriol active metabolite) involved in UTI	Delivery	Quantitative chemiluminescent immunoassay in the LIAISON platform	50	Serum sample	—

Abbreviations: EIA, enzyme immunoassay; ELISA, enzyme-linked immunosorbent assay; IUGR, intrauterine growth retardation; GDM, gestational diabetes mellitus; GH, gestational hypertension; PE, preeclampsia; RFLP, restriction fragment length polymorphism; UTI, urinary tract infection; VDR, vitamin D receptor.

2.5. Study Quaity Assessment

The quality of the studies is assessed using the following tools: The Eight Star Newcastle–Ottawa Scale (NOS) for observational studies (cohorts and case-controls) [47,48] was used to evaluate the methodological quality—specifically, the risk of bias—of the original studies. Assessment with the Newcastle–Ottawa Scale produces a score ranging from 0 to 9, with the overall score based on three sub-scores based on the subject selection (0–4), the comparability of the subject (0–2), and the clinical outcome (0–3). The study assessment was carried out independently by two individuals (NU and IPC), and discrepancies were brought to a third individual (MMSV) if a compromise could not be reached among the two original individuals after discussion.

Further assessment of the quality of the included studies was carried out using the Scottish Intercollegiate Guidelines Network (SIGN) [49]. Using the SIGN ensures that the validity—including key factors such as bias and confounding—of a study is robustly assessed. The SIGN system in based on the principles of evidence-based medicine, an approach that ensures the use of the most up-to-date, reliable, and scientifically solid evidence available in making decisions about a particular situation being studied [64].

The SIGN system establishes levels of evidence and recommendations to describe a given study and its results. The levels of evidence are based on the study design and the methodological quality of individual studies and are scored from best to worst using the numbers 1, 2, 3, and 4. These scores are further ranked using the ++, +, and—signs. The grades of recommendation, rated from best to worst as A, B, C, and D, are based on the strength of the evidence on which the recommendation is based, and they do not reflect the clinical importance of the recommendation.

3. Results

3.1. Study Characteristics

Our search approach yielded up 215 studies identified through database searching; a total of 14 original research studies and 13 review studies remained. After consideration, it was decided to include only the 14 original studies in this review.

Considering the SIGN and NOS scores, the 14 original studies could be regarded as good (high) quality. The important methodological features and the general characteristics of all the review studies are summarized in Table 1. The chosen studies were analyzed according to the design, location and year, source of data, sample size, factor, vitamin D level assessment, and major findings. Meanwhile, the vitamin D analysis details and vitamin D cutoff values of the included articles are listed in Table 2.

The studies were published between 2012 and 2021. The original research studies used data from India [57,58,61], Denmark [59], the United States [54,56], Germany [55], Nigeria [53], the Northern Hemisphere [51], Canada [52], Brazil [50], Egypt [60], Sweden [62], and Mexico [63]. The review research studies included data from Brazil, India, the United States, Puerto Rico, Spain, Iran, and Australia [65–77].

All but six observational studies of vitamin D were conducted in high-income country settings, and most populations had either a presumed risk or a high prevalence of deficiency at baseline (Table 1). The dosing approaches and assay methods in the trials varied: one trial contained multiple intervention arms testing the daily dietary intake of Vitamin D, vitamin D supplementation, and the frequency of UV exposure in the first trimester, in the second trimester, and at the time of delivery. One recent trial tested daily 4400 vs. 400 IU D3. In other studies, the relationship between disease risks was evaluated by measuring serum vitamin D levels with different assay methods (Table 2). This trial [65] showed that a significant effect of sufficient vitamin D status (25OHD $\geq$ 30 ng/mL) was observed in both early and late pregnancy compared with insufficient levels (25OHD < 30 ng/mL) (OR, 0.28; 95% CI, 0.10–0.96).

Vitamin D supplementation appeared to improve maternal vitamin D levels in the two trials for which data were available [65]. In addition, the results of trials by Christine Rohr Thomsen indicate a seasonal variation effect of the risk of gestational hypertension

($p = 0.01$), PE ($p = 0.001$), and early-onset PE ($p = 0.014$) [51,59]. Women with an estimated date of conception in June had the highest risk of preeclampsia, while women with an estimated date of conception in August had the highest risk of gestational hypertension.

Observational studies of vitamin D status during pregnancy and the risk of pre-eclampsia have not shown consistent associations. Vitamin D levels were lower ($p < 0.01$) in women with PE [50–52,57,58,60,61]. The investigators of a study from the USA [54] observed that vitamin D supplementation initiated in weeks 10–18 of pregnancy did not reduce preeclampsia incidence in the intention-to-treat paradigm. However, vitamin D levels of 30 ng/mL or higher at trial entry and in late pregnancy were associated with a lower risk of preeclampsia (8.08% vs. 8.33%, respectively; relative risk: 0.97; 95% CI, 0.61–1.53). A nested case control study from North Carolina reported that women with vitamin D levels <50 nmol/L had a nearly fourfold greater risk of severe preeclampsia compared with those with levels $\geq$ 75 nmol/L [78]. In contrast, a nested case-control study in Massachusetts found no statistically significant differences in the risk of pre-eclampsia for women with vitamin D levels < 37.5 nmol/L (AOR 1.35 [0.40, 4.50]) [71]. Another prospective cohort study of pregnancies at a high risk for pre-eclampsia in Canada found no effect of vitamin D during early pregnancy on pre-eclampsia risk [72].

A group of studies relate the vitamin D status with the alteration of different metabolic pathways such as carbon and peptide metabolism. The imbalance of long-chain polyunsaturated fatty acid metabolites produced by a vitamin D deficiency contributes to inflammation and endothelial dysfunction [61]. This deficiency also contributes to a low antimicrobial peptide metabolism [63], resulting in several urinary infections.

3.2. Original Research Studies

Nandi and colleagues [58] published a cross-sectional study in 2019. The study included 119 pregnant women (69 normotensive controls [NC] and 50 women with PE). The women with PE had lower maternal and cord serum vitamin D levels ($p < 0.01$ for both) than the NC women. A total of 94% of women in the PE group and 76% in the NC group were deficient in maternal vitamin D levels, while for cord vitamin D levels, 98% of women with PE and 85.2% of NC women were deficient. In 2020, this group reported [61] how the imbalance in the long-chain polyunsaturated fatty acid (LCPUFA) metabolites derived from vitamin D deficiency contributes to placental inflammation and endothelial dysfunction in PE.

Rohr Thomsen and colleagues [59] published a cohort study based on data from the Aarhus Birth Cohort (ABC). Of the 50,665 women included, 4285 (8.5%) were diagnosed with a hypertensive disorder of pregnancy, 1999 (3.9%) were diagnosed with PE, and 2386 were diagnosed (4.7%) with gestational hypertension (GH). The hypertensive disorders of pregnancy, including GH, PE, and early-onset PE, increased the risk for women conceiving during spring and early summer, peaking in midsummer, and later decreasing steadily during late summer and fall to reach the nadir by winter. Seasonal variation was found for GH ($p = 0.01$), PE ($p = 0.001$) and early-onset PE ($p = 0.01$). In another prospective comparative study [68], a significant negative correlation was observed between vitamin D and systolic and diastolic blood pressure in the PE group ($p < 0.05$), whereas no significant correlation was observed between vitamin D and systolic/diastolic blood pressure in the control group. The mean vitamin D level was significantly lower in the PE group than that in the control group (9 $\pm$ 5 and 14 $\pm$ 8 ng/mL, respectively), with a statistically significant $p < 0.05$. A vitamin D level < 5 ng/mL was associated with a 14.58-fold (95% CI; 12.16–17.55) increase in the odds ratio of PE, whereas a vitamin D level of 5–10 ng/mL was associated with an 11.42-fold (95% CI; 8.26–13.6) increase in the odds ratio of PE.

In 2017, Accortt and colleagues [56] found an association between a higher postpartum allostatic load and an index of multisystem physiological wear and tear, operationalizing emergent chronic disease risk and predicting morbidity and vitamin D. Adding vitamin D deficiency to the allostatic load index produced a stronger association with adverse outcome. Brodowski and colleagues [55] assessed the effect of vitamin D supplementation

(4400 vs. 400 IU/day) initiated early in pregnancy (10–18 weeks) on the development of PE. When started at weeks 10–18 of pregnancy, vitamin D supplementation did not reduce the incidence of PE. However, vitamin D levels of ≥30 ng/mL at trial entry and in late pregnancy were associated with a lower risk of PE.

Lawal and colleagues [53] showed that no relationship exists between vitamin D deficiency and GDM. That case-control study had 200 pregnant women; the proportion of cases (n = 100) and controls (n = 100) with vitamin D insufficiency was 62% and 54%, respectively. Lechtermann and colleagues [51] indicated that patients with PE had lower serum levels of vitamin D in response to seasonal changes.

In 2020, Schoenmakers and colleagues [62] found a correlation between a relatively high concentration of 1,2(OH)2D and hypercalcemia in pregnant women during the third trimester. The retrospective and explorative study investigated the prevalence of hypercalcemia in a cohort of 2121 women—1827 screened for hypercalcemia in T3. The prevalence was 1.7% higher than that in the general population.

Olmos-Ortiz and colleagues suggest [64] cardiovascular risk and perinatal infections due to vitamin D3 (calcitriol) deficiency, especially in male-carrying pregnancies due to the lower calcitriol-activating enzyme. The placental calcitriol was significantly elevated in women with urinary tract infections, and it was negatively correlated with blood pressure. Regarding newborns' sex, the calcitriol-activating enzyme showed a higher expression in female-carrying mothers.

The level of evidence is relatively high—2++ or 2+, according to SIGN, which belong to a great level of recommendation: B. The systematic review about the importance of the maintenance of a good level of vitamin D could be used as a recommendation guide in the studied population: pregnant women.

4. Discussion

Overall, this systematic review suggests that maternal low levels of vitamin D during pregnancy lead to a greater risk of gestational diabetes, preeclampsia, early labor, and other complications. However, due to the variability in numerous elements of the study design (e.g., vitamin D assessment methods, pregnant mobility assessment methods, and the timing of the data collection), it remains a challenge to synthesize the findings. This data suggest that low maternal vitamin D appears to have a negative impact or detrimental impact on the health status of pregnant women, which is an important conclusion that prevents many women from getting adequate nutrition with the adequate support of vitamin D, and it is not possible to use supplementation during the pregnancy period.

Recently, vitamin D has been recognized as interacting with a nuclear receptor in various organs [71–76]. Vitamin D deficiency is associated with increased risks of morbidity and mortality in cardiovascular, malignant, and autoimmune diseases [72,77,78]. In recent years, the interest in the consequences of maternal vitamin D deficiency and its effect on pregnancy has increased. Vitamin D insufficiency is considered common in pregnant women, and deficiencies have been linked to adverse pregnancy outcomes [78–80].

Considering whether prenatal vitamin D deficiency is associated with maternal morbidity seems reasonable. The findings from several studies suggest an increasing prevalence of vitamin D deficiency in pregnancy and its associated adverse outcomes [81–85]. To further understand the role of vitamin D in pregnancy and the seemingly associated adverse outcomes, interventional and observational studies are needed.

Furthermore, a current systematic review described the overall mean prevalence rates of vitamin D deficiency in pregnant women and newborns as 54% and 75%, respectively [86]. In postpartum periods, the prevalence of vitamin D deficiency in women is also high: 63% [86,87]. Although evidence points to the high prevalence of deficiency, there exist strategies to raise maternal vitamin D concentrations, including supplementation, advice for sun exposure (15–20% of the body surface area), and the intake of vitamin D–fortified foods. The vitamin D status during pregnancy varies around the world as a function of maternal sunlight exposure, the degree of skin pigmentation, latitude, lifestyle, BMI, and the intake of vitamin D supplements. People with darker skin pigmentation and limited sunlight exposure are at the greatest risk for deficiency [88].

Supplement intake can also play an important role in improving vitamin D status among pregnant women. Taking vitamin D-enriched food and supplements can be advised in order to maintain optimum serum levels during pregnancy. The recommendations for vitamin D intake during pregnancy range from 200 to 4000 IU/day worldwide. The current WHO guideline recommends 200 IU/day of vitamin D supplement intake among pregnant women with vitamin D deficiency in order to reduce the risk of PE, a low birth weight, and a preterm birth [89]. The American Pregnancy Association recommends 100 μg/day of vitamin D intake, a considerably larger amount of vitamin D than the recommended intake of 10 μg/day for women [90]. In China, a daily intake of 600 IU is suggested during pregnancy [91]. In the United Kingdom, it is advised to have a maternal vitamin D intake of 400 IU/day. The United Kingdom Health Department provides free vitamin D supplementation to pregnant women and newborn children [92]. Switzerland follows the Institute of Medicine-recommended nutrient intake: 1500–2000 IU/day for women at risk of vitamin D deficiency and 600 IU for women without such risk [93]. In Canada, pregnant women are suggested to take 400–600 IU/day [94]. In Turkey, free supplementation of vitamin D (1200 IU/day) is provided to all women from early pregnancy to 6 months after delivery [95]. A similar approach to vitamin D supplementation (400 IU/day) is followed in New Zealand for pregnant women identified as being at risk of vitamin D deficiency [96]. Meanwhile, for women not at risk, the ministry of health of New Zealand recommends 200 IU/day [97–99].

After many years of study, researchers at the Medical University of South Carolina College of Medicine suggested 4000 IU/day of vitamin D for pregnant women. The findings suggest that, starting at 12–16 weeks of gestation, vitamin D supplementation at a rate of 4000 IU/day is most effective in achieving vitamin D sufficiency in order to attain an optimal nutritional and hormonal vitamin D status throughout pregnancy [88]. A treatment (<37 weeks) goal > 40 ng/mL was associated with a reduction in preterm birth risk [31].

Further, no trials or observational studies specifically regarding vitamin D supplementation/intake and maternal morbidity during pregnancy were identified. Nevertheless, vitamin D requirements are higher among pregnant women, and maintaining optimum serum levels of vitamin D during maternity and for fetus growth is important. Adequate levels of vitamin D seem to be a determinant at the time of implantation and placentation for the development of preeclampsia. There is not a consensus regarding the vitamin D blood concentration value that predisposes women to maternal morbidity; hence, is not easy to recommend a specific supplementation treatment. The present systematic review lacks the experimental data needed to establish a general cutoff value of vitamin D in order to settle how important it could be to improving the maternal diet with vitamin D supplements. Further exploration of vitamin D's role in pregnancy and its potential role in maternal morbidity would be worthwhile, including maternal age and sexual dimorphism.

5. Strengths and Limitations of This Review

This study has limitations. First, there were limited data on maternal vitamin D supplementation during pregnancy regarding long-term outcomes. Second, the studies included here show significant methodological differences, which problematizes the obtention of a consensus on the evidence currently available on the relationship between vitamin D and maternal morbidity during pregnancy. In addition, we may not have been able to access all publications on the relationship between vitamin D and maternal morbidity during pregnancy because the area of analysis is limited to studies that are published in English and Spanish and that are available through the PubMed, Cochrane, and Embase databases.

6. Conclusions

Despite the inherent limitations discussed above that limit the ability to draw conclusions across studies, some important findings were noted. Collectively, the studies suggest that appropriate levels of vitamin D during pregnancy are associated with less mobility during pregnancy. Pregnant women should be counselled to maintain an adequate intake of vitamin D, with suitable nutritional support to adequately control their levels. In this systematic review of the literature, we found evidence relating vitamin D to maternal morbidity-related outcomes. However, well-designed, randomized vitamin D supplementation trials in pregnant women carried out to determine the optimal vitamin D status and dosing and evaluate the potential effectiveness of supplementation with respect to the risk of maternal morbidity are still greatly needed.

Author Contributions: Conceptualization: W.B.G. Methodology: N.U., W.B.G., I.P.-C., J.M.S., A.L.-M. and M.M.M.-S.-V.; Writing—Original Draft Preparation: N.U., M.F.H. and M.M.M.-S.-V.; Writing—Review & Editing: N.U., W.B.G., I.P.-C., J.M.S., A.L.-M. and M.M.M.-S.-V. All authors have read and agreed to the published version of the manuscript.

Funding: This research received no external funding.

Conflicts of Interest: WBG receives funding from Bio-Tech Pharmacal Inc. (Fayetteville, AR, USA). The other authors declare no conflict of interest.

References

1. Weinert, L.S.; Reichelt, A.J.; Schmitt, L.R.; Boff, R.; Oppermann, M.L.; Camargo, J.L.; Silveiro, S.P. Vitamin D Deficiency Increases the Risk of Adverse Neonatal Outcomes in Gestational Diabetes. *PLoS ONE* **2016**, *11*, e0164999. [CrossRef] [PubMed]
2. Holick, M.F. The vitamin D deficiency pandemic: Approaches for diagnosis, treatment and prevention. *Rev. Endocr. Metab. Disord.* **2017**, *18*, 153–165. [CrossRef] [PubMed]
3. Feleke, Y.; Abdulkadir, J.; Mshana, R.; Mekbib, T.A.; Brunvand, L.; Berg, J.P.; Falch, J.A. Low levels of serum calcidiol in an African population compared to a North European population. *Eur. J. Endocrinol. Eur. Fed. Endocr. Soc.* **1999**, *141*, 358–360. [CrossRef] [PubMed]
4. Farrant, H.J.; Krishnaveni, G.V.; Hill, J.C.; Boucher, B.J.; Fisher, D.J.; Noonan, K.; Osmond, C.; Veena, S.R.; Fall, C.H. Vitamin D insufficiency is common in Indian mothers but is not associated with gestational diabetes or variation in newborn size. *Eur. J. Clin. Nutr.* **2009**, *63*, 646–652. [CrossRef] [PubMed]
5. Marcenes, W.; Kassebaum, N.J.; Bernabé, E.; Flaxman, A.; Naghavi, M.; Lopez, A.; Murray, C.J. Global Burden of Oral Conditions in 1990–2010: A Systematic Analysis. *J. Dent. Res.* **2013**, *92*, 592–597. [CrossRef] [PubMed]
6. Hill, K.; Thomas, K.; AbouZahr, C.; Walker, N.; Say, L.; Inoue, M.; Suzuki, E.; Maternal Mortality Working Group. Estimates of Maternal Mortality Worldwide between 1990 and 2005: An Assessment of Available Data. *Lancet* **2007**, *370*, 1311–1319. [CrossRef]
7. Kilpatrick, S.J. Next Steps to Reduce Maternal Morbidity and Mortality in the USA. *Women's Health* **2015**, *11*, 193–199. [CrossRef]
8. Callaghan, W.M.; Creanga, A.A.; Kuklina, E.V. Severe Maternal Morbidity among Delivery and Postpartum Hospitalizations in the United States. *Obstet. Gynecol.* **2012**, *120*, 1029–1036. [CrossRef]
9. Hirshberg, A.; Srinivas, S.K. Epidemiology of Maternal Morbidity and Mortality. *Semin Perinatol.* **2017**, *41*, 332–337. [CrossRef]
10. Barrera, D.; Díaz, L.; Noyola-Martínez, N.; Halhali, A. Vitamin D and Inflammatory Cytokines in Healthy and Preeclamptic Pregnancies. *Nutrients* **2015**, *7*, 6465–6490. [CrossRef]
11. Ross, A.C.; Manson, J.E.; Abrams, S.A.; Aloia, J.F.; Brannon, P.M.; Clinton, S.K.; Durazo-Arvizu, R.A.; Gallagher, J.C.; Gallo, R.L.; Jones, G. The 2011 Report on Dietary Reference Intakes for Calcium and Vitamin D from the Institute of Medicine: What Clinicians Need to Know. *J. Clin. Endocr.* **2011**, *96*, 53–58. [CrossRef] [PubMed]
12. Sempos, C.T.; Vesper, H.W.; Phinney, K.W.; Thienpont, L.M.; Coates, P.M.; Vitamin D Standardization Program (VDSP). Vitamin D Status as an International Issue: National Surveys and the Problem of Standardization. *Scand. J. Clin. Lab. Investig.* **2012**, *72*, 32–40.
13. Varsavsky, M.; Moreno, P.R.; Fernández, A.B.; Fernández, I.L.; Gómez, J.M.Q.; Rubio, V.Á.; Martín, A.G.; Berdonces, M.C.; Cortés, S.N.; Muñoz, M.R. Recommended Vitamin D Levels in the General Population. *Endocrinol. Diabetes Nutr.* **2017**, *64*, 7–14. [CrossRef] [PubMed]
14. Wang, W.; Li, G.; He, X.; Gao, J.; Wang, R.; Wang, Y.; Zhao, W. Serum 25-Hydroxyvitamin D Levels and Prognosis in Hematological Malignancies: A Systematic Review and Meta-Analysis. *Cell. Physiol. Biochem.* **2015**, *35*, 1999–2005. [CrossRef] [PubMed]
15. Cashman, K.D.; Dowling, K.G.; Škrabáková, Z.; Gonzalez-Gross, M.; Valtueña, J.; De Henauw, S.; Moreno, L.; Damsgaard, C.T.; Michaelsen, K.F.; Mølgaard, C. Vitamin D Deficiency in Europe: Pandemic? *Am. J. Clin. Nutr.* **2016**, *103*, 1033–1044. [CrossRef] [PubMed]
16. Grzegorzewska, A.E.; Izdebska, A.; Niepolski, L.; Warchol, W.; Jagodzinski, P.P. Self-Reported Physical Activity, Quality of Life, and Psychological Status in Relation to Plasma 25-Hydroxyvitamin D Concentration in Patients Treated with Hemodialysis. *Kidney Blood Press. Res.* **2016**, *41*, 886–900. [CrossRef]

17. Yuan, Y.; Liu, H.; Ji, C.; Guo, X.; Hu, L.; Wen, J.; Cai, M. Association of Maternal Serum 25-Hydroxyvitamin D Concentrations in Second Trimester with Delivery Mode in A Chinese Population. *Int. J. Med. Sci.* **2017**, *14*, 1008–1014. [CrossRef] [PubMed]
18. Dawson-Hughes, B.; Heaney, R.P.; Holick, M.F.; Lips, P.; Meunier, P.J.; Vieth, R. Estimates of Optimal Vitamin D Status. *Osteoporos. Int.* **2005**, *16*, 713–716. [CrossRef] [PubMed]
19. Rosen, C.J.; Adams, J.S.; Bikle, D.D.; Black, D.M.; Demay, M.B.; Manson, J.E.; Murad, M.H.; Kovacs, C.S. The Nonskeletal Effects of Vitamin D: An Endocrine Society Scientific Statement. *Endocr. Rev.* **2012**, *33*, 456–492. [CrossRef] [PubMed]
20. Vieth, R. Why the Minimum Desirable Serum 25-Hydroxyvitamin D Level should be 75 Nmol/L (30 Ng/mL). *Best Pract. Res. Clin. Endocrinol. Metab.* **2011**, *25*, 681–691. [CrossRef]
21. Bischoff-Ferrari, H.A.; Giovannucci, E.; Willett, W.C.; Dietrich, T.; Dawson-Hughes, B. Estimation of Optimal Serum Concentrations of 25-Hydroxyvitamin D for Multiple Health Outcomes. *Am. J. Clin. Nutr.* **2006**, *84*, 18–28. [CrossRef] [PubMed]
22. Gertner, J.M.; Glassman, M.S.; Coustan, D.R.; Goodman, D.B. Fetomaternal Vitamin D Relationships at Term. *J. Pediatr.* **1980**, *97*, 637–640. [CrossRef]
23. Delvin, E.E.; Glorieux, F.H.; Salle, B.L.; David, L.; Varenne, J.P. Control of Vitamin D Metabolism in Preterm Infants: Feto-Maternal Relationships. *Arch. Dis. Child.* **1982**, *57*, 754–757. [CrossRef] [PubMed]
24. O'Brien, K.O.; Li, S.; Cao, C.; Kent, T.; Young, B.V.; Queenan, R.A.; Pressman, E.K.; Cooper, E.M. Placental CYP27B1 and CYP24A1 Expression in Human Placental Tissue and their Association with Maternal and Neonatal Calcitropic Hormones. *J. Clin. Endocrinol. Metab.* **2014**, *99*, 1348–1356. [CrossRef]
25. Hollis, B.W.; Pittard, W.B., III. Evaluation of the Total Fetomaternal Vitamin D Relationships at Term: Evidence for Racial Differences. *J. Clin. Endocrinol. Metab.* **1984**, *59*, 652–657. [CrossRef]
26. Young, B.E.; McNanley, T.J.; Cooper, E.M.; McIntyre, A.W.; Witter, F.; Harris, Z.L.; O'Brien, K.O. Vitamin D Insufficiency is Prevalent and Vitamin D is Inversely Associated with Parathyroid Hormone and Calcitriol in Pregnant Adolescents. *J. Bone Miner. Res.* **2012**, *27*, 177–186. [CrossRef]
27. Institute of Medicine (IOM). *Dietary Reference Intakes for Calcium and Vitamin D*; National Academies Press: Washington, DC, USA, 2011.
28. Heaney, R.P.; Holick, M.F. Why the IOM recommendations for vitamin D are deficient. *J. Bone Miner. Res.* **2011**, *26*, 455–457. [CrossRef]
29. Holick, M.F. Vitamin D: A D-Lightful Solution for Health. *J. Investig. Med.* **2011**, *59*, 872–880. [CrossRef]
30. Singh, S.; Garg, R.; Meena, A.; Kumar, D. Perinatal Outcome in Vitamin D Deficiency and Effect of Oral and Intramuscular Vitamin D3 Supplementation in Antenatal Women on Pregnancy Outcomes. *J. South Asian Feder. Obst. Gyne.* **2021**, *13*, 86–89. [CrossRef]
31. Hollis, B.W.; Wagner, C.L.; Howard, C.R.; Ebeling, M.; Shary, J.R.; Smith, P.G.; Taylor, S.N.; Morella, K.; Lawrence, R.A.; Hulsey, T.C. Maternal Versus Infant Vitamin D Supplementation During Lactation: A Randomized Controlled Trial. *Pediatrics* **2015**, *136*, 625–634. [CrossRef]
32. McDonnell, S.L.; Baggerly, K.A.; Baggerly, C.A.; Aliano, J.L.; French, C.B.; Baggerly, L.L.; Ebeling, M.D.; Rittenberg, C.S.; Goodier, C.G.; Niño, J.F.; et al. Maternal 25(OH)D concentrations $\geq$40 ng/mL associated with 60% lower preterm birth risk among general obstetrical patients at an urban medical center. *PLoS ONE* **2017**, *12*, e0180483. [CrossRef] [PubMed]
33. Olmos-Ortiz, A.; Avila, E.; Durand-Carbajal, M.; Díaz, L. Regulation of Calcitriol Biosynthesis and Activity: Focus on Gestational Vitamin D Deficiency and Adverse Pregnancy Outcomes. *Nutrients* **2015**, *7*, 443–480. [CrossRef] [PubMed]
34. Tamblyn, J.; Hewison, M.; Wagner, C.; Bulmer, J.; Kilby, M. Immunological Role of Vitamin D at the Maternal–fetal Interface. *J. Endocrinol.* **2015**, *224*, R107–R121. [CrossRef] [PubMed]
35. Ma, R.; Gu, Y.; Zhao, S.; Sun, J.; Groome, L.J.; Wang, Y. Expressions of Vitamin D Metabolic Components VDBP, CYP2R1, CYP27B1, CYP24A1, and VDR in Placentas from Normal and Preeclamptic Pregnancies. *Am. J. Physiol. Endocrinol. Metab.* **2012**, *303*, E928–E935. [CrossRef] [PubMed]
36. Díaz, L.; Arranz, C.; Avila, E.; Halhali, A.; Vilchis, F.; Larrea, F. Expression and Activity of 25-Hydroxyvitamin D-1α-Hydroxylase are Restricted in Cultures of Human Syncytiotrophoblast Cells from Preeclamptic Pregnancies. *J. Clin. Endocrinol. Metab.* **2002**, *87*, 3876–3882. [PubMed]
37. Pospechova, K.; Rozehnal, V.; Stejskalova, L.; Vrzal, R.; Pospisilova, N.; Jamborova, G.; May, K.; Siegmund, W.; Dvorak, Z.; Nachtigal, P. Expression and Activity of Vitamin D Receptor in the Human Placenta and in Choriocarcinoma BeWo and JEG-3 Cell Lines. *Mol. Cell. Endocrinol.* **2009**, *299*, 178–187. [CrossRef]
38. Slominski, A.T.; Kim, T.; Shehabi, H.Z.; Semak, I.; Tang, E.K.; Nguyen, M.N.; Benson, H.A.; Korik, E.; Janjetovic, Z.; Chen, J. In Vivo Evidence for a Novel Pathway of Vitamin D3 Metabolism Initiated by P450scc and Modified by CYP27B1. *FASEB J.* **2012**, *26*, 3901–3915. [CrossRef]
39. Aguilar-Cordero, M.; Lasserrot-Cuadrado, A.; Mur-Villar, N.; León-Ríos, X.; Rivero-Blanco, T.; Pérez-Castillo, I. Vitamin D, Preeclampsia and Prematurity: A Systematic Review and Meta-Analysis of Observational and Interventional Studies. *Midwifery* **2020**, *87*, 102707. [CrossRef]
40. Marino, R.; Misra, M. Extra-Skeletal Effects of Vitamin D. *Nutrients* **2019**, *11*, 1460. [CrossRef]
41. Larque, E.; Morales, E.; Leis, R.; Blanco-Carnero, J.E. Maternal and Foetal Health Implications of Vitamin D Status during Pregnancy. *Ann. Nutr. Metab.* **2018**, *72*, 179–192. [CrossRef] [PubMed]

42. Wei, S.; Qi, H.; Luo, Z.; Fraser, W.D. Maternal Vitamin D Status and Adverse Pregnancy Outcomes: A Systematic Review and Meta-Analysis. *J. Matern. Fetal. Neonatal. Med.* **2013**, *26*, 889–899. [CrossRef]
43. Dawodu, A.; Akinbi, H. Vitamin D Nutrition in Pregnancy: Current Opinion. *Int. J. Woman's Health* **2013**, *5*, 333–343. [CrossRef] [PubMed]
44. Schöttker, B.; Ball, D.; Gellert, C.; Brenner, H. Serum 25-Hydroxyvitamin D Levels and overall Mortality. A Systematic Review and Meta-Analysis of Prospective Cohort Studies. *Ageing Res. Rev.* **2013**, *12*, 708–718. [CrossRef] [PubMed]
45. Moher, D.; Shamseer, L.; Clarke, M.; Ghersi, D.; Liberati, A.; Petticrew, M.; Shekelle, P.; Stewart, L.A. Preferred Reporting Items for Systematic Review and Meta-Analysis Protocols (PRISMA-P) 2015 Statement. *Syst. Rev.* **2015**, *151*, 264–269. [CrossRef]
46. Moher, D.; Liberati, A.; Tetzlaff, J.; Altman, D.G. Preferred Reporting Items for Systematic Reviews and Meta-Analyses: The PRISMA Statement. *Ann. Intern. Med.* **2009**, *151*, 264–269. [CrossRef] [PubMed]
47. Wells, G.A.; Tugwell, P.; O'Connell, D.; Welch, V.; Peterson, J.; Shea, B.; Losos, M. The Newcastle-Ottawa Scale (NOS) for Assessing the Quality of Nonrandomized Studies in Meta-Analyses. *Syst. Rev.* Available online: https://www.ohri.ca/programs/clinical_epidemiology/oxford.asp. (accessed on 1 June 2022).
48. Stang, A. Critical Evaluation of the Newcastle-Ottawa Scale for the Assessment of the Quality of Nonrandomized Studies in Meta-Analyses. *Eur. J. Epidemiol.* **2010**, *25*, 603–605. [CrossRef] [PubMed]
49. Scottish Intercollegiate Guidelines Network (SIGN). *SIGN 50: A Guideline Developer's Handbook*; Scottish Intercollegiate Guidelines Network: Edinburgh, UK, 2015.
50. Sackett, D.L. Evidence-Based Medicine. *Semin. Perinatol.* **1997**, *21*, 3–5. [CrossRef]
51. Nandi, A.; Wadhwani, N.; Randhir, K.; Wagh, G.; Joshi, S. Association of Vitamin D with Fatty Acids in Pregnancy. *Prostaglandins Leukot. Essent. Fat. Acids* **2020**, *157*, 102030. [CrossRef] [PubMed]
52. Jeyakumar, A.; Shinde, V.; Ravindran, R. Pooled estimate of Vitamin D deficiency among pregnant women in India: A systematic review and meta-analysis. *J. Heath Popul. Nutr.* **2021**, *40*, 48. [CrossRef]
53. Nandi, A.A.; Wadhwani, N.S.; Randhir, K.N.; Madiwale, S.D.; Deshpande, J.S.; Wagh, G.N.; Joshi, S.R. Maternal vitamin D deficiency influences long-chain polyunsaturated fatty acids and pregnancy outcome in association with alterations in one-carbon metabolism. *Nutr. Res.* **2021**, *86*, 37–49. [CrossRef] [PubMed]
54. Rohr Thomsen, C.; Brink Henriksen, T.; Uldbjerg, N.; Milidou, I. Seasonal Variation in the Hypertensive Disorders of Pregnancy in Denmark. *Acta Obstet. Gynecol. Scand.* **2020**, *99*, 623–630. [CrossRef] [PubMed]
55. Mirzakhani, H.; Litonjua, A.A.; McElrath, T.F.; O'Connor, G.; Lee-Parritz, A.; Iverson, R.; Macones, G.; Strunk, R.C.; Bacharier, L.B.; Zeiger, R. Early Pregnancy Vitamin D Status and Risk of Preeclampsia. *J. Clin. Investig.* **2016**, *126*, 4702–4715. [CrossRef] [PubMed]
56. Accortt, E.E.; Mirocha, J.; Schetter, C.D.; Hobel, C.J. Adverse Perinatal Outcomes and Postpartum Multi-Systemic Dysregulation: Adding Vitamin D Deficiency to the Allostatic Load Index. *Matern. Child Health J.* **2017**, *21*, 398–406. [CrossRef] [PubMed]
57. Brodowski, L.; Burlakov, J.; Hass, S.; von Kaisenberg, C.; von Versen-Höynck, F. Impaired Functional Capacity of Fetal Endothelial Cells in Preeclampsia. *PLoS ONE* **2017**, *12*, e0178340. [CrossRef] [PubMed]
58. Lawal, O.A. A Study of Vitamin D Status of Women with Gestational Diabetes Mellitus in Abuja, Nigeria. *Fac. Pathol.* **2016**, *62* (Suppl. S1), 527.
59. Lechtermann, C.; Hauffa, B.P.; Herrmann, R.; Schündeln, M.M.; Gellhaus, A.; Schmidt, M.; Grasemann, C. Maternal Vitamin D Status in Preeclampsia: Seasonal Changes are Not Influenced by Placental Gene Expression of Vitamin D Metabolizing Enzymes. *PLoS ONE* **2014**, *9*, e105558. [CrossRef] [PubMed]
60. Achkar, M.; Dodds, L.; Giguère, Y.; Forest, J.; Armson, B.A.; Woolcott, C.; Agellon, S.; Spencer, A.; Weiler, H.A. Vitamin D Status in Early Pregnancy and Risk of Preeclampsia. *Obstet. Gynecol.* **2015**, *212*, 511.e1–511.e7. [CrossRef] [PubMed]
61. Rezende, V.B.; Sandrim, V.C.; Palei, A.C.; Machado, L.; Cavalli, R.C.; Duarte, G.; Tanus-Santos, J.E. Vitamin D Receptor Polymorphisms in Hypertensive Disorders of Pregnancy. *Mol. Biol. Rep.* **2012**, *39*, 10903–10906. [CrossRef] [PubMed]
62. Osman, O.M.; Gaafar, T.; Eissa, T.S.; Abdella, R.; Ebrashy, A.; Ellithy, A. Prevalence of Vitamin D Deficiency in Egyptian Patients with Pregnancy-Induced Hypertension. *J. Perinat. Med.* **2020**, *48*, 583–588. [CrossRef]
63. Schoenmakers, I.; Piec, I.; Baban, S.; Barebring, L.; Green, D.; Washbourne, C.J.; Tang, J.C.Y.; Fraser, W.D.; Augustin, H. Gestational hypercalcemia: Prevalence and biochemical profile. *J. Steroid Biochem. Mol. Biol.* **2020**, *199*, 105611. [CrossRef]
64. Olmos-Ortiz, A.; Olivares-Huerta, A.; García-Quiroz, J.; Avila, E.; Halhali, A.; Quesada-Reyna, B.; Larrea, F.; Zaga-Clavellina, V.; Díaz, L. Cord serum calcitriol inversely correlates with maternal blood pressure in urinary tract infection-affected pregnancies: Sex-dependent immune implications. *Nutrients* **2021**, *13*, 3114. [CrossRef]
65. de Souza, E.A.; Pisani, L.P. The Relationship among Vitamin D, TLR4 Pathway and Preeclampsia. *Mol. Biol. Rep.* **2020**, *47*, 6259–6267. [CrossRef] [PubMed]
66. Nandi, A.A.; Wadhwani, N.S.; Joshi, S.R. Altered Metabolic Homeostasis between Vitamin D and Long Chain Polyunsaturated Fatty Acids in Preeclampsia. *Med. Hypotheses* **2017**, *100*, 31–36. [CrossRef] [PubMed]
67. Agarwal, S.; Kovilam, O.; Agrawal, D.K. Vitamin D and its Impact on Maternal-Fetal Outcomes in Pregnancy: A Critical Review. *Crit. Rev. Food Sci. Nutr.* **2018**, *58*, 755–769. [CrossRef]
68. Palacios, C.; De-Regil, L.M.; Lombardo, L.K.; Peña-Rosas, J.P. Vitamin D Supplementation during Pregnancy: Updated Meta-Analysis on Maternal Outcomes. *J. Steroid Biochem. Mol. Biol.* **2016**, *164*, 148–155. [CrossRef] [PubMed]

69. Pérez-López, F.R.; Pasupuleti, V.; Mezones-Holguin, E.; Benites-Zapata, V.A.; Thota, P.; Deshpande, A.; Hernandez, A.V. Effect of Vitamin D Supplementation during Pregnancy on Maternal and Neonatal Outcomes: A Systematic Review and Meta-Analysis of Randomized Controlled Trials. *Fertil. Steril.* **2015**, *103*, 1278–1288.e4. [CrossRef] [PubMed]
70. Weinert, L.S.; Silveiro, S.P. Maternal–fetal Impact of Vitamin D Deficiency: A Critical Review. *Matern. Child Health J.* **2015**, *19*, 94–101. [CrossRef] [PubMed]
71. Girling, J.; Sykes, L. Thyroid Disorders and Other Endocrinological Disorders in Pregnancy. *Obstet. Gynaecol. Reprod. Med.* **2013**, *23*, 171–179. [CrossRef]
72. Tabesh, M.; Salehi-Abargouei, A.; Tabesh, M.; Esmaillzadeh, A. Maternal Vitamin D Status and Risk of Pre-Eclampsia: A Systematic Review and Meta-Analysis. *J. Clin. Endocrinol. Metab.* **2013**, *98*, 3165–3173. [CrossRef]
73. Alzaim, M.; Wood, R.J. Vitamin D and Gestational Diabetes Mellitus. *Nutr. Rev.* **2013**, *71*, 158–167. [CrossRef] [PubMed]
74. Senti, J.; Thiele, D.K.; Anderson, C.M. Maternal Vitamin D Status as a Critical Determinant in Gestational Diabetes. *J. Obstet. Gynecol. Neonatal. Nurs.* **2012**, *41*, 328–338. [CrossRef] [PubMed]
75. Barrett, H.; McElduff, A. Vitamin D and Pregnancy: An Old Problem Revisited. *Best Pract. Res. Clin. Endocrinol. Metabo.* **2010**, *24*, 527–539. [CrossRef] [PubMed]
76. Mulligan, M.L.; Felton, S.K.; Riek, A.E.; Bernal-Mizrachi, C. Implications of Vitamin D Deficiency in Pregnancy and Lactation. *Obstet. Gynecol.* **2010**, *202*, 429.e1–429.e9. [CrossRef] [PubMed]
77. Thorne-Lyman, A.; Fawzi, W.W. Vitamin D during Pregnancy and Maternal, Neonatal and Infant Health Outcomes: A Systematic Review and Meta-Analysis. *Paediatr. Perinat. Epidemiol.* **2012**, *26*, 75–90. [CrossRef] [PubMed]
78. Baker, A.M.; Haeri, S.; Camargo, C.A., Jr.; Espinola, J.A.; Stuebe, A.M. A nested case-control study of midgestation vitamin D deficiency and risk of se-vere preeclampsia. *J. Clin. Endocrinol. Metab.* **2010**, *95*, 5105–5109. [CrossRef] [PubMed]
79. Powe, C.E.; Seely, E.W.; Rana, S.; Bhan, I.; Ecker, J.; Karumanchi, S.A.; Thadhani, R. First trimester vitamin D, vitamin D binding protein, and subsequent preeclampsia. *Hypertension* **2010**, *56*, 758–763. [CrossRef] [PubMed]
80. Shand, A.; Nassar, N.; Von Dadelszen, P.; Innis, S.; Green, T. Maternal vitamin D status in pregnancy and adverse pregnancy outcomes in a group at high risk for pre-eclampsia. *BJOG Int. J. Obstet. Gynaecol.* 2010; 117, 1593–1598.
81. Holick, M.F.; Chen, T.C. Vitamin D Deficiency: A Worldwide Problem with Health Consequences. *Am. J. Clin. Nutr.* **2008**, *87*, 1080S–1086S. [CrossRef]
82. Holick, M.F. Vitamin D Deficiency. *N. Engl. J. Med.* **2007**, *357*, 266–281. [CrossRef]
83. Baggerly, C.A.; Cuomo, R.E.; French, C.B.; Garland, C.F.; Gorham, E.D.; Grant, W.B.; Heaney, R.P.; Holick, M.F.; Hollis, B.W.; McDonnell, S.L. Sunlight and Vitamin D: Necessary for Public Health. *J. Am. Coll. Nutr.* **2015**, *34*, 359–365. [CrossRef]
84. Bodnar, L.M.; Catov, J.M.; Simhan, H.N.; Holick, M.F.; Powers, R.W.; Roberts, J.M. Maternal Vitamin D Deficiency Increases the Risk of Preeclampsia. *J. Clin. Endocrinol. Metab.* **2007**, *92*, 3517–3522. [CrossRef] [PubMed]
85. Bui, T.; Christin-Maitre, S. Vitamin D and Pregnancy. *Ann. Endocrinol.* **2011**, *72* (Suppl. S1), S23–S28. [CrossRef]
86. Brannon, P.M. Vitamin D and Adverse Pregnancy Outcomes: Beyond Bone Health and Growth. *Proc. Nutr. Soc.* **2012**, *71*, 205–212. [CrossRef]
87. Palaniswamy, S.; Williams, D.; Järvelin, M.; Sebert, S. Vitamin D and the Promotion of Long-Term Metabolic Health from a Programming Perspective. *Nutr. Metab. Insights* **2015**, *8*, S29526.
88. Saraf, R.; Morton, S.M.; Camargo, C.A., Jr.; Grant, C.C. Global Summary of Maternal and Newborn Vitamin D Status—A Systematic Review. *Matern. Child Nutr.* **2016**, *12*, 647–668. [CrossRef] [PubMed]
89. De-Regil, L.M.; Palacios, C.; Lombardo, L.K.; Peña-Rosas, J.P. Vitamin D Supplementation for Women during Pregnancy. *Cochrane Database Syst. Rev.* **2016**, *1*, CD008873.
90. Hollis, B.W.; Johnson, D.; Hulsey, T.C.; Ebeling, M.; Wagner, C.L. Vitamin D Supplementation during Pregnancy: Double-blind, Randomized Clinical Trial of Safety and Effectiveness. *J. Bone Miner. Res.* **2011**, *26*, 2341–2357. [CrossRef]
91. World Health Organization (WHO). *Vitamin D Supplementation during Pregnancy*; WHO: Geneva, Switzerland, 2021; ISBN 978-92-4-150493-5.
92. American Pregnancy Association. *Vitamin D and Pregnancy*; American Pregnancy Association: Irving, TX, USA, 2021.
93. Hong-Bi, S.; Yin, X.; Xiaowu, Y.; Ying, W.; Yang, X.; Ting, C.; Na, W. High Prevalence of Vitamin D Deficiency in Pregnant Women and its Relationship with Adverse Pregnancy Outcomes in Guizhou, China. *J. Int. Med. Res.* **2018**, *46*, 4500–4505. [CrossRef]
94. Hynes, C.; Jesurasa, A.; Evans, P.; Mitchell, C. Vitamin D Supplementation for Women before and during Pregnancy: An Update of the Guidelines, Evidence, and Role of GPs and Practice Nurses. *Br. J. Gen. Pract.* **2017**, *67*, 423–424. [CrossRef] [PubMed]
95. Lötscher, Q.; l'Alemand, D.; Bischoff-Ferrari, H.; Burckhardt, P. Vitamin D Deficiency: Evidence, Safety, and Recommendations for the Swiss Population. 2012. Available online: https://www.zora.uzh.ch/id/eprint/73029/ (accessed on 1 June 2022).
96. Health Canada. Vitamin D and Calcium: Updated Dietary Reference Intakes. 2021. Available online: https://www.canada.ca/en/health-canada/services/food-nutrition/healthy-eating/vitamins-minerals/vitamin-calcium-updated-dietary-reference-intakes-nutrition.html (accessed on 1 June 2022).
97. Özdemir, A.A.; Gündemir, Y.E.; Küçük, M.; Sarıcı, D.Y.; Elgörmüş, Y.; Çağ, Y.; Bilek, G. Vitamin D Deficiency in Pregnant Women and their Infants. *J. Clin. Pediatr. Endocrinol.* **2018**, *10*, 44. [CrossRef] [PubMed]

98. Keshavarz, P.; Jandaghi, P.; Shafiee, M.; Islam, N.; Vatanparast, H. Maternal Vitamin D Status among Different Ethnic Groups and its Potential Contribution to Adverse Pregnancy and Child Outcomes. In *Vitamin D Deficiency*; IntechOpen: London, UK, 2020.
99. Ministry of Health. Companion Statement on Vitamin D and Sun Exposure in Pregnancy and Infancy in New Zealand: A Supplement to the Consensus Statement on Vitamin D and Sun Exposure in New Zealand. 2021. Available online: https://www.health.govt.nz/publication/companion-statement-vitamin-d-and-sun-exposure-pregnancy-and-infancy-new-zealand (accessed on 20 June 2022).

nutrients

MDPI

Article

Urbanization and Unfavorable Changes in Metabolic Profiles: A Prospective Cohort Study of Indonesian Young Adults

Farid Kurniawan [1,2,3,*], Mikhael D. Manurung [3], Dante S. Harbuwono [1,2], Em Yunir [1,2], Roula Tsonaka [4], Tika Pradnjaparamita [2], Dhanasari Vidiawati [5,6], Angelica Anggunadi [7], Pradana Soewondo [1,2], Maria Yazdanbakhsh [3], Erliyani Sartono [3] and Dicky L. Tahapary [1,2]

1 Division of Endocrinology, Metabolism, and Diabetes, Department of Internal Medicine, Dr. Cipto Mangunkusumo National General Hospital and Faculty of Medicine Universitas Indonesia, Jakarta 10430, Indonesia
2 Metabolic, Cardiovascular, and Aging Research Cluster, The Indonesian Medical Educational and Research Institute, Faculty of Medicine Universitas Indonesia, Jakarta 10430, Indonesia
3 Department of Parasitology, Leiden University Medical Center, 2333 ZA Leiden, The Netherlands
4 Department of Biomedical Data Science, Leiden University Medical Center, 2333 ZC Leiden, The Netherlands
5 Division of Family Medicine, Department of Community Medicine, Faculty of Medicine Universitas Indonesia, Jakarta 10310, Indonesia
6 Makara UI Satellite Clinic, Universitas Indonesia, Depok 16425, Indonesia
7 Center for Sport and Exercise Studies Cluster, The Indonesian Medical Educational and Research Institute, Faculty of Medicine Universitas Indonesia, Jakarta 10430, Indonesia
* Correspondence: f.kurniawan@lumc.nl or farid.kurniawan01@ui.ac.id; Tel.: +31-715269278

Citation: Kurniawan, F.; Manurung, M.D.; Harbuwono, D.S.; Yunir, E.; Tsonaka, R.; Pradnjaparamita, T.; Vidiawati, D.; Anggunadi, A.; Soewondo, P.; Yazdanbakhsh, M.; et al. Urbanization and Unfavorable Changes in Metabolic Profiles: A Prospective Cohort Study of Indonesian Young Adults. *Nutrients* **2022**, *14*, 3326. https://doi.org/10.3390/nu14163326

Academic Editors: William B. Grant and Ronan Lordan

Received: 8 July 2022
Accepted: 11 August 2022
Published: 14 August 2022

Abstract: The substantial increase in the prevalence of non-communicable diseases in Indonesia might be driven by rapid socio-economic development through urbanization. Here, we carried out a longitudinal 1-year follow-up study to evaluate the effect of urbanization, an important determinant of health, on metabolic profiles of young Indonesian adults. University freshmen/women in Jakarta, aged 16–25 years, who either had recently migrated from rural areas or originated from urban settings were studied. Anthropometry, dietary intake, and physical activity, as well as fasting blood glucose and insulin, leptin, and adiponectin were measured at baseline and repeated at one year follow-up. At baseline, 106 urban and 83 rural subjects were recruited, of which 81 urban and 66 rural were followed up. At baseline, rural subjects had better adiposity profiles, whole-body insulin resistance, and adipokine levels compared to their urban counterparts. After 1-year, rural subjects experienced an almost twice higher increase in BMI than urban subjects (estimate (95%CI): 1.23 (0.94; 1.52) and 0.69 (0.43; 0.95) for rural and urban subjects, respectively, $P_{int} < 0.01$). Fat intake served as the major dietary component, which partially mediates the differences in BMI between urban and rural group at baseline. It also contributed to the changes in BMI over time for both groups, although it does not explain the enhanced gain of BMI in rural subjects. A significantly higher increase of leptin/adiponectin ratio was also seen in rural subjects after 1-year of living in an urban area. In conclusion, urbanization was associated with less favorable changes in adiposity and adipokine profiles in a population of young Indonesian adults.

Keywords: urbanization; adiposity; dietary intake; adipokines; young adults; prospective cohort

1. Introduction

As a low-middle income country, Indonesia is facing two major health problems. On the one hand, an increasing prevalence of non-communicable diseases such as cardiovascular diseases (CVD), obesity, and type 2 diabetes (T2D) is becoming rampant. While on the other, infectious diseases such as helminth infections, malaria, and tuberculosis are still highly prevalent in some rural areas, resulting in stark differences of these disease patterns between urban and rural settings [1,2].

People residing in urban areas are characterized by relatively high caloric and fat intake compared to their rural counterparts [3]. Moreover, urban people tend to be less physically active [4]. These factors can cause a disruption in energy homeostasis, with a surplus stored in the body as fat [5]. Increasing body fat increases the chance of obesity [6]. Previous studies have shown that higher adipose tissue mass is associated with higher inflammation and insulin resistance [7], which eventually could lead to T2D [8] and CVD [9].

Rapid socio-economic development in Indonesia has promoted the migration of people from rural to urban areas to seek a better life [10]. Previous studies have shown that urbanization is associated with new environmental and lifestyle changes that have the potential to put rural individuals at risk of deteriorating metabolic health [11–13]. The limitation of these previous studies evaluating the effect of urban-rural environment on metabolic health is their cross-sectional design, which lacks the power to show causality.

The worldwide increase of obesity is not only observed in older populations but also in young adults [14]. Based on the Indonesian National Basic Health Survey 2018, there is a high burden of obesity and prediabetes in the young adult population [15]. As this population constitutes a significant proportion of Indonesians [16], the increase in the prevalence of these diseases may become a major health burden.

Early problem identification and intervention targeted towards this economically active young adult population in the context of metabolic health could have a great impact on decreasing the incidence rate, or even lowering the prevalence of non-communicable diseases. To this end, we conducted a prospective cohort study to assess the effect of urbanization over time and its contributing factors on the metabolic health profiles of the Indonesian young adult population.

2. Methods

2.1. Study Design and Population

This prospective cohort study was conducted on the Depok campus of the University of Indonesia (UI). Freshmen/women UI bachelor students were recruited in this study. Baseline data were collected in the first three months of the start of the academic year, between August-November 2018, while the follow-up sample collection was performed one year later. Subjects' recruitment was started by providing information about the study during the medical examination of newly arrived students, via social media, and by spreading flyers/leaflets after classes, as well as in student dormitories. A short interview was performed to collect information regarding the areas where the students originated from. Afterwards, a detailed explanation of the study was given to the subjects who agreed to participate and fulfilled the criteria set in this study. After written informed consent, subjects were invited to visit the Makara UI Satellite Clinic to undergo clinical assessment, measurements, and blood sampling. The subjects were classified into the urban group if they were born and lived in urban areas, such as in Jakarta metropolitan areas or in one of the provincial capital cities. The rural group comprised subjects that were originally born and lived in rural areas, defined as the villages that are located at the district levels across Indonesia. Pregnancy and students with previously known diabetes, prediabetes, severe liver or kidney dysfunction, cardiovascular and autoimmune diseases were excluded from the study. This study was approved by the Ethical Committee of Faculty of Medicine Universitas Indonesia (No. 1181/UN2.F1/ETIK/2017).

2.2. Anthropometric Measurements

Body height was measured using a portable stadiometer (SECA Model 213, Seca Gmbh Co., Hamburg, Germany), while body weight and body composition were measured using a Tanita body impedance analyzer (TBF-300A, Tanita Corp, Tokyo, Japan). Body mass index (BMI) was calculated in kg divided by squared height in meters. Three measurements of waist circumference were taken for each subject using an ergonomic circumference measuring tape (SECA Model 201, Seca Gmbh Co., Hamburg, Germany) and according to the WHO standardized protocol. The average of all three measurements was then used for analysis.

2.3. Fasting Blood Glucose, HbA1c, Fasting Insulin, and HOMA-IR Measurement

All clinical measurements and blood samples collection were performed after overnight fasting. Finger prick blood was used for measurement of fasting blood glucose (Accu-Check Performa, Roche Diagnostic GmBH, Germany) and HbA1c (A1c EZ 2.0 HbA1c Analyzer, BioHermes, Wuxi, China) levels. The results of fasting blood glucose (FBG) and HbA1c were used to detect subjects with undiagnosed diabetes and prediabetes that had to be excluded from the study. Serum fasting insulin levels were measured in a certified commercial laboratory (Prodia Lab) by a solid-phase, enzyme-labeled chemiluminescent immunometric assay (Siemens IMMULITE 2000XPi) with an assay range of 2–300 mU/L. For the levels below 2 mU/L, a standardized formula from the instrument manufacturer was used to interpolate the concentrations. Homeostatic model assessment for insulin resistance (HOMA-IR) as a validated measure for whole-body insulin resistance (IR) in humans was calculated using the formula: HOMA-IR = fasting serum insulin × fasting glucose/22.5) [17].

2.4. Leptin, Adiponectin, and Leptin/Adiponectin Ratio

Serum leptin and adiponectin levels were measured by ELISA using commercial reagents (DuoSet ELISA R&D System) according to the manufacturer's protocol. Leptin to adiponectin (L/A) ratio, a more sensitive marker for adipose tissue dysfunction, was calculated by L/A = leptin level (ng/mL)/adiponectin level (μg/mL) [18].

2.5. Dietary Intake Analysis

One week before the intended measurement date, each subject was informed and instructed on how to make a 3-day food record consisting of two working days and one day during the weekend. For each recording day, all participants were required to write down all of the food and drink they consumed throughout the day. The household servings portion for each meal, food preparation methods, brand name of the foods or beverages if applicable, as well as the addition of sugar, were recorded, as described previously [19]. On the study subjects' clinical measurement and blood sampling day, a certified dietician performed an interview with the subjects to review the completeness and validity of the food record data. These dietary intake data were then analyzed using NutriSurvey 2007 (EBISpro, Willstatt, Germany) software. The amount of total calorie, carbohydrate, fat, and protein intake for each day were obtained and then averaged for further analysis, as published [20].

2.6. Physical Activity Analysis

Physical activity was assessed using the adapted Global Physical Activity Questionnaire (GPAQ), which was developed by the World Health Organization [21] and validated for the Indonesian population [22]. This self-reported questionnaire comprised 16 questions that were grouped to collect information regarding physical activity over a typical week in three domains: activity at work, transportation (travel to and from places), and recreational activity [21]. All subjects were asked to fill in the questionnaire based on their one-week activities before the measurement date. According to GPAQ analysis guidelines [23], an estimation of the total weekly volume of moderate and vigorous physical activities (MVPA) was given as Metabolic Equivalent-minutes/week (MET. minutes/week), along with the total time spent on MVPA (minutes/week) and total time of sedentary activities in one week (minutes/week) [24]. Furthermore, based on their total volume and time spent on MVPA, the subject's physical activity level was classified into three categories (low, moderate, and high) [23].

2.7. Statistical Analysis

Continuous variables with normal distribution were presented as mean and standard deviation [mean (SD)]. Meanwhile, non-normally distributed data were presented as geometric mean and 95% confidence interval (geomean (95%CI)) and were log-transformed (log2) for analysis. Linear regression (IBM SPSS Statistics ver. 25) was performed to compare the mean differences of independent variables between two groups at baseline when adjustment for covariates was needed. The chi-square test was used to compare categorical data. Mediation analysis for evaluating the effect of dietary intake components on anthropometry parameter differences between rural and urban group at baseline was performed using PROCESS macro ver. 4.0 for SPSS, as described previously [25].

The changes in parameters measured at baseline and 1-year follow-up for each group, and the differences of these changes between urban and rural subjects, were analyzed using linear-mixed model as implemented in the lme4 R package [26]. For each parameter, the covariates used in the linear mixed model were origin (urban/rural), time, and their interaction. The within subject correlation was accounted for using a random-intercepts term. The statistical significance of the effects (i.e., changes from baseline within each group and between groups) were tested using the F-test with Satterthwaite's degree-of-freedom as implemented in lmerTest [27]. Mediation analysis for the BMI and adipokines changes was performed using 5000 bootstrap samples to obtain the 95% confidence interval for the indirect effect of the covariates. In particular, we evaluated the statistical significance in the decrease/increase of the estimate of the outcome variables after correcting for the changes in certain covariates. Linear mixed model analyses and bootstrapping were performed using R version 4.1.2 in RStudio version 1.4. For all tests, statistical significance was considered at the two-sided 5% level.

3. Results

3.1. Study Population

A total of 189 (106 urban; 83 rural) subjects were recruited at baseline. For urban subjects, 87.7% originated from Jakarta metropolitan areas, while the rest were from other provincial capital cities. The overall loss to follow-up was 22.1%, leaving 81 urban and 66 rural subjects at the one-year assessment time point. The main reasons for loss to follow-up were refusal to continue (18 subjects/9.4%), could not be contacted (22 subjects/11.6%), and moved to study at another university (2 subjects/1.1%). The proportion of loss to follow-up was similar between rural and urban groups (see flow-chart of the study in Figure S1).

3.2. Metabolic Profiles of Urban vs. Rural Subjects at Baseline

Age and proportion of males and females were similar between the rural and urban groups. Adiposity indices (BMI, waist circumference, and fat percentage) were significantly higher in urban compared to rural subjects (mean differences (95%CI) after adjustment for age and sex: 2.81 (1.55; 4.07) kg/m^2, $p < 0.001$; 6.37 (3.25; 9.50) cm, $p < 0.001$; and 5.07 (2.70; 7.44) %, $p < 0.001$; for BMI, waist circumference and fat percentage, respectively). Moreover, if BMI was grouped based on the WHO cut-off for Asian populations [28], we observed a higher proportion of overweight/obese in urban compared to rural subjects. Conversely, the proportion of underweight subjects was almost three times higher in the rural than in the urban group (Table 1).

Table 1. Baseline characteristics of the study population.

Variables	Urban N = 106	Rural N = 83	*p* Values # (Adjusted for Age and Sex)	*p* Values # (Adjusted for Age, Sex, and BMI)
Age, yrs old (mean, SD)	18.4 (0.7)	18.6 (0.7)	0.09	
Sex, n male (%)	39 (36.8)	31 (37.3)	0.94	
BMI, kg/m^2 (mean, SD)	22.9 (5.0)	20.0 (3.2)	**<0.001**	
BMI grouping, n (%)				
- Underweight (<18.5)	14 (13.2)	28 (33.7)		
- Normoweight (18.5–22.9)	50 (47.2)	41 (49.4)	**0.001**	
- Overweight (23–24.9)	17 (16.0)	7 (8.4)		
- Obese (≥25.0)	25 (23.6)	7 (8.4)		
Waist circumference, cm (mean, SD)	78.5 (12.8)	72.1 (8.2)	**<0.001**	
Fat percentage, % (mean, SD)	28.2 (9.1)	22.8 (8.3)	**<0.001**	
FBG, mg/dL (mean, SD)	87.1 (8.2)	86.7 (7.8)	0.54	
HbA1c, % (mean, SD)	5.1 (0.4)	5.1 (0.3)	0.24	
Fasting insulin †, IU/mL	5.3 (4.3–6.6)	2.9 (2.2–3.8)	**0.001**	0.06
HOMA-IR †	1.1 (0.9–1.4)	0.6 (0.5–0.8)	**0.001**	0.06
Leptin †, ng/mL	11.6 (9.7–13.8)	6.9 (5.3–9.1)	**<0.001**	0.07
Adiponectin †, µg/mL	4.1 (3.7–4.5)	4.9 (4.4–5.3)	**0.02**	0.19
Leptin-Adiponectin (L/A) Ratio †	2.9 (2.3–3.5)	1.4 (1.1–1.9)	**<0.001**	0.03
Dietary intake, mean (SD)				
- Total calories, kcal	1444 (335)	1289 (422)	**0.002**	**0.009**
- Fat, gram	52 (15)	44 (16)	**<0.001**	**0.01**
- Protein, gram	50 (14)	41 (13)	**<0.001**	**0.001**
- Carbohydrate, gram	193 (55)	179 (73)	0.08	0.06

† Not normally distributed continuous variables, presented as geomean (95%CI) and log transformed for analysis. # Analyzed with linear regression for continuous variables and Chi-square test for categorical variables. The *p*-values shown in bold represent the statistically significant differences with $p<0.05$. BMI: body mass index; FBG: fasting blood glucose; HOMA-IR: homeostatic model assessment for insulin resistance.

There was no difference in the fasting blood glucose and HbA1c levels between the two groups. Urban subjects had double the HOMA-IR, leptin levels, and L/A ratio than their rural counterparts. The opposite was observed for adiponectin levels. Further adjustment for BMI revealed that the differences remained significant for L/A ratio, while for HOMA-IR, leptin, and adiponectin became not statistically significant (Table 1).

3.3. Dietary Intake and Physical Activity at Baseline

Regarding dietary intake, we observed that urban subjects had significantly higher total calorie, fat, and protein intake compared to their rural counterparts (mean differences (95%CI) after adjustment for age and sex: 162.0 (59.4; 264.7) kcal, $p = 0.002$; 8.2 (3.7; 12.6) gram, $p < 0.001$, and 8.4 (4.7; 12.2) gram, $p < 0.001$), for total calorie, fat, and protein intake, respectively) (Table 1). Additionally, the differences in BMI, waist circumference, and fat percentage between the two groups were slightly attenuated after further adjustment for fat and protein intake, despite remaining statistically significant ((2.22 (0.92; 3.52) kg/m^2, $p = 0.001$ for BMI; 4.95 (1.74; 8.16) cm, $p = 0.003$ for waist circumference; and 4.38 (1.91; 6.85)%, $p = 0.001$ for fat percentage)). Moreover, mediation analysis showed that fat intake, compared to the other dietary intake components, might be the major driver of the differences in the adiposity profiles between urban and rural subjects at baseline (Table S1).

Next, we compared the physical activity profiles between the two groups at baseline based on the GPAQ analysis. The results showed that urban subjects had higher total volume and total time spent on MVPA compared to their rural counterparts. However, if these parameters were categorized as low, moderate, or high physical activity levels, no statistically significant differences were observed between the two groups. Meanwhile, for the total time of sedentary activities, we observed lower values for urban compared to rural subjects. (Table S2).

3.4. *Effect of Urbanization over Time on Adiposity Profiles, Insulin Resistance, and Adipokines*

At follow-up, after one year, both groups experienced an increase in their BMI. When we compared the degree of changes over time, we found that the increase of BMI in rural subjects was almost double what was seen in their urban counterparts (estimate (95%CI) after adjustment for age and sex: 1.23 (0.94; 1.52), $p < 0.001$ and 0.69 (0.43; 0.95), $p < 0.001$, for rural and urban subjects, respectively, $P_{int} < 0.01$). Although a similar pattern was observed for fat percentage, the difference between the groups did not reach statistical significance (2.18 (1.39; 2.97), $p < 0.001$ in rural subjects vs. 1.33 (0.62; 2.04), $p < 0.001$ in urban subjects, $P_{int} = 0.12$). Meanwhile, HOMA-IR at one-year follow-up did not change significantly compared to baseline in either rural or urban groups (Figure 1).

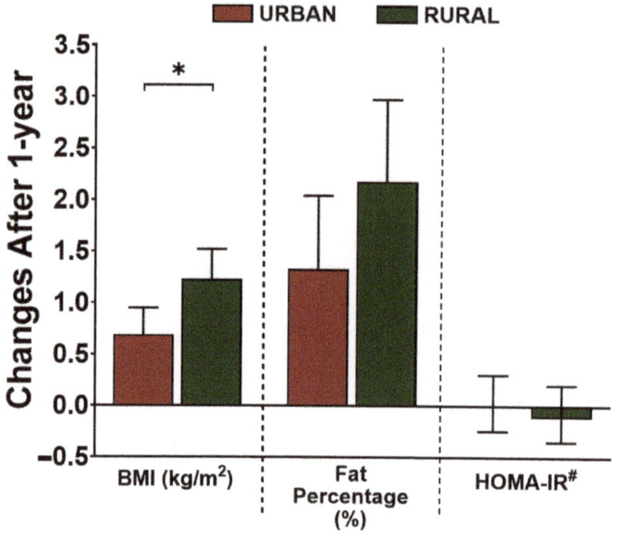

Figure 1. Changes of BMI, fat percentage, and whole-body insulin resistance (HOMA-IR) in urban and rural subjects after 1-year of living in an urban environment. The changes are presented as estimate and 95% confidence interval (95%CI). The changes in each group and the differences of changes between the urban and rural group for each parameter were analyzed using a linear-mixed model, adjusted for age and sex. The *p*-value depicted in the figure represents the *p*-value for interaction (P_{int}), the level of significance in the differences of changes between the two groups. * $p < 0.05$. # HOMA-IR was log-transformed (base 2) for analysis. The estimates (95%CI) were back-transformed (2^{β}) and presented as a multiplicative scale compared to baseline. BMI: body mass index; HOMA-IR: homeostatic model assessment for insulin resistance.

Similar analysis was performed for adipokines data, which revealed that both groups had increased leptin levels at 1-year follow-up, with a trend towards a higher increase in rural than urban subjects (Figure 2A, Table 2). Additionally, no changes in the adiponectin levels were observed in urban subjects at the follow-up time point, but a significant decrease was found in the rural subjects (Figure 2B, Table 2). These changes caused no differences in the adiponectin levels between the two groups at 1-year follow-up time point (Table S3). When L/A ratio was considered, a significant three times higher increase was seen in the rural compared to urban group (Figure 2C, Table 2). After further adjustment with the changes in BMI over time, these changes of leptin, adiponectin, and L/A ratio were attenuated and became non-significant for urban subjects (Table 2).

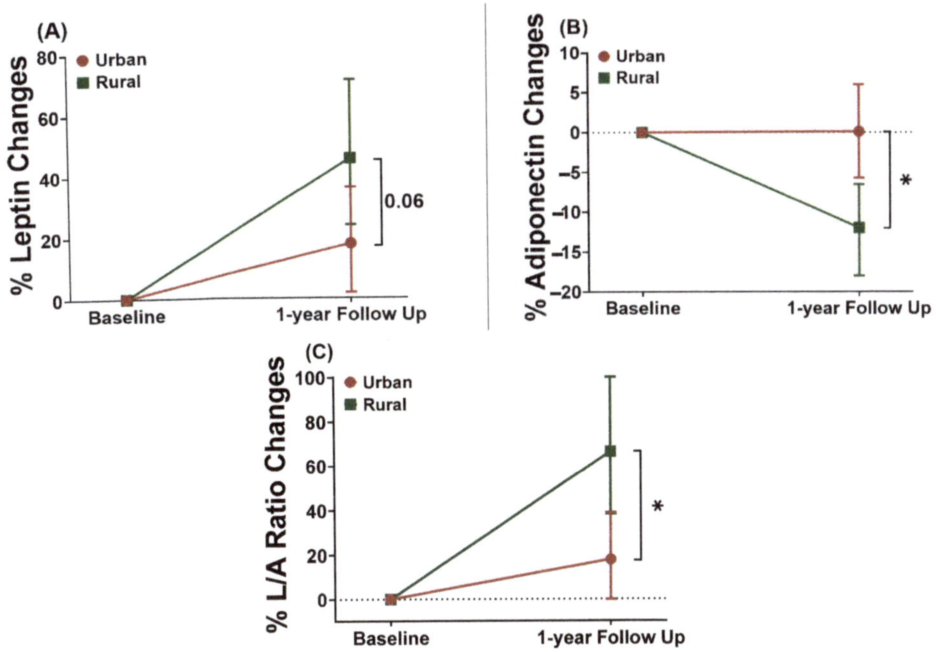

Figure 2. Changes of leptin levels (**A**), adiponectin levels (**B**), and leptin-adiponectin (L/A) ratio (**C**) in urban and rural subjects after 1-year of living in an urban environment. The changes are presented as estimate and 95% confidence interval (95%CI). The changes in each group and the differences of changes between urban and rural group for each parameter were analyzed using a linear-mixed model, adjusted for age and sex. All parameters were log-transformed (base 2) for analysis. The estimates (95%CIs) were back-transformed (2^{β}) and presented as percent changes compared to baseline. The *p*-value depicted in the figure represents the *p*-value for interaction (P_{int}), the level of significance in the differences of changes between the two groups. * $p < 0.05$.

Table 2. Mediation analysis of the effect of changes in BMI overtime on the leptin, adiponectin, and L/A ratio in urban and rural subjects at 1-year follow-up.

Variables †	Adjusted for Age and Sex		P_{int}	Adjusted for Age, Sex, and BMI				P_{int}
	Urban	Rural		Urban	Indirect Effect #	Rural	Indirect Effect #	
Leptin	0.24 (0.03; 0.45) $p = 0.03$	0.54 (0.31; 0.78) $p < 0.001$	0.06	0.09 (−0.11; 0.29) $p = 0.38$	−0.25; −0.07	0.33 (0.10; 0.55) $p = 0.005$	−0.29; −0.12	0.12
Adiponectin	0.002 (−0.08; 0.09) $p = 0.97$	−0.19 (−0.29; −0.10) $p < 0.001$	0.003	0.04 (−0.04; 0.12) $p = 0.34$	0.01; 0.06	−0.12 (−0.22; −0.03) $p = 0.008$	0.03; 0.10	0.008
L/A ratio	0.23 (−0.003; 0.47) $p = 0.05$	0.73 (0.47; 1.00) $p < 0.001$	0.006	0.06 (−0.16; 0.28) $p = 0.60$	−0.30, −0.08	0.45 (0.20; 0.70) $p < 0.001$	−0.39; −0.17	0.02

† All variables were analyzed using a linear mixed model on log transformed data, presented as estimate and 95% confidence interval. # Indirect effect of BMI on the variables analyzed, obtained by performing bootstrapping with 5000 iterations and presented as its 95% confidence interval. BMI: body mass index; L/A ratio: leptin/adiponectin ratio; P_{int}: *p*-value for interaction.

3.5. *Effect of Urbanization over Time on Dietary Intake and Physical Activity*

The changes over time in two important factors associated with urbanization-related lifestyle, namely, dietary intake and physical activity, were considered next. At the follow-up time point, a significant increase in total calorie, fat, and protein intake was seen in both groups. However, only the increase in protein consumption was significantly higher in rural than in urban subjects ((7.99 (4.42; 11.56), $p < 0.001$ vs. 14.03 (9.95; 18.10), $p < 0.001$, for urban and rural, respectively, $P_{int} = 0.03$) (see Figure 3). These changes also resulted in the loss of differences in the protein intake between the two groups at the 1-year follow-up time point (Table S3). Similar to the findings at baseline, adjustment for the increase in fat intake after one year contributed to the largest attenuation of the BMI increase in both groups (in urban: 29.0% vs. 5.8% vs. 1.4%, and in rural: 19.5% vs. 8.9% vs. 7.3%, for fat, protein, and carbohydrate intake, respectively) (Table 3). Although the increase in protein intake was almost twice as high in the rural group compared to the urban group after 1-year, adjustment for protein intake changes did not attenuate the differences in the increase in BMI between the two groups (Table S4).

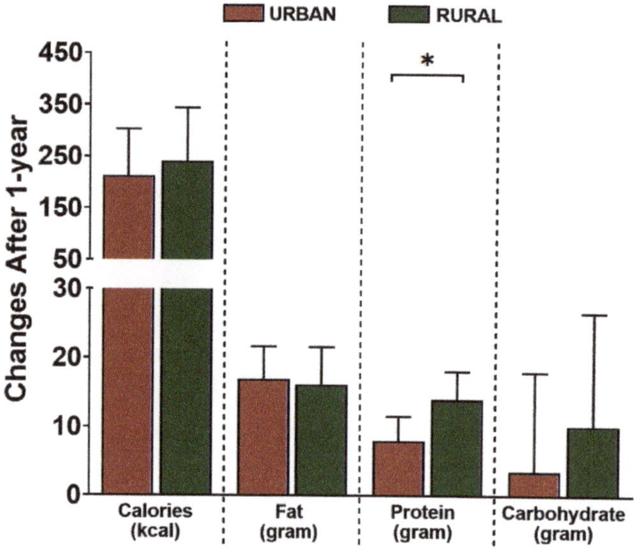

Figure 3. Changes of calorie-, fat-, protein-, and carbohydrate intake in urban and rural subjects after 1-year of living in an urban environment. The changes are presented as estimate and 95% confidence interval (95%CI). The changes in each group and the differences of changes between urban and rural group for each parameter were analyzed using a linear-mixed model, adjusted for age and sex. The *p*-value depicted in the figure represents the *p*-value for interaction (P_{int}), the level of significance in the differences of changes between the two groups. * $p < 0.05$.

Table 3. Mediation analysis of the effect of changes in dietary intake and physical activity over time on the changes of BMI at 1-year follow-up in both urban and rural subjects.

Model †	Urban				Rural				P_{int}
	Estimate (95%CI)	*p* Values	% Changes ††	Indirect Effect # (95%CI)	Estimate (95%CI)	*p* Values	% Changes ††	Indirect Effect # (95%CI)	
Adjusted for age and sex	0.69 (0.43, 0.95)	<0.001			1.23 (0.94; 1.52)	<0.001			0.007

Table 3. *Cont.*

Model †	Urban				Rural				P_{int}
	Estimate (95%CI)	*p* Values	% Changes ††	Indirect Effect # (95%CI)	Estimate (95%CI)	*p* Values	% Changes ††	Indirect Effect # (95%CI)	
Model with changes in dietary intake									
(+) Total calories intake	0.55 (0.28; 0.28)	<0.001	−20.3	−0.13 (−0.33; −0.02)	1.02 (0.71; 1.32)	<0.001	−17.1	−0.15 (−0.52; −0.04)	0.02
(+) Carbohydrate intake	0.68 (0.42; 0.93)	<0.001	−1.4	−0.01 (−0.12; 0.03)	1.14 (0.85; 1.43)	<0.001	−7.3	−0.09 (−0.34; 0.01)	0.02
(+) Fat intake	0.49 (0.20; 0.78)	<0.001	−29.0	−0.20 (−0.47; −0.04)	0.99 (0.67; 1.31)	<0.001	−19.5	−0.24 (−0.54; −0.06)	0.01
(+) Protein intake	0.65 (0.38; 0.93)	<0.001	−5.8	−0.04 (−0.22; 0.08)	1.12 (0.78; 1.45)	<0.001	−8.9	−0.11 (−0.43; 0.14)	0.02
(+) Fat and protein intake	0.50 (0.21; 0.79)	<0.001	−27.5	−0.19 (−0.48; 0.02)	1.04 (0.70; 1.37)	<0.001	−15.4	−0.19 (−0.53; 0.05)	0.007
Model with changes in physical activity									
(+) Total volume of MVPA	0.68 (0.42; 0.95)	<0.001	−1.4	−0.01 (−0.14; 0.05)	1.23 (0.94; 1.52)	<0.001	0.0	0.0 (−0.10; 0.04)	0.007
(+) Total minutes of MVPA	0.68 (0.42; 0.95)	<0.001	−1.4	−0.01 (−0.15; 0.06)	1.23 (0.94; 1.52)	<0.001	0.0	0.0 (−0.10; 0.05)	0.007
(+) Total sedentary time	0.70 (0.44; 0.96)	<0.001	1.4	0.01 (−0.03; 0.10)	1.26 (0.94; 1.58)	<0.001	2.4	0.03 (−0.11; 0.22)	0.007
Model with changes in dietary intake and physical activity									
(+) Fat and protein intake and total volume of MVPA	0.48 (0.19; 0.78)	0.001	−30.4	−0.21 (−0.50; −0.01)	1.10 (0.75; 1.44)	<0.001	−10.6	−0.13 (−0.49; 0.15)	0.003
(+) Fat and protein intake and total minutes of MVPA	0.49 (0.19; 0.78)	0.001	−29.0	−0.20 (−0.51; −0.01)	1.09 (0.75; 1.44)	<0.001	−11.4	−0.14 (−0.52; 0.14)	0.003
(+) Fat and protein intake and total sedentary time	0.51 (0.22; 0.80)	<0.001	−26.1	−0.18 (−0.47; 0.001)	1.15 (0.79; 1.50)	<0.001	−6.5	−0.08 (−0.47; 0.21)	0.002

† All variables as an additional adjustment for age and sex, and all analyses were performed using linear-mixed model. The group of covariates used for model adjustment were shown in bold. †† Proportion of changes in the estimate of the model compared to the model adjusted for age and sex only. # Indirect effect of covariate(s) on BMI, obtained by performing bootstrapping with 5000 iterations and presented as its 95% confidence interval. BMI: body mass index; MVPA: moderate-vigorous physical activity; P_{int}: *p*-value for interaction.

With respect to physical activity, we found a significant decrease in the total volume of MVPA after one year in the urban group only. However, the difference of changes between the two groups was not statistically significant. A similar pattern was also observed for the total time spent on MVPA. Moreover, there was a significantly higher decrease in total sedentary time after one year in the rural group, with a trend for a decrease in the urban group (Figure S2). Furthermore, addition of the physical activity parameters to the model with fat and protein intake did not significantly further attenuate the estimated changes of BMI in either group (Table 3).

4. Discussion

Here, we report the first prospective cohort study in an Indonesian young adult population that evaluated the effect of urbanization on metabolic health profiles. Our study showed that rural subjects had overall better adiposity, insulin resistance, and adipokine profiles compared to their urban counterparts. Importantly, we observed a significantly higher increase in BMI and leptin/adiponectin ratios in the rural subjects migrating to an urban area compared to subjects originating from urban areas.

The higher adiposity indices, proportion of overweight/obese, and whole-body insulin resistance in urban compared to rural residents of Indonesia have been reported before [29].

Unhealthy dietary behavior, such as high intake of calories and fat-dense foods associated with urban living, is thought to contribute to the higher adiposity profiles [3]. Indeed, we confirmed this pattern of dietary intake in our study. Although all further adjustments with total calorie, fat, or protein intake attenuated the anthropometric differences between rural and urban groups, our study showed that fat intake contributed the most. Additionally, the longitudinal follow-up to see how urban lifestyle affects metabolic health in those migrating from rural areas compared to urban residents, first confirmed a significant increase of BMI after one-year follow-up in both groups, as seen in previous studies, showing that the majority of freshmen gain weight during their first year of university life [30,31]. The increase of total calorie, fat, and protein intake after one year in both groups might partially explain these changes in BMI. Our study also implicated fat intake changes as the dietary intake component that might be the major contributor for BMI increase after one year in both groups. Significantly, the rural group experienced almost a twice higher increase in BMI than the urban group. Although a significantly higher protein intake was observed in rural compared to urban group at one year follow-up, this could not explain the differences in the BMI increase between the two groups. Interestingly, previous studies have shown that higher meat or meat-products intake, which mostly consist of protein and fat, is associated with more weight gain independent of the total calorie intake [32].

Another factor contributing to adiposity profiles is the level of physical activity [33], as this promotes burning of calories, leading to negative energy balance and subsequently less probability for fat deposition [34]. In our study, at baseline, we found that rural group had lower total volume and time spent on MVPA with a higher sedentary time compared to urban group. This suggests that physical activity does not explain the differences observed in BMI. However, it has to be noted that studies of physical activity in rural and urban areas can be influenced by factors such as the level of education, ethnicity, and tools utilized [35,36]. In our study, the questionnaires used at baseline, which took place when the study subjects had already arrived in urban area, might not truly reflect the subjects' level of physical activity during their residence in rural areas. At the 1-year follow-up time point, we found no significant differences in the changes of total volume and time spent on MVPA between the two groups. As for fat or protein intake, physical activity did not explain the higher gain in BMI seen in the rural group.

The addition of the physical activity parameters into the model with the adjustment of dietary intake also could not explain the higher increase of BMI in rural compared to urban group. This result implies that with similar changes of dietary intake and physical activity within one year, rural subjects experienced bigger changes in BMI than their urban counterparts. Hence, other factors, such as the gut microbiome [37] or epigenetic changes [38], could potentially influence the adiposity changes in the rural population upon migration to urban areas. Other factors that could potentially influence the changes of weight or BMI in our study subjects, as shown in previous studies, are psychological stress [39] and socioeconomic-cultural backgrounds [40,41]. These factors were not evaluated in our study.

The increase of BMI in both groups, if continued for the long term, could potentially cause obesity and induce other metabolic and cardiovascular diseases. In other cases, if the BMI increase does not lead to obesity, the distribution of body fat caused by the weight gain also needs to be considered. Previous studies have shown that Asian populations tend to have higher cardiometabolic risks compared to Caucasian populations with the same levels of BMI, in particular, due to central obesity or visceral adiposity [42,43]. These risks would potentially be higher in the rural group as a substantial number of subjects are underweight. Several studies showed that individuals with previous malnourished condition have an increased risk of obesity later in life, especially if they adopt unhealthy lifestyles [44,45]. Moreover, individuals that have experienced a double burden of malnutrition or undernutrition in early life followed by later overweight/obesity, also pose a substantially enhanced risk of non-communicable diseases (NCDs) [46].

The observed differences in insulin resistance and adipokine levels at baseline between urban and rural groups aligned with the findings from previous studies, showing a higher

HOMA-IR and lower adiponectin levels in urban compared to rural population [47,48]. Moreover, after adjustment for BMI, the differences in HOMA-IR, leptin, and adiponectin were no longer statistically significant, indicating a major contribution by BMI. Interestingly, after 1-year living in an urban area, a significant decrease in adiponectin levels, along with a significant increase in L/A ratio, was observed in rural groups compared to the urban group. These changes were attenuated after adjustment for the changes in BMI. This finding shows that rural subjects also experienced worse changes in the adipokine profiles, which was partially mediated by the changes in BMI. It also indicates that there might be other factors than BMI, potentially contributing to the changes in adipokines in the rural subjects after 1-year living in an urban area. As shown from previous studies, gut microbiota has been associated with changes in leptin and adiponectin levels in response to a high-fat diet [49,50].

In our study, although urban subjects had a significant increase in BMI levels after one year, this did not cause changes in adiponectin levels compared to the significant decrease observed in the rural group. These changes even led to the loss of the differences in adiponectin levels between the two groups at 1-year follow-up time point. These findings showed that rural individuals tend to be more vulnerable in their metabolic parameters upon changes in BMI.

Leptin and adiponectin have opposite effects on subclinical inflammation and insulin resistance. Leptin upregulates pro-inflammatory cytokines such as tumor necrosis factor-α and interleukin-6, while adiponectin has anti-inflammatory properties [18]. Adipose tissue dysfunction, marked by higher leptin and lower adiponectin levels, has been associated with insulin resistance and the incidence of T2D [51]. However, in our study, there were no significant changes in insulin resistance in the groups studied after one year of residence in an urban area. The relatively short follow-up period and preserved pancreatic beta-cell function in the young adult population might potentially explain this [52].

The longitudinal study design, inclusion of several metabolic health parameters, and incorporation of dietary intake and physical activity measurements were several strong points of our study. There is only one previous prospective cohort study known to the author that has evaluated the effect of urbanization on CVD risk factors and major NCDs [12]. However, this study did not incorporate dietary intake analysis and measurement of biological metabolic markers, such as insulin resistance index, leptin, and adiponectin. Our study also observed the importance of fat intake contribution in the increase of BMI in both the freshmen urban group and the rural individuals who recently migrated to an urban area. Previous study evaluating this freshmen weight gain only took into account eating behavior changes but did not perform detailed dietary intake analysis [53].

However, the relatively small number of subjects and short duration of follow-up could be considered as limitations in our study. The addition of tools to evaluate the quality of dietary intake, such as the Healthy Eating Index and the utilization of health technology devices like an accelerometer to assess physical activity more objectively, could provide more accurate data in future studies. Additionally, the inclusion of psychological stress assessment and questionnaires or tools to accommodate the evaluation of the socio-economic and cultural aspects would also result in a more comprehensive data for future research. Moreover, investigation of the gut microbiome, epigenetic changes, as well as immunological factors, might shed more light on the mechanisms that underlie rapid changes in the metabolic profiles upon urbanization.

In conclusion, the findings in our study show that adoption of an urban lifestyle could potentially cause poorer metabolic health changes in rural individuals who migrate to an urban area. Our findings, in part, complement a previous study that showed the rising BMI in residents of increasingly urbanizing rural areas in low-middle income countries is due to an increase in low-quality diet [54]. However, it also indicates that there is a more rapid increase in BMI of subjects arriving from rural areas that could not be explained by either diet or physical activity. Therefore, further studies are needed, as it is important for policymakers to design innovative approaches to prevent this negative effect of urbanization in young adult population, with particular attention to those migrating from rural areas.

Supplementary Materials: The following supporting information can be downloaded at: https://www.mdpi.com/article/10.3390/nu14163326/s1. Figure S1. Flow Chart of The Study. Figure S2. The changes of physical activity levels (A & B) and sedentary time (C) in urban and rural subjects after 1-year of living in an urban environment. Table S1. Mediation analysis of the effect of dietary intake on the differences of adiposity profiles in urban and rural subjects at baseline. Table S2. The levels of physical activity and sedentary time measured with GPAQ in urban and rural subjects at baseline. Table S3. Characteristics of study population at 1-year follow-up time. Table S4. The effect of the differences in protein intake changes on the differences of BMI increase after 1-year between urban and rural subjects.

Author Contributions: Conceptualization, F.K. and M.Y.; methodology, F.K., A.A., E.S. and D.L.T.; investigation, F.K., D.L.T., D.V. and T.P.; resources, E.S., M.D.M. and T.P.; formal analysis, F.K. and M.D.M.; funding acquisition, D.L.T., D.S.H., E.Y., M.Y. and P.S.; supervision, M.Y., D.S.H., E.Y. and P.S.; validation, R.T.; writing—original draft, F.K. and E.S.; writing—review and editing, D.L.T., M.Y., M.D.M., A.A., R.T., D.S.H., E.Y. and P.S. All authors have read and agreed to the published version of the manuscript.

Funding: The study was supported by the grant from Ministry of Research and Technology Republic of Indonesia (Grant No. NKB-1555/UN2.R3.1/HKP.05.00/2019) and PUTI Universitas Indonesia (Grant No. NKB-762/UN2.RST/HKP.05.02/2020). The doctoral study of F.K. was funded by a scholarship from The Indonesian Endowment Fund for Education (Lembaga Pengelola Dana Pendidikan/LPDP) Ministry of Finance the Republic of Indonesia, Ref S-364/LPDP.3/2019. The funders had no role in study design, data collection and analysis, decision to publish, or preparation of the manuscript.

Institutional Review Board Statement: The study was conducted in accordance with the Declaration of Helsinki and approved by Ethical Committee for Health Research of Faculty of Medicine Universitas Indonesia (No. 1181/UN2.F1/ETIK/2017).

Informed Consent Statement: Informed consent was obtained from all subjects involved in the study.

Data Availability Statement: The data that support the findings of this study are available from the corresponding author (F.K.) upon reasonable request.

Acknowledgments: The authors would like to thank all study participants in this study. Thanks are due to all research assistants and secretaries for their help during the field work. The authors would also like to thank Makara UI Satellite Clinic for providing the space and permission to perform all the study subject's recruitment and measurements there. The authors would like to thank Emma Houlder for proofreading the manuscript.

Conflicts of Interest: The authors declare that they have no conflict of interest.

References

1. Htet, A.S.; Bjertness, M.B.; Sherpa, L.Y.; Kjollesdal, M.K.; Oo, W.M.; Meyer, H.E.; Stigum, H.; Bjertness, E. Urban-rural differences in the prevalence of non-communicable diseases risk factors among 25–74 years old citizens in Yangon Region, Myanmar: A cross sectional study. *BMC Public Health* **2016**, *16*, 1225. [CrossRef]
2. Neiderud, C.J. How urbanization affects the epidemiology of emerging infectious diseases. *Infect Ecol Epidemiol.* **2015**, *5*, 27060. [CrossRef] [PubMed]
3. Mayen, A.L.; Marques-Vidal, P.; Paccaud, F.; Bovet, P.; Stringhini, S. Socioeconomic determinants of dietary patterns in low- and middle-income countries: A systematic review. *Am. J. Clin. Nutr.* **2014**, *100*, 1520–1531. [CrossRef] [PubMed]
4. Mashili, F.L.; Kagaruki, G.B.; Mbatia, J.; Nanai, A.; Saguti, G.; Maongezi, S.; Magimba, A.; Mghamba, J.; Kamugisha, M.; Mgina, E.; et al. Physical activity and associated socioeconomic determinants in rural and urban Tanzania: Results from the 2012 WHO-STEPS Survey. *Int. J. Popul. Res.* **2018**, 4965193. [CrossRef]
5. Hall, K.D.; Heymsfield, S.B.; Kemnitz, J.W.; Klein, S.; Schoeller, D.A.; Speakman, J.R. Energy balance and its components: Implications for body weight regulation. *Am. J. Clin. Nutr.* **2012**, *95*, 989–994. [CrossRef] [PubMed]
6. Hill, J.O.; Wyatt, H.R.; Peters, J.C. Energy balance and obesity. *Circulation* **2012**, *126*, 126–132. [CrossRef] [PubMed]
7. Schuster, D.P. Obesity and the development of type 2 diabetes: The effects of fatty tissue inflammation. *Diabetes Metab Syndr Obes* **2010**, *3*, 253–262. [CrossRef] [PubMed]

8. Chen, P.; Hou, X.; Hu, G.; Wei, L.; Jiao, L.; Wang, H.; Chen, S.; Wu, J.; Bao, Y.; Jia, W. Abdominal subcutaneous adipose tissue: A favorable adipose depot for diabetes? *Cardiovasc. Diabetol* **2018**, *17*, 93. [CrossRef] [PubMed]
9. Ha, E.E.; Bauer, R.C. Emerging roles for adipose tissue in cardiovascular disease. *Arter. Thromb. Vasc. Biol.* **2018**, *38*, e137–e144. [CrossRef]
10. Bloom, D.E.; Canning, D.; Fink, G. Urbanization and the wealth of nations. *Science* **2008**, *319*, 772–775. [CrossRef]
11. Ebrahim, S.; Kinra, S.; Bowen, L.; Andersen, E.; Ben-Shlomo, Y.; Lyngdoh, T.; Ramakrishnan, L.; Ahuja, R.C.; Joshi, P.; Das, S.M.; et al. The effect of rural-to-urban migration on obesity and diabetes in India: A cross-sectional study. *PLoS Med.* **2010**, *7*, e1000268. [CrossRef] [PubMed]
12. Ruiz-Alejos, A.; Carrillo-Larco, R.M.; Miranda, J.J.; Anderson, C.A.M.; Gilman, R.H.; Smeeth, L.; Bernabe-Ortiz, A. Addressing the impact of urban exposure on the incidence of type 2 diabetes mellitus: The PERU MIGRANT Study. *Sci. Rep.* **2018**, *8*, 5512. [CrossRef] [PubMed]
13. Sobngwi, E.; Mbanya, J.C.; Unwin, N.C.; Porcher, R.; Kengne, A.P.; Fezeu, L.; Minkoulou, E.M.; Tournoux, C.; Gautier, J.F.; Aspray, T.J.; et al. Exposure over the life course to an urban environment and its relation with obesity, diabetes, and hypertension in rural and urban Cameroon. *Int. J. Epidemiol.* **2004**, *33*, 769–776. [CrossRef] [PubMed]
14. Poobalan, A.; Aucott, L. Obesity Among Young Adults in Developing Countries: A Systematic Overview. *Curr. Obes. Rep.* **2016**, *5*, 2–13. [CrossRef] [PubMed]
15. National Institute for Health Research and Development (NIHRD), Ministry of Health, Republic of Indonesia. Laporan Nasional RISKESDAS. Jakarta. 2018. Available online: http://labdata.litbang.kemkes.go.id/riset-badan-litbangkes/menu-riskesnas/menu-riskesdas/426-rkd-2018 (accessed on 24 April 2021).
16. Indonesian Central Bureau of Statistics, *Hasil Sensus Penduduk 2020. Report No.: Berita Resmi Statistik No. 7/01/Th. XXIV*; Indonesian Central Bureau of Statistics: Jakarta, Indonesia, 2021.
17. Radziuk, J. Homeostastic model assessment and insulin sensitivity/resistance. *Diabetes* **2014**, *63*, 1850–1854. [CrossRef]
18. Lopez-Jaramillo, P.; Gomez-Arbelaez, D.; Lopez-Lopez, J.; Lopez-Lopez, C.; Martinez-Ortega, J.; Gomez-Rodriguez, A.; Triana-Cubillos, S. The role of leptin/adiponectin ratio in metabolic syndrome and diabetes. *Horm. Mol. Biol. Clin. Investig.* **2014**, *18*, 37–45. [CrossRef]
19. Ortega, R.M.; Perez-Rodrigo, C.; Lopez-Sobaler, A.M. Dietary assessment methods: Dietary records. *Nutr. Hosp.* **2015**, *31*, 38–45.
20. Rupasinghe, W.S.; Perera, H.; Wickramaratne, N. A comprehensive review on dietary assessment methods in epidemiological research. *J. Pub. Health Nutri.* **2020**, *3*, 204–211.
21. Armstrong, T.; Bull, F. Development of the World Health Organization Global Physical Activity Questionnaire (GPAQ). *J. Public Health-Heid* **2006**, *14*, 66–70. [CrossRef]
22. Bull, F.C.; Maslin, T.S.; Armstrong, T. Global physical activity questionnaire (GPAQ): Nine country reliability and validity study. *J. Phys. Act. Health* **2009**, *6*, 790–804. [CrossRef]
23. Global Physical Activity Questionnaire (GPAQ) Analysis Guide. World Health Organization. 2012. Available online: http://www.who.int/chp/steps/resources/GPAQ_Analysis_Guide.pdf (accessed on 2 April 2021).
24. Wanner, M.; Hartmann, C.; Pestoni, G.; Martin, B.W.; Siegrist, M.; Martin-Diener, E. Validation of the Global Physical Activity Questionnaire for self-administration in a European context. *BMJ Open Sport Exerc. Med.* **2017**, *3*, e000206. [CrossRef] [PubMed]
25. Hayes, A.H. Introduction to Mediation, Moderation, and Conditional Process Analysis. In *A Regression-Based Approach*, 3rd ed.; Guilford Press: New York, NY, USA, 2022.
26. Bates, D.; Mächler, M.; Bolker, B.; Walker, S. Fitting linear mixed-effects models using lme4. *J. Stat. Softw.* **2015**, *67*, 1–48. [CrossRef]
27. Kuznetsova, A.; Brockhoff, P.B.; Christensen, R.H.B. lmerTest package: Tests in linear mixed effects models. *J. Stat. Softw.* **2017**, *82*, 1–26. [CrossRef]
28. World Health Organization. Regional Office for the Western Pacific Region. In *The Asia-Pacific perspective: Redefining Obesity and Its Treatment*; Health Communications: Sydney, Australia, 2000.
29. Nurwanti, E.; Hadi, H.; Chang, J.S.; Chao, J.C.J.; Paramashanti, B.A.; Gittelsohn, J.; Bai, C.H. Rural-urban differences in dietary behavior and obesity: Results of the Riskesdas study in 10-18-year-old Indonesian children and adolescents. *Nutrients* **2019**, *11*, 2813. [CrossRef]
30. Beaudry, K.M.; Ludwa, I.A.; Thomas, A.M.; Ward, W.E.; Falk, B.; Josse, A.R. First-year university is associated with greater body weight, body composition and adverse dietary changes in males than females. *PLoS ONE* **2019**, *14*, e0218554. [CrossRef]
31. Vadeboncoeur, C.; Foster, C.; Townsend, N. Freshman 15 in England: A longitudinal evaluation of first year university student's weight change. *BMC Obes.* **2016**, *3*, 45.
32. Vergnaud, A.C.; Norat, T.; Romaguera, D.; Mouw, T.; May, A.M.; Travier, N.; Luan, J.; Wareham, N.; Slimani, N.; Rinaldi, S.; et al. Meat consumption and prospective weight change in participants of the EPIC-PANACEA study. *Am. J. Clin. Nutr.* **2010**, *92*, 398–407. [CrossRef]
33. Ojiambo, R.M.; Easton, C.; Casajus, J.A.; Konstabel, K.; Reilly, J.J.; Pitsiladis, Y. Effect of urbanization on objectively measured physical activity levels, sedentary time, and indices of adiposity in Kenyan adolescents. *J. Phys. Act. Health* **2012**, *9*, 115–123. [CrossRef]
34. Bowen, L.; Taylor, A.E.; Sullivan, R.; Ebrahim, S.; Kinra, S.; Krishna, K.V.R.; Kulkarni, B.; Ben-Shlomo, Y.; Ekelund, U.; Wells, J.C.K.; et al. Associations between diet, physical activity and body fat distribution: A cross sectional study in an Indian population. *BMC Public Health* **2015**, *15*, 281. [CrossRef]

35. Martin, S.L.; Kirkner, G.J.; Mayo, K.; Matthews, C.E.; Durstine, J.L.; Hebert, J.R. Urban, rural, and regional variations in physical activity. *J. Rural Health* **2005**, *21*, 239–244. [CrossRef]
36. Mumu, S.J.; Ali, L.; Barnett, A.; Merom, D. Validity of the global physical activity questionnaire (GPAQ) in Bangladesh. *BMC Public Health* **2017**, *17*, 650. [CrossRef] [PubMed]
37. Murphy, E.F.; Cotter, P.D.; Healy, S.; Marques, T.M.; O'Sullivan, O.; Fouhy, F.; Clarke, S.F.; O'Toole, P.W.; Quigley, E.M.; Stanton, C.; et al. Composition and energy harvesting capacity of the gut microbiota: Relationship to diet, obesity and time in mouse models. *Gut* **2010**, *59*, 1635–1642. [CrossRef] [PubMed]
38. Cuevas-Sierra, A.; Ramos-Lopez, O.; Riezu-Boj, J.I.; Milagro, F.I.; Martinez, J.A. Diet, gut microbiota, and obesity: Links with host genetics and epigenetics and potential applications. *Adv. Nutr.* **2019**, *10*, S17–S30. [CrossRef] [PubMed]
39. Boyce, J.A.; Kuijer, R.G. Perceived stress and freshman weight change: The moderating role of baseline body mass index. *Physiol. Behav.* **2015**, *139*, 491–496. [CrossRef]
40. Arnetz, B.B.; Templin, T.N.; Jen, K.C.; Sudan, S.; Arnetz, J.E. Lifestyle and psychosocial factors associated with maintenance of normal body mass index in college students: A cross sectional study. *BMC Res. Notes* **2020**, *13*, 516.
41. Baum, C.L. The effects of college on weight: Examining the 'Freshman 15' myth and other effects of college over the life cycle. *Demography* **2017**, *54*, 311–336. [CrossRef]
42. Haldar, S.; Chia, S.C.; Henry, C.J. Body composition in Asians and Caucasians: Comparative analyses and influences on cardiometabolic outcomes. *Adv. Food Nutr. Res.* **2015**, *75*, 97–154.
43. Wulan, S.N.; Westerterp, K.R.; Plasqui, G. Ethnic differences in body composition and the associated metabolic profile: A comparative study between Asians and Caucasians. *Maturitas* **2010**, *65*, 315–319. [CrossRef]
44. Grillol, L.P.; Siqueira, A.F.; Silva, A.C.; Martins, P.A.; Verreschi, I.T.; Sawaya, A.L. Lower resting metabolic rate and higher velocity of weight gain in a prospective study of stunted vs nonstunted girls living in the shantytowns of Sao Paulo, Brazil. *Eur. J. Clin. Nutr.* **2005**, *59*, 835–842. [CrossRef]
45. Hoffman, D.J.; Sawaya, A.L.; Verreschi, I.; Tucker, K.L.; Roberts, S.B. Why are nutritionally stunted children at increased risk of obesity? Studies of metabolic rate and fat oxidation in shantytown children from Sao Paulo, Brazil. *Am. J. Clin. Nutr.* **2000**, *72*, 702–707. [CrossRef]
46. Wells, J.C.; Sawaya, A.L.; Wibaek, R.; Mwangome, M.; Poullas, M.S.; Yajnik, C.S.; Demaio, A. The double burden of malnutrition: Aetiological pathways and consequences for health. *Lancet* **2020**, *395*, 75–88. [CrossRef]
47. Thanikachalam, M.; Fuller, C.H.; Lane, K.J.; Sunderarajan, J.; Harivanzan, V.; Brugge, D.; Thanikachalam, S. Urban environment as an independent predictor of insulin resistance in a South Asian population. *Int. J. Health Geogr.* **2019**, *18*, 5. [CrossRef] [PubMed]
48. Wang, J.; Li, H.; Franco, O.H.; Yu, Z.; Liu, Y.; Lin, X. Adiponectin and metabolic syndrome in middle-aged and elderly Chinese. *Obesity* **2008**, *16*, 172–178. [CrossRef] [PubMed]
49. Yao, H.; Fan, C.; Fan, X.; Lu, Y.; Wang, Y.; Wang, R.; Tang, T.; Qi, K. Effects of gut microbiota on leptin expression and body weight are lessened by high-fat diet in mice. *Br. J. Nutr.* **2020**, *124*, 396–406. [CrossRef]
50. Yao, H.Y.; Fan, C.N.; Lu, Y.Y.; Fan, X.Q.; Xia, L.L.; Li, P.; Wang, R.; Tang, T.T.; Wang, Y.Y.; Qi, K.M. Alteration of gut microbiota affects expression of adiponectin and resistin through modifying DNA methylation in high-fat diet-induced obese mice. *Genes Nutr.* **2020**, *15*, 12. [CrossRef]
51. Bidulescu, A.; Dinh, P.C.; Sarwary, S.; Forsyth, E.; Luetke, M.C.; King, D.B.; Liu, J.K.; Davis, S.K.; Correa, A. Associations of leptin and adiponectin with incident type 2 diabetes and interactions among African Americans: The Jackson heart study. *BMC Endocr. Disord.* **2020**, *20*, 31. [CrossRef]
52. Aguayo-Mazzucato, C. Functional changes in beta cells during ageing and senescence. *Diabetologia* **2020**, *63*, 2022–2029. [CrossRef]
53. Wengreen, H.J.; Moncur, C. Change in diet, physical activity, and body weight among young-adults during the transition from high school to college. *Nutr. J.* **2009**, *8*, 32. [CrossRef]
54. Bixby, H.; Bentham, J.; Zhou, B.; Di Cesare, M.; Paciorek, C.J.; Bennett, J.E.; Taddei, C.; Stevens, G.A.; Rodriguez-Martinez, A.; Carrillo-Larco, R.M.; et al. Rising rural body-mass index is the main driver of the global obesity epidemic in adults. *Nature* **2019**, *569*, 260–264.

nutrients

Systematic Review

Vitamin D-Related Risk Factors for Maternal Morbidity and Mortality during Pregnancy: Systematic Review and Meta-Analysis

María Morales-Suárez-Varela [1,2], Nazlı Uçar [1], José Miguel Soriano [3], Agustín Llopis-Morales [1], Beth S. Sanford [4] and William B. Grant [5,*]

1 Area of Preventive Medicine and Public Health, Department of Preventive Medicine and Public Health, Food Sciences, Toxicology, and Legal Medicine, School of Pharmacy, University de Valencia, 46100 Burjassot, Spain
2 Biomedical Research Center Network on Epidemiology and Public Health (CIBERESP), Institute of Health Carlos III, 28029 Madrid, Spain
3 Food & Health Lab, Institute of Materials Science, University of Valencia, 46980 Paterna, Spain
4 College of Nursing, Rasmussen University, 4012 19th Ave S, Fargo, ND 58103, USA
5 Sunlight, Nutrition, and Health Research Center, P.O. Box 641603, San Francisco, CA 94164, USA
* Correspondence: wbgrant@infionline.net

Abstract: Vitamin D deficiency (serum 25-hydroxyvitamin D [25(OH)D] levels <20 ng/mL in serum) is a common health condition among pregnant women, especially in high-risk groups. Evidence has connected vitamin D levels with many health-related problems during pregnancy, including gestational diabetes and preeclampsia. Because of vitamin D's effect on both mother and fetus, we systematically review the association between 25(OH)D level and its health effects. From a total of 143 studies, 43 came from PubMed, 4 from Cochrane, and 96 from EMBASE. After screening, we identified 38 studies as candidates for inclusion. Ultimately, we limited this review to 23 articles originating from 12 countries, written in English or Spanish, and conducted between 2010 and 2022. We conducted this review according to the Preferred Reporting Items for Systematic Review and Meta-Analysis (PRISMA) guidelines and evaluated the quality and strength of the evidence by using the Navigation Guide Systematic Review Methodology (SING). These systematic reviews summarize findings that support vitamin D's role in reducing risks of multiple outcomes and the possible contribution of adequate vitamin D levels to a healthy pregnancy.

Keywords: maternal mortality; maternal morbidity; preeclampsia; pregnancy; vitamin D deficiency; supplementation; vitamin D; 25-hydroxyvitamin D

Citation: Morales-Suárez-Varela, M.; Uçar, N.; Soriano, J.M.; Llopis-Morales, A.; Sanford, B.S.; Grant, W.B. Vitamin D-Related Risk Factors for Maternal Morbidity and Mortality during Pregnancy: Systematic Review and Meta-Analysis. *Nutrients* **2022**, *14*, 4124. https://doi.org/10.3390/nu14194124

Academic Editor: Andrea Fabbri

Received: 14 September 2022
Accepted: 30 September 2022
Published: 4 October 2022

1. Introduction

Vitamin D is a fat-soluble vitamin critical within the body for many functions, including cell proliferation, differentiation, apoptosis, and immune modulation [1]. Vitamin D is transmitted from the mother to the fetus via the placenta and is fundamental at all stages of embryonic and fetal development, from implantation to general growth, including skeletal maturation and placental function [2–4].

Worldwide, about 1 million people suffer from vitamin D deficiency (serum 25-hydroxyvitamin D [25(OH)D] <20 ng/mL, referring to vitamin D_2 and/or D_3) [5]. Due to the increased physiological demand for vitamin D during pregnancy, pregnant women are considered a high-risk group for developing vitamin D deficiency (VDD), with prevalence ranging from 51.3% [6] to 100% [7]. VDD in pregnant women increases maternal mortality and morbidity rates. Worldwide, the highest prevalence (>80%) of deficiency in pregnancy was observed among Chinese women (100%) [7] and pregnant Turkish women (95.6%) [8]. In Middle Eastern countries, VDD among pregnant women is an estimated 60–80% [9,10]. Among Iranian pregnant women, studies have reported prevalence rates of

78%, 76%, 70.4%, and 69.2% [11]. The estimated prevalence in pregnant women in the USA and Canada was reported to be 42–72% [12]. In Sweden, a longitudinal study reported 37% of first-trimester pregnant women had 25(OH)D concentrations <20 ng/mL [13], in comparison with 23% of Canadian women [14]. In Mexico, a previous cross-sectional study reported VDD among 61% of women in the third trimester, and 98% of their newborns had vitamin D deficiency [15].

There are a number of reasons why maternal VDD rates are high during pregnancy. One reason is sun avoidance. In the Middle East, that can be due to wearing concealing clothing as well as not going outdoors during the hot summers. Another reason is that diet provides only a small amount of vitamin D, and then only from animal products, including eggs, fish and meat, unless the food is fortified. Thus, low-latitude countries, which have largely plant-based diets, obtain little vitamin D from food. A third reason is that health care providers do not generally recommend enough vitamin D supplementation in general and during pregnancy in particular.

Vitamin D may impact maternal, fetal, and postnatal growth by affecting calcium absorption [16], parathyroid hormone expression [17], phosphate metabolism [18], growth plate function [19], and possibly regulating the insulin-like growth factor axis [20]. VDD during pregnancy has therefore been associated with adverse health outcomes in the mother, including increased risk of preeclampsia, glucose intolerance, gestational diabetes, preterm birth, and hypocalcemia crisis [21], as well as poor fetal skeletal development [22]. Through this review's comprehensive meta-analysis, we aim to determine the effect of vitamin D supplementation in preventing maternal mortality and morbidity.

2. Materials and Methods

2.1. PICO Strategy

We used the PICO strategy (Pregnancy, Intake/level, C/D, mOrtality/mOrbidity) to identify potentially relevant studies. In PICO, the question needs to identify the patient or population problem we intend to study, the planned intervention or treatment, the comparison of one intervention with another (if applicable), and the anticipated outcome. Our PICO framework was "Is there more mortality or morbidity in pregnant women with low levels of vitamin D than in those with adequate levels of vitamin D?", in which P is pregnant women, I is a low intake/level of vitamin D, C is adequate intake/level of vitamin D, and O is pregnancy mortality and morbidity.

2.2. Literature Search

We searched the PubMed, Cochrane and Embase databases; keywords included "pregnancy," "gestation," "vitamin D," "mortality," "morbidity," and "review." First, we performed a literature search to identify publications eligible for inclusion in PubMed, Cochrane and Embase. Keywords included "pregnancy" OR "gestation" AND "vitamin D" AND "review" AND "mortality" OR "morbidity." The search was limited to human subjects and English- and Spanish-language articles published between 2010 and January 2022. We recovered 43 studies from PubMed, 4 from Cochrane, and 96 from Embase, for a total of 143 studies.

Results were screened in a three-stage process based on title, abstract, and full-text review in duplicate by reviewers at each stage. Study selections were compared and discrepancies were resolved by discussion with N.U. and M.M.-S.-V. Duplicates and studies not meeting selection criteria were removed at each round. Search results were uploaded in Mendeley to remove duplicates, and the reference list was entered in Excel for study selection. The initial screening identified 38 candidates, of which 23 articles met inclusion and exclusion criteria.

The Preferred Reporting Items for Systematic reviews and Meta-Analyses (PRISMA) flowchart (Figure 1) shows the number of articles at each stage of the screening process.

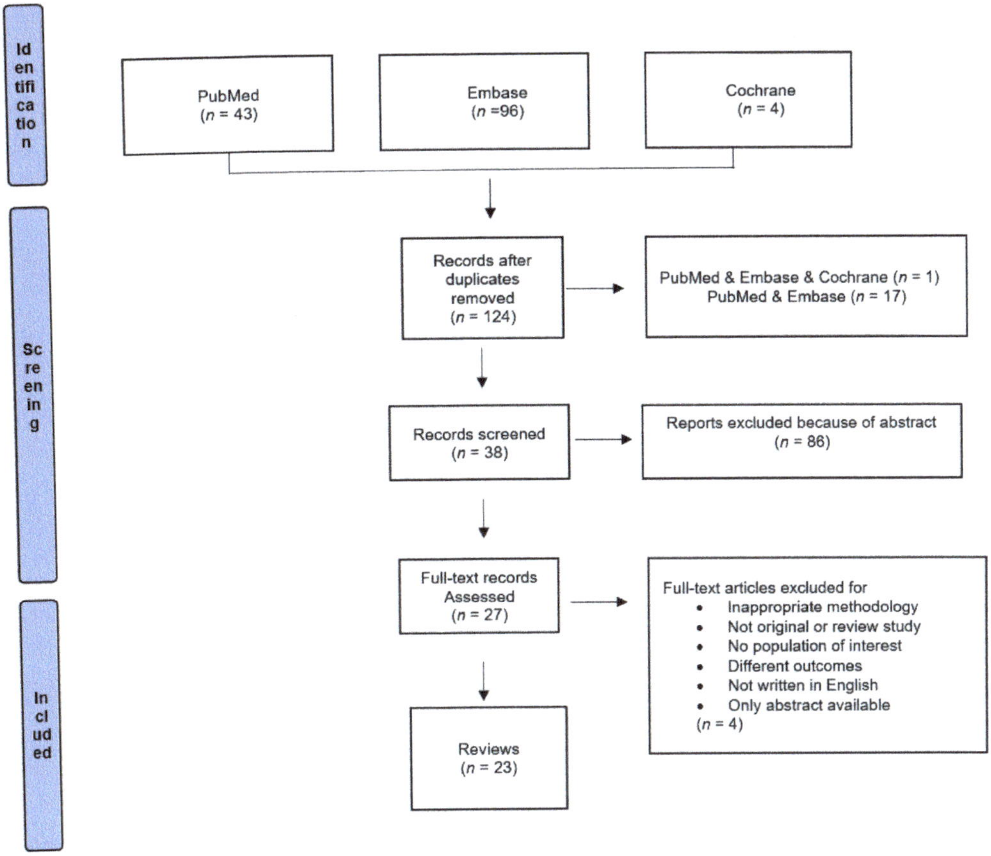

Figure 1. Search and screening strategy for candidate studies.

2.3. Study Inclusion/Exclusion Criteria and Data Extraction

Studies included in this review met the following criteria: reviews, narrative reviews, clinical review, systematic review, and meta-analysis studies to look at the effects of vitamin D on maternal mortality and morbidity. All studies focused on how vitamin D levels in pregnancy related to maternal mortality and morbidity and were longitudinal in nature. Specific inclusion/exclusion criteria were developed, and only published works meeting all criteria were included. The selection criteria were the following:

1. Review, narrative review, clinical review, systematic review or meta-analysis;
2. Available in English or Spanish;
3. Published between 2010 and January 2022;
4. Study carried out on humans;
5. Exposure of interest is vitamin D status or supplementation during pregnancy;
6. Data on vitamin D or metabolite concentration in maternal blood during pregnancy are available;
7. Main outcomes of interest are the incidence of maternal mortality and morbidity.

After we thoroughly assessed the candidate studies, 23 were included in this meta-analysis. We examined articles to tabulate data, which we summarized under the headings of design, location, vitamin D status, and main findings.

2.4. Quality Assessment

This systematic review and meta-analysis was conducted according to the Preferred Reporting Items for Systematic Reviews and Meta-Analyses (PRISMA) guidelines [23,24]. The systematic review has been registered in the International prospective register of systematic reviews (PROSPERO) (CRD42022343174).

In addition, to ensure that studies reflected the most current science and evidence-based practice, the Scottish Intercollegiate Guidelines Network (SIGN) was used to ensure rigorous assessment of study quality, validity, bias, and possible confounding variables [25]. Using SIGN ensures a robust assessment of a study's validity, including key factors such as bias and confounding. SIGN is based on the principles of evidence-based medicine, an approach that ensures using the most up-to-date, reliable, and scientifically solid evidence available in making decisions about a situation being studied [26].

SIGN establishes levels of evidence and recommendations to describe a given study and its results. Levels of evidence are based on study design and the methodological quality of individual studies. Scores are ranked best to worst using 1, 2, 3, and 4, with those scores further ranked with ++, +, and − signs. Grades of recommendation, rated best to worst as A, B, C, and D, are based on the strength of the evidence on which the recommendation is based and do not reflect the recommendation's clinical importance.

3. Results

3.1. Study Characteristics

Our search yielded 143 studies; 23 review studies remained after further screening. Studies were published between 2010 and 2022. They used data from Chile [27], Canada [28,29], Spain [30,31], Pakistan [32], Brazil [33,34], the United States [35–42], Germany [43], Iran [44,45], India [46], Puerto Rico [47], Poland [48], and Australia [49]. The chosen studies were analyzed according to design, location, vitamin D concentration and supplementation, major findings, and SIGN scores. Table 1 summarizes the study characteristics.

Overall, most included studies were deemed high quality. With the SIGN scores, the 23 review articles could be regarded as good quality. The risk of bias was generally low, with at least 75% of judgments assessed as low risk for four domains: blinding of participants, personnel and outcome assessment, selective reporting, and other biases.

3.2. Review and Meta-Analysis Studies

Twenty-three reviews reported analyses for vitamin D supplementation [27–49]. Vitamin D supplementation is is protective for pregnant women, having a significant effect on the incidence of preeclampsia [27,28,30,33,35,44,46] and reduced risk of preterm births [29], prematurity [30], gestational diabetes [31,39,41], and both maternal and infant infections [40].

Through a meta-analysis of observational studies, we suggest that vitamin D supplementation acts as a protective factor for preeclampsia and prematurity [30]. One meta-analysis [30] reviewed cohort studies evaluating the association between vitamin D and prematurity. Four studies used concentrations <75 nmol/L, and six studies considered concentrations <50 nmol/L as vitamin D cutoff points. Results of the subgroup analysis of cohort studies showed a significant association between maternal vitamin D and preterm birth only for concentrations <75 nmol/L: pooled odds ratio (OR) = 1.56; 95% confidence interval (CI), 1.25–1.94; $I^2 = 70\%$; $p = 0.02$. For concentrations <50 nmol/L, pooled OR = 1.09; 95% CI, 0.91–1.30; $I^2 = 91\%$; $p < 0.00001$.

Table 1. Reviews that show vitamin D-related risk factors for maternal morbidity and mortality during pregnancy.

Ref.	Design	Location	Vitamin Concentration and Supplementation	Findings	SIGN	
					LE	GR
González-Wong et al. 2021 [27]	Narrative review	Chile	Supplementation during pregnancy	Vitamin D may lower risk of PE.	3	C
Kinshella et al. 2021 [28]	Review	Canada	Supplementation during pregnancy	Half of reviews reported that vitamin D supplementation had a significant protective effect on PE incidence; little evidence that vitamin D supplementation affected PTB or stillbirth. Pooled outcomes showed 38% reduced risk of developing PE among pregnant women who received vitamin D supplementation in comparison with those who did not, without heterogeneity between studies (RR = 0.62; 95% CI, 0.43–0.91; $I^2 = 0\%$; 12 studies, $n = 1353$), which did not change in direction or significance with sensitivity analyses.	1 ++	A
Aguliar-Cordero et al. 2020 [30]	Systematic review and meta-analysis of observational and interventional studies	Spain	Maternal concn during pregnancy	Interventional studies indicated that vitamin D supplementation acts as a prevention factor for PE and prematurity. Observational studies showed that vitamin D insufficiency and deficiency are associated with higher risk of developing PE. However, prematurity and vitamin D were associated only for maternal vitamin D concentrations < 75 nmol/L. Random-effects meta-analysis indicated no significant association between vitamin D, PE, and prematurity in either observational or interventional studies.	2 ++	A
Oh, Keats, and Bhutta 2020 [29]	Systematic review and meta-analysis	Canada	Micronutrient and vitamin supplementation during pregnancy	Vitamin D supplementation may have reduced risk of PTB by 36% (average RR = 0.64; 95% CI, 0.40–1.04; studies = 7), though the upper limit of the confidence interval just crossed the line of no effect.	1 ++	A
Sibtain et al. 2020 [32]	Systematic review	Pakistan	Status during pregnancy and fetoplacental unit	Studies recommended substantial vitamin D supplementation during pregnancy. High-quality randomized controlled trials (RCTs) required to see optimal level of vitamin D.	2 ++	A
De Souza and Pisani 2020 [33]	Narrative review	Brazil	Status and risk of PE	Although studies showing relation between vitamin D and lower risk of PE are limited, maternal status of vitamin D seems to influence risk of developing PE. Therefore, vitamin D supplementation in women may improve pregnancy outcomes.	3	C

Table 1. *Cont.*

Ref.	Design	Location	Vitamin Concentration and Supplementation	Findings	SIGN	
					LE	GR
Palacios, Kostiuk, and Peña-Rosas 2019 [35]	Review	USA	Supplementation during pregnancy	Supplementation with vitamin D alone during pregnancy probably reduces risk of PE (RR = 0.48; 95% CI, 0.30–0.79; 4 trials, n = 499) and gestational diabetes (RR = 0.51; 95% CI, 0.27–0.97; 4 trials, n = 446) and probably reduces risk of low birthweight (<2500 g; RR = 0.55, 95% CI, 0.35–0.87; 5 trials, n = 697) compared with women who received placebo or no intervention. Vitamin D supplementation may make little or no difference in risk of PTB (<37 weeks) compared with no intervention or placebo (RR = 0.66; 95% CI, 0.34–1.30; 7 trials, n = 1640), and vitamin D supplementation may reduce risk of severe postpartum bleeding (RR = 0.68; 95% CI, 0.51–0.91; 1 trial, n = 1134).	2 ++	A
Pilz et al. 2018 [43]	Review of clinical data	Germany	Status during pregnancy and lactation	Based on available evidence derived from RCTs on vitamin D supplementation in pregnancy, this study reported that vitamin D is safe and improves vitamin D and calcium status, thereby protecting skeletal health. Data from RCTs and meta-analyses of RCTs suggest other beneficial effects but are inconsistent on whether vitamin D supplementation improves clinical neonatal or maternal outcomes such as SGA, fetal/infant growth, infant/neonatal mortality, asthma/wheeze, PE, or GDM.	2 ++	A
Akbari et al. 2018 [44]	Systematic review and updated meta-analysis	Iran	Status during pregnancy	Based on the forest plot, lower levels of 25(OH)D were significantly associated with risk of PE (fixed and random $p < 0.001$). Women with vitamin D deficiency (<20 ng/mL) at higher risk of PE. Association can be specific up to 90% at 10.60-ng/mL cutoff.	2 ++	A
Nandi, Wadhwani, and Joshi 2017 [46]	Narrative review	India	Vitamin D and LCPUFAs and their role in PE development	Vitamin D [1,25$(OH)_2D_3$] induces CBS gene expression while it can suppress the oxidative stress–induced COX-2 upregulation and thromboxane production. On that basis, it is hypothesized that disturbed vitamin D and LCPUFA metabolism influences regulation of the one-carbon cycle, which will trigger inflammation through oxidative stress in PE. That may lead to altered fetoplacental growth and development of PE.	3	C

Table 1. *Cont.*

Ref.	Design	Location	Vitamin Concentration and Supplementation	Findings	SIGN	
					LE	GR
Wagner et al. 2017 [36]	Review	USA	Status and supplementation during pregnancy	A growing body of observational studies indicated that maternal hypovitaminosis D (<20 ng/mL or <50 nmol/L) is a significant risk factor for adverse neonatal outcomes, including asthma, multiple sclerosis, and other neurological disorders. Results of RCTs of vitamin D supplementation during pregnancy recently showed decreased complications of pregnancy/birth and GDM.	2 ++	A
Agarwal, Kovilam, and Agrawal 2018 [37]	Critical review	USA	PE, PTL, GDM, GH	In pregnancy, vitamin D deficiency is associated with increased incidence of adverse maternal and fetal outcomes, primarily PE, GDM, low birth weight, and PTB. Other outcomes still under study; no definite conclusions drawn yet.	2 ++	A
Palacios et al. 2016 [47]	Meta-analysis	Puerto Rico	Oral supplementation (alone or with other vitamins and minerals); maternal levels and risk of developing PE, GDM, PTB, impaired glucose tolerance, Cesarean delivery, GH, and other adverse conditions	Data suggest that pregnant women supplemented with vitamin D had significantly higher vitamin D levels than controls (mean difference, 54.7 nmol/L; 95% CI, 36.6–72.9). Two trials showed lower risk of PE (8.9% vs. 15.5%; average risk ratio, 0.52; 95% CI, 0.25–1.05) and two others showed no difference in risk of GDM with vitamin D supplementation. Additionally, three trials showed that supplementation with vitamin D plus calcium reduced risk of PE (5% vs. 9%; average risk ratio, 0.51; 95% CI, 0.32–0.80).	2 ++	A
Pérez-López et al. 2015 [31]	Systematic review and meta-analysis of RCTs	Spain	Circulating levels, PE, GDM, SGA, low birth weight, PTB, birth weight, birth length, Cesarean delivery	Circulating vitamin D levels significantly higher at term than in control group (mean difference, 66.5 nmol/L; 95% CI, 66.2–66.7). Birth weight and birth length were significantly greater in vitamin D group; mean difference, 107.6 g (95% CI, 59.9–155.3) and 0.3 cm (95% CI, 0.10–0.41), respectively. Incidence of PE, GDM, SGA, low birth weight, PTB, and Cesarean delivery not influenced by vitamin D supplementation. Across RCTs, doses and types of vitamin D supplements, gestational age at first administration, and outcomes were diverse.	2 ++	A

Table 1. *Cont.*

Ref.	Design	Location	Vitamin Concentration and Supplementation	Findings	SIGN	
					LE	GR
Weinert and Silveiro 2015 [34]	Critical review	Brazil	Low birth weight, growth restriction, respiratory tract infection, altered glucose homeostasis, increased incidence of GDM, PE, bacterial vaginosis	Current state of evidence is controversial for some other endpoints, and actual benefit of vitamin D supplementation in pregnancy remains unclear.	3	D
Pludowski et al. 2013 [48]	Review of recent evidence	Poland	Status during pregnancy	Various health effects of vitamin D deficiency during pregnancy continue to be reported, notably with increased risk of PE, infection, PTL and PTB, Cesarean delivery, GDM.	2 ++	A
Girling and Sykes 2013 [38]	Narrative review	USA	Physiology and current management of thyroid dysfunction and the rarer endocrine disorders in pregnancy and includes current guidance on supplementation	Over recent years, awareness of potential adverse effects of vitamin D deficiency has driven guidance for vitamin D supplementation for pregnant and lactating women.	3	D
Tabesh, Salehi-Abargouei, and Esmaillzadeh 2013 [45]	Systematic review and meta-analysis	Iran	Maternal serum and risk of PE	Overall significant association between vitamin D deficiency and risk of PE; however, significant between-study heterogeneity evident ($I^2 = 52.7\%$; $p = 0.039$). In subgroup analysis, overall effect was significant for studies defining vitamin D deficiency as ≤50 nmol/L (20 ng/mL) but not for those that used <38 nmol/L (15.2 ng/mL). Association was seen for "cohort or nested case–control studies" as well as for "cross-sectional or case–control studies" (2.78; 95% CI, 1.45–5.33; $p = 0.002$). For analysis by study location, associations remained significant only for U.S. studies.	2 ++	A
Alzaim and Wood 2013 [39]	Critical review	USA	Status during pregnancy and GDM	Suggested that vitamin D deficiency in pregnant women increases risk for GDM. However, that determination is based largely on only six published observational studies and one short-term intervention study with an active analog form of 1,25(OH)$_2$ vitamin D. Effect of treating existing vitamin D deficiency on later development of GDM in pregnant women unknown.	2 ++	A

Table 1. *Cont.*

Ref.	Design	Location	Vitamin Concentration and Supplementation	Findings	SIGN	
					LE	GR
Thorne-Lyman and Fawzi 2012 [40]	Systematic review and meta-analysis	USA	Supplementation during pregnancy for maternal, perinatal, or infant health outcomes	Only low-level evidence relates vitamin D supplementation or intake during pregnancy to perinatal and infant health-related outcomes. Emerging evidence suggesting plausible effects on intrauterine growth restriction, PE, and both maternal and infant infections as important outcomes in need of further research in low-income settings.	2 ++	A
Senti et al. 2012 [41]	Systematic review	USA	Status during pregnancy and GDM	Study findings consist solely of level 2 evidence for associating maternal vitamin D deficiency with risk of GDM. Five (83%) studies reported inverse relationship between circulating vitamin D levels and markers of glucose homeostasis associated with GDM or increased risk for GDM associated with reduced maternal levels of vitamin D. In one study, researchers did not identify association between vitamin D and GDM but did identify association between higher vitamin D levels and lower fasting glucose and insulin levels.	3	D
Barrett and McElduff 2010 [50]	Narrative review	Australia	Supplementation during pregnancy	Vitamin D's role in multiple nonclassical metabolic processes, though shown primarily by association studies in human populations, has a possible physiological basis. Considerable evidence associates low maternal vitamin D levels with worse outcomes for both mother and fetus in pregnancy and for the neonate. Whether association between vitamin D status and a wide range of adverse health outcomes is because vitamin D acts as a marker for some other health parameter such as obesity or occurs because of a direct causal relationship usually remains to be determined. Optimal concentration of vitamin D is unclear or at least controversial. RCTs of vitamin D supplementation with measurement of vitamin D to determine baseline status, level achieved on supplementation with appropriate documentation of possible confounders, and assessment of various health outcomes are required. Trying to achieve a vitamin D concentration of >50 nmol/L seems reasonable in most populations, including pregnant women.	3	B

Table 1. *Cont.*

Ref.	Design	Location	Vitamin Concentration and Supplementation	Findings	SIGN	
					LE	GR
Mulligan et al. 2010 [42]	Narrative review	USA	Metabolism and implications of deficiency in pregnancy and lactation	Vitamin D deficiency is associated with increased prevalence of PE, a common cause of increased mortality rates in pregnancy. Current recommendations for daily vitamin D intake (200 IU) are inadequate to maintain serum levels of vitamin D in the recommended range during pregnancy and lactation. More studies are needed to determine serum levels and degree of supplementation necessary to optimize maternal and fetal outcomes. However, because vitamin D supplementation is simple and cost-effective with a low likelihood of toxicity, we recommend increased supplementation in all pregnant women to keep serum levels of vitamin D in the reference range for adults (>32 ng/mL).	4	D

Abbreviations: 25(OH)D, 25-hydroxyvitamin D; 95% CI, 95% confidence interval; CBS, cystathionine beta synthase; COX-2, cyclooxygenase 2; GDM, gestational diabetes mellitus; LCPUFA, long-chain polyunsaturated fatty acid; PE, preeclampsia; PTL, preterm labor; PTB, preterm birth; RCT, randomized controlled trial; RR = relative risk; SGA, small for gestational age.

In Oh, Keats, and Bhutta [29], the risk for preterm birth may have been reduced by 36% through vitamin D supplementation (average relative risk (RR) = 0.64; 95% CI, 0.40–1.04; studies = 7), though the upper limit of the confidence interval just crossed the line of no effect. In Tabesh and colleagues [45], a significant association was found between VDD and risk of preeclampsia; however, significant between-study heterogeneity was found (I^2 = 52.7%; p = 0.03). In Palacios and colleagues [47], data from original studies involving 446 women showed probable gestational diabetes risk reduction with vitamin D supplementation in comparison with no intervention or placebo groups (RR = 0.51; 95% CI, 0.27–0.97; moderate-certainty evidence). In addition, vitamin D supplementation probably makes little or no difference in risk of gestational hypertension in comparison with no intervention or placebo (n = 1130; RR = 0.78; 95% CI, 0.41–1.49).

4. Discussion

The most common causes of maternal mortality are severe bleeding, infections/sepsis, embolisms (blockage in the heart or lungs), stroke (can be a blown blood vessel or an embolus), blood pressure disorders (preeclampsia/eclampsia), and gestational diabetes. According to the WHO, the leading causes of maternal mortality (nearly 75%) are severe bleeding, postnatal infections, blood pressure disorders, and complications of labor and delivery [49]. In the US, the top causes of maternal mortality are hemorrhage, cardiovascular complications, infections/sepsis, embolism (pulmonary embolism or other embolisms), cerebrovascular accidents, and noncardiovascular medical conditions such as gestational diabetes mellitus (GDM) [51].

In this review, we aimed to evaluate the current evidence on the effect of vitamin D and vitamin D supplementation on maternal mortality and morbidity. Interventions yielded significant effects, albeit with sparse evidence in some areas.

In recent years, scientific interest has increasingly focused on the consequences of VDD on pregnant women, in particular, the impact of its deficiency on adverse maternal health outcomes. With all significant effects taken together, vitamin D supplementation was associated with a reduced risk of maternal mortality and morbidity-related outcomes.

Many observational studies did report that vitamin D levels were associated with adverse maternal, fetal, and neonatal outcomes, including increased risk of developing preeclampsia, preterm labor, gestational diabetes, being small for gestational age, low birth weight, an increased rate of Cesarean delivery, and infertility [50]. Since vitamin D RCTs have not yielded much useful information regarding the role and requirements for vitamin D for many health outcomes while observational studies have [52], results from observational studies are highlighted in the following paragraphs.

4.1. Hemorrhage

Low maternal 25(OH)D concentrations have been found to be associated with an increased risk of postpartum hemorrhage. An observational study from Taiwan involving 600 pregnant women with 25(OH)D concentrations measured in the 36th week of pregnancy found that 25(OH)D below 30 ng/mL was associated with a factor of four-to-five increased risk of postpartum hemorrhage [53].

4.2. Gestational Diabetes

In a meta-analysis of 31 observational trials, low vitamin D levels increased the risk of gestational diabetes by 49 (OR = 1.49; 95% CI, 1.18–1.89) [54]. Another meta-analysis of 24 observational studies showed similar results [22]. Observational studies also have shown VDD in pregnancy increases the risk of preeclampsia and that vitamin D supplementation, with or without calcium, may reduce that risk [55].

4.3. Pulmonary Embolism or Other Embolism

The incidence of pulmonary embolism, a common cause of maternal mortality, has been found in many studies to be increased in a state of maternal VDD, and the risk decreased with supplementation [27,28,30,33,35,37,42–46]. Other studies concluded pulmonary embolism was not influenced by supplementation [31] or that the connection was unclear [34].

4.4. Preterm Birth Risk

Reports have conflicting findings on the role of vitamin D in reducing preterm birth risk. Some studies identified the association of VDD and the inflammatory response with premature rupture of the amniotic membrane and preterm delivery [56,57]. Pooled analysis of four randomized controlled trials in that study showed no significant effect of vitamin D supplementation in preventing preterm birth. Other publications have reported alterations in the cervicovaginal fluid content of vitamin D and vitamin D binding protein as biomarkers of vaginal inflammation and preterm birth risk several weeks before delivery [58]. A review published in 2017 reported that based on 6 vitamin D RCTs, vitamin D supplementation could significantly reduce the risk of preterm birth (pooled RR = 0.57 (95% CI, 0.36–0.91)) and from 18 observational studies that maternal 25(OH)D <20 ng/mL was associated with a pooled OR = 1.25 (95% CI, 1.13–1.38) [59]. The best observational study on preterm delivery to date was conducted at the Medical University of South Carolina [60]. A total of 1064 consecutive pregnant women were enrolled at their first prenatal visit around the 12th to 14th week of pregnancy. The participants included 488 whites, 395 African Americans, 117 Hispanics, 19 Asians, and 29 multiple or other ethnicities. Their serum 25(OH)D concentration was measured, and they were given bottles of 5000 IU vitamin D3 and counseled on how to achieve 25(OH)D >40 ng/mL. Achieved 25(OH)D was also measured during pregnancy. Those who achieved >40 ng/mL had a 62% lower risk of PTB compared to those <20 ng/mL ($p < 0.0001$). There was no effect of race/ethnicity on the outcomes.

The journal literature on vitamin D and maternal mortality is relatively limited. However, there is a reasonable body of literature on the role of vitamin D in reducing the risk of adverse pregnancy and birth outcomes for both the developing fetus and mother, e.g., [61]. The review by Wagner et al. [62] outlines important findings regarding complications, including preterm birth, preeclampsia, and gestational diabetes, as well as adverse effects that appear in early childhood, such as asthma and neurological development. This review also points out that vitamin D regulates gene expression through DNA methylation, which has profound effects on fetal development and life after birth. They point out that pregnant women should supplement with 4000–5000 IU/d vitamin D_3 and achieve 25(OH)D concentrations >40 ng/mL.

5. Strengths and Limitations

An important strength of this review is that it presents an overview of reviews of the effect of vitamin D on the risk of many risk factors for maternal morbidity and mortality during pregnancy. Table 1 can serve as a starting point for those wanting to know the results to date and can help guide future research efforts. The studies included here show significant methodological differences, including mixed ethnicities and genetic reservoirs, countries, times and conditions of vitamin D evaluation, and different brands and qualities of vitamin D supplements among studies. Those factors all contributed to the heterogeneity of the included studies. A limitation is that we may not have been able to access all publications on the relationship between vitamin D and maternal mortality and morbidity during pregnancy because we limited our analysis to studies published in English and Spanish and available through the PubMed, Cochrane, and Embase databases.

6. Conclusions

Our meta-analysis showed evidence to support vitamin D supplementation as a cost-effective public health strategy to minimize adverse maternal health outcomes. Whenever possible, supplementation should be based on initial vitamin D serum levels with the intent to obtain and maintain optimal levels of a minimum of 40 ng/mL throughout the pregnancy for maximum impact [61]. In venues where testing is not affordable or convenient, innovative evidence-based technologies such as the *Vitamin D Deficiency Risk Assessment Quiz* (beta) and the *Vitamin D*Calculator* can aid providers of prenatal care in assessing individual VDD risk and calculating an individualized evidence-based loading and maintenance doses based on target optimal blood levels of 40 ng/mL, respectively (GrassrootsHealth.net, accessed 15 September 2022) In light of the results of the present review, further studies should be conducted. Randomized, controlled, blinded vitamin D supplementation trials must be conducted with pregnant women using standard nutrient physiological design criteria to ensure homogeneity of study design [62], including vitamin D levels (baseline and at time of birth) for all participants, to facilitate future systematic review and meta-analyses, In addition, RCT design may not be ideal for vitamin D outcomes studies because vitamin D intake is difficult to quantify from other sources, as well as lack of compliance, which can lead to unclear study results. Alternatively, they could be observational studies with vitamin D supplementation as done at the Medical University of South Carolina [59]. These studies must include large enough sample sizes to permit evaluating the prevalence of maternal mortality and morbidity.

Author Contributions: Conceptualization: W.B.G. Methodology: N.U., W.B.G., J.M.S., A.L.-M. and M.M.-S.-V. Writing—original draft preparation: N.U., W.B.G. and M.M.-S.-V.; writing—review and editing: N.U., W.B.G., J.M.S., A.L.-M., B.S.S. and M.M.-S.-V. All authors have read and agreed to the published version of the manuscript.

Funding: This research received no external funding.

Institutional Review Board Statement: Not applicable.

Informed Consent Statement: Not applicable.

Data Availability Statement: Not applicable.

Conflicts of Interest: W.B.G. receives funding from Bio-Tech Pharmacal Inc. (Fayetteville, AR, USA). The other authors declare no conflict of interest.

References

1. Kim, I.; Kim, S.S.; Song, J.I.; Yoon, S.H.; Park, G.Y.; Lee, Y.-W. Association between vitamin D level at birth and respiratory morbidities in very-low-birth-weight infants. *Korean J. Pediatr.* **2019**, *62*, 166–172. [CrossRef] [PubMed]
2. Eremkina, A.; Mokrysheva, N.; Pigarova, E.; Mirnaya, S. Vitamin D: Effects on pregnancy, maternal, fetal and postnatal outcomes. *Ter. Arkhiv.* **2018**, *90*, 115–127. [CrossRef]
3. Baqai, S.; Siraj, A.; Imran, R. Association of vitamin-d insufficiency during pregnancy with maternal & perinatal morbidity and mortality. *Pak. Armed. Forces Med. J.* **2020**, *70*, 323–327.
4. Thomas, D.J.; Khan, H.U.; Jaidev, S.P.; Hegde, P. A study on vitamin D levels in preterm and term neonates and their mothers. *Int. J. Contemp. Pediatrics* **2020**, *7*, 387–392. [CrossRef]
5. Sankar, J.; Lotha, W.; Ismail, J.; Anubhuti, C.; Meena, R.S.; Sankar, M.J. Vitamin D deficiency and length of pediatric intensive care unit stay: A prospective observational study. *Ann. Intensiv. Care* **2016**, *6*, 3. [CrossRef] [PubMed]
6. Gur, E.B.; Gokduman, A.; Turan, G.A.; Tatar, S.; Hepyilmaz, I.; Zengin, E.B.; Eskicioglu, F.; Guclu, S. Mid-pregnancy vitamin D levels and postpartum depression. *Eur. J. Obstet. Gynecol. Reprod. Biol.* **2014**, *179*, 110–116. [CrossRef]
7. Song, S.J.; Si, S.; Liu, J.; Chen, X.; Zhou, L.; Jia, G.; Liu, G.; Niu, Y.; Wu, J.; Zhang, W.; et al. Vitamin D status in Chinese pregnant women and their newborns in Beijing and their relationships to birth size. *Public Health Nutr.* **2013**, *16*, 687–692. [CrossRef]
8. Ates, S.; Sevket, O.; Ozcan, P.; Ozkal, F.; Kaya, M.O.; Dane, B. Vitamin D status in the first-trimester: Effects of Vitamin D deficiency on pregnancy outcomes. *Afr. Health Sci.* **2016**, *16*, 36–43. [CrossRef]
9. Behjat Sasan, S.; Zandvakili, F.; Soufizadeh, N.; Baybordi, E. The Effects of Vitamin D Supplement on Prevention of Recurrence of Preeclampsia in Pregnant Women with a History of Preeclampsia. *Obstet. Gynecol. Int.* **2017**, *2017*, 8249264. [CrossRef]
10. Rostami, M.; Tehrani, F.R.; Simbar, M.; Hosseinpanah, F.; Majd, H.A.; Khan, F.R. Rationale and Design of Khuzestan Vitamin D Deficiency Screening Program in Pregnancy: A Stratified Randomized Vitamin D Supplementation Controlled Trial. *JMIR Res. Protoc.* **2017**, *6*, e54. [CrossRef]
11. Sepandi, M.; Esmailzadeh, S.; Hosseini, M.S.; Hashemi, S.R.; Abbaszadeh, S.; Alimohamadi, Y.; Taghdir, M. Prevalence of Vitamin D Deficiency Among Iranian Pregnant Women. *Nutr. Diet. Suppl.* **2020**, *12*, 97–102. [CrossRef]
12. Saraf, R.; Morton, S.M.; Camargo, C.A., Jr.; Grant, C.C. Global summary of maternal and newborn vitamin D status–a systematic review. *Matern. Child Nutr.* **2016**, *12*, 647–668. [CrossRef] [PubMed]

13. Lundqvist, A.; Sandström, H.; Stenlund, H.; Johansson, I.; Hultdin, J. Vitamin D Status during Pregnancy: A Longitudinal Study in Swedish Women from Early Pregnancy to Seven Months Postpartum. *PLoS ONE* **2016**, *11*, e0150385. [CrossRef]
14. Perreault, M.; Moore, C.J.; Fusch, G.; Teo, K.K.; Atkinson, S.A. Factors Associated with Serum 25-Hydroxyvitamin D Concentration in Two Cohorts of Pregnant Women in Southern Ontario, Canada. *Nutrients* **2019**, *11*, 123. [CrossRef]
15. Ochoa-Correa, E.D.C.; Garcia-Hernandez, P.A.; Villarreal-Perez, J.Z.; Treviño-Garza, C.; Villarreal, L.E.M.-D.; Zapata-Castilleja, C.A.; De La O-Cavazos, M.E. Vitamin D deficiency in Mexican mothers and their newborns. *Gac. Médica México* **2017**, *153*, 559–565. [CrossRef]
16. Christakos, S. Mechanism of action of 1,25-dihydroxyvitamin D3 on intestinal calcium absorption. *Rev. Endocr. Metab. Disord.* **2012**, *13*, 39–44. [CrossRef]
17. Kumar, R.; Thompson, J.R. The Regulation of Parathyroid Hormone Secretion and Synthesis. *J. Am. Soc. Nephrol.* **2011**, *22*, 216–224. [CrossRef] [PubMed]
18. Lederer, E. Regulation of serum phosphate. *J. Physiol.* **2014**, *592*, 3985–3995. [CrossRef]
19. Eisman, A.J.; Bouillon, R. Vitamin D: Direct effects of vitamin D metabolites on bone: Lessons from genetically modified mice. *BoneKEy Rep.* **2014**, *3*, 499. [CrossRef]
20. Ciresi, A.; Giordano, C. Vitamin D across growth hormone (GH) disorders: From GH deficiency to GH excess. *Growth Horm. IGF Res.* **2017**, *33*, 35–42. [CrossRef]
21. Larqué, E.; Morales, E.; Leis, R.; Blanco-Carnero, J.E. Maternal and Foetal Health Implications of Vitamin D Status during Pregnancy. *Ann. Nutr. Metab.* **2018**, *72*, 179–192. [CrossRef] [PubMed]
22. Wei, S.-Q.; Qi, H.-P.; Luo, Z.-C.; Fraser, W.D. Maternal vitamin D status and adverse pregnancy outcomes: A systematic review and meta-analysis. *J. Matern. Fetal Neonatal Med.* **2013**, *26*, 889–899. [CrossRef] [PubMed]
23. Moher, D.; Shamseer, L.; Clarke, M.; Ghersi, D.; Liberati, A.; Petticrew, M.; Shekelle, P.; Stewart, L.A. Preferred reporting items for systematic review and meta-analysis protocols (prisma-p) 2015 statement. *Syst. Rev.* **2015**, *4*, 1. [CrossRef] [PubMed]
24. Moher, D.; Liberati, A.; Tetzlaff, J.; Altman, D.G.; PRISMA Group. Preferred Reporting Items for Systematic Reviews and Meta-Analyses: The PRISMA Statement. *Ann. Intern. Med.* **2009**, *151*, 264–269. [CrossRef]
25. *SIGN 50: A Guideline Developer's Handbook*; Scottish Intercollegiate Guidelines Network: Edinburgh, UK, 2015; Available online: https://www.sign.ac.uk/assets/sign50_2015.pdf (accessed on 29 September 2022).
26. Sackett, D.L. Evidence-based medicine. *Semin. Perinatol.* **1997**, *21*, 3–5. [CrossRef]
27. González-Wong, C.; Fuentes-Barría, H.; Aguilera-Eguía, R.; Urbano-Cerda, S.; Vera-Aguirre, V. The role of vitamin D in preeclampsia risk: A narrative review. *Rev. Chil. Nutr.* **2021**, *48*, 118–125. [CrossRef]
28. Kinshella, M.-L.; Omar, S.; Scherbinsky, K.; Vidler, M.; Magee, L.; von Dadelszen, P.; Moore, S.; Elango, R.; The PRECISE Conceptual Framework Working Group. Effects of Maternal Nutritional Supplements and Dietary Interventions on Placental Complications: An Umbrella Review, Meta-Analysis and Evidence Map. *Nutrients* **2021**, *13*, 472. [CrossRef]
29. Oh, C.; Keats, E.C.; Bhutta, Z.A. Vitamin and Mineral Supplementation During Pregnancy on Maternal, Birth, Child Health and Development Outcomes in Low- and Middle-Income Countries: A Systematic Review and Meta-Analysis. *Nutrients* **2020**, *12*, 491. [CrossRef]
30. Aguilar-Cordero, M.; Lasserrot-Cuadrado, A.; Mur-Villar, N.; León-Ríos, X.; Rivero-Blanco, T.; Pérez-Castillo, I. Vitamin D, preeclampsia and prematurity: A systematic review and meta-analysis of observational and interventional studies. *Midwifery* **2020**, *87*, 102707. [CrossRef]
31. Pérez-López, F.R.; Pasupuleti, V.; Mezones-Holguin, E.; Benites-Zapata, V.A.; Thota, P.; Deshpande, A.; Hernandez, A.V. Effect of vitamin D supplementation during pregnancy on maternal and neonatal outcomes: A systematic review and meta-analysis of randomized controlled trials. *Fertil. Steril.* **2015**, *103*, 1278–1288.e1274. [CrossRef]
32. Sibtain, S.; Sinha, P.; Manoharan, M.; Azeez, A. Controversies related to vitamin D deficiency effect on the maternal and feto-placental unit–an update. *J. Obstet. Gynaecol.* **2020**, *40*, 759–766. [CrossRef] [PubMed]
33. De Souza, E.A.; Pisani, L.P. The relationship among vitamin D, TLR4 pathway and preeclampsia. *Mol. Biol. Rep.* **2020**, *47*, 6259–6267. [CrossRef] [PubMed]
34. Weinert, L.S.; Silveiro, S.P. Maternal–Fetal Impact of Vitamin D Deficiency: A Critical Review. *Matern. Child Health J.* **2015**, *19*, 94–101. [CrossRef] [PubMed]
35. Palacios, C.; Kostiuk, L.K.; Peña-Rosas, J.P. Vitamin D supplementation for women during pregnancy. *Cochrane Database Syst. Rev.* **2019**, *7*, CD008873. [CrossRef]
36. Wagner, C.L.; Hollis, B.W.; Kotsa, K.; Fakhoury, H.; Karras, S.N. Vitamin D administration during pregnancy as prevention for pregnancy, neonatal and postnatal complications. *Rev. Endocr. Metab. Disord.* **2017**, *18*, 307–322. [CrossRef]
37. Agarwal, S.; Kovilam, O.; Agrawal, D.K. Vitamin D and its impact on maternal-fetal outcomes in pregnancy: A critical review. *Crit. Rev. Food Sci. Nutr.* **2018**, *58*, 755–769. [CrossRef]
38. Girling, J.; Sykes, L. Thyroid disorders and other endocrinological disorders in pregnancy. *Obstet. Gynaecol. Reprod. Med.* **2013**, *23*, 171–179. [CrossRef]
39. Alzaim, M.; Wood, R.J. Vitamin D and gestational diabetes mellitus. *Nutr. Rev.* **2013**, *71*, 158–167. [CrossRef]
40. Thorne-Lyman, A.; Fawzi, W.W. Vitamin D During Pregnancy and Maternal, Neonatal and Infant Health Outcomes: A Systematic Review and Meta-analysis. *Paediatr. Périnat. Epidemiol.* **2012**, *26*, 75–90. [CrossRef]

41. Senti, J.; Thiele, D.K.; Anderson, C.M. Maternal Vitamin D Status as a Critical Determinant in Gestational Diabetes. *J. Obstet. Gynecol. Neonatal Nurs.* **2012**, *41*, 328–338. [CrossRef]
42. Mulligan, M.L.; Felton, S.K.; Riek, A.E.; Bernal-Mizrachi, C. Implications of vitamin D deficiency in pregnancy and lactation. *Am. J. Obstet. Gynecol.* **2010**, *202*, 429.e1–429.e9. [CrossRef] [PubMed]
43. Pilz, S.; Zittermann, A.; Obeid, R.; Hahn, A.; Pludowski, P.; Trummer, C.; Lerchbaum, E.; Pérez-López, F.R.; Karras, S.N.; März, W. The Role of Vitamin D in Fertility and during Pregnancy and Lactation: A Review of Clinical Data. *Int. J. Environ. Res. Public Health* **2018**, *15*, 2241. [CrossRef] [PubMed]
44. Akbari, S.; Khodadadi, B.; Ahmadi, S.A.Y.; Abbaszadeh, S.; Shahsavar, F. Association of vitamin D level and vitamin D deficiency with risk of preeclampsia: A systematic review and updated meta-analysis. *Taiwan. J. Obstet. Gynecol.* **2018**, *57*, 241–247. [CrossRef] [PubMed]
45. Tabesh, M.; Salehi-Abargouei, A.; Tabesh, M.; Esmaillzadeh, A. Maternal Vitamin D Status and Risk of Pre-Eclampsia: A Systematic Review and Meta-Analysis. *J. Clin. Endocrinol. Metab.* **2013**, *98*, 3165–3173. [CrossRef] [PubMed]
46. Nandi, A.A.; Wadhwani, N.S.; Joshi, S.R. Altered metabolic homeostasis between vitamin D and long chain polyunsaturated fatty acids in preeclampsia. *Med. Hypotheses* **2017**, *100*, 31–36. [CrossRef] [PubMed]
47. Palacios, C.; De-Regil, L.M.; Lombardo, L.K.; Peña-Rosas, J.P. Vitamin D supplementation during pregnancy: Updated meta-analysis on maternal outcomes. *J. Steroid Biochem. Mol. Biol.* **2016**, *164*, 148–155. [CrossRef] [PubMed]
48. Pludowski, P.; Holick, M.F.; Pilz, S.; Wagner, C.L.; Hollis, B.W.; Grant, W.B.; Shoenfeld, Y.; Lerchbaum, E.; Llewellyn, D.J.; Kienreich, K.; et al. Vitamin D effects on musculoskeletal health, immunity, autoimmunity, cardiovascular disease, cancer, fertility, pregnancy, dementia and mortality—A review of recent evidence. *Autoimmun. Rev.* **2013**, *12*, 976–989. [CrossRef]
49. Barrett, H.; McElduff, A. Vitamin D and pregnancy: An old problem revisited. *Best Pract. Res. Clin. Endocrinol. Metab.* **2010**, *24*, 527–539. [CrossRef]
50. CDC. Pregnancy Mortality Surveillance System. 22 June 2022. Available online: https://www.cdc.gov/reproductivehealth/maternalinfanthealth/pregnancy-mortality-surveillance-system.htm (accessed on 28 September 2022).
51. WHO. Maternal Mortality. 19 September 2019. Available online: https://www.who.int/en/news-room/fact-sheets/detail/maternal-mortality (accessed on 28 September 2022).
52. Grant, W.B.; Boucher, B.J.; Al Anouti, F.; Pilz, S. Comparing the Evidence from Observational Studies and Randomized Controlled Trials for Nonskeletal Health Effects of Vitamin D. *Nutrients* **2022**, *14*, 3811. [CrossRef]
53. Li, W.-J.; Chen, K.-H.; Huang, L.-W.; Tsai, Y.-L.; Seow, K.-M. Low Maternal Serum 25-Hydroxyvitamin D Concentration Is Associated with Postpartum Hemorrhage: A Retrospective Observational Study. *Front. Endocrinol.* **2022**, *13*, 816480. [CrossRef]
54. Aghajafari, F.; Nagulesapillai, T.; Ronksley, P.E.; Tough, S.C.; O'Beirne, M.; Rabi, D.M. Association between maternal serum 25-hydroxyvitamin D level and pregnancy and neonatal outcomes: Systematic review and meta-analysis of observational studies. *BMJ* **2013**, *346*, f1169. [CrossRef] [PubMed]
55. Baker, A.M.; Haeri, S.; Camargo, A.C.; Stuebe, A.M.; Boggess, A.K. A Nested Case-Control Study of First-Trimester Maternal Vitamin D Status and Risk for Spontaneous Preterm Birth. *Am. J. Perinatol.* **2011**, *28*, 667–672. [CrossRef] [PubMed]
56. Bodnar, L.M.; Rouse, D.J.; Momirova, V.; Peaceman, A.; Sciscione, A.; Spong, C.Y.; Varner, M.; Malone, F.D.; Iams, J.D.; Mercer, B.M.; et al. Maternal 25-Hydroxyvitamin D and Preterm Birth in Twin Gestations. *Obstet. Gynecol.* **2013**, *122*, 91–98. [CrossRef] [PubMed]
57. Thota, C.; Menon, R.; Fortunato, S.J.; Brou, L.; Lee, J.-E.; Al-Hendy, A. 1,25-Dihydroxyvitamin D Deficiency Is Associated with Preterm Birth in African American and Caucasian Women. *Reprod. Sci.* **2014**, *21*, 244–250. [CrossRef] [PubMed]
58. Liong, S.; Di Quinzio, M.K.W.; Fleming, G.; Permezel, M.; Georgiou, H.M. Is Vitamin D Binding Protein a Novel Predictor of Labour? *PLoS ONE* **2013**, *8*, e76490. [CrossRef]
59. Zhou, S.-S.; Tao, Y.-H.; Huang, K.; Zhu, B.-B.; Tao, F.-B. Vitamin D and risk of preterm birth: Up-to-date meta-analysis of randomized controlled trials and observational studies. *J. Obstet. Gynaecol. Res.* **2017**, *43*, 247–256. [CrossRef]
60. McDonnell, S.L.; Baggerly, K.A.; Baggerly, C.A.; Aliano, J.L.; French, C.B.; Baggerly, L.L.; Ebeling, M.D.; Rittenberg, C.S.; Goodier, C.G.; Niño, J.F.M.; et al. Maternal 25(OH)D concentrations $\geq$40 ng/mL associated with 60% lower preterm birth risk among general obstetrical patients at an urban medical center. *PLoS ONE* **2017**, *12*, e0180483. [CrossRef]
61. Morales-Suárez-Varela, M.M.; Uçar, N.; Peraita-Costa, I.; Huertas, M.F.; Soriano, J.M.; Llopis-Morales, A.; Grant, W.B. Vitamin D-Related Risk Factors for Maternal Morbidity during Pregnancy: A Systematic Review. *Nutrients* **2022**, *14*, 3166. [CrossRef]
62. Heaney, R.P. Guidelines for optimizing design and analysis of clinical studies of nutrient effects. *Nutr. Rev.* **2014**, *72*, 48–54. [CrossRef]

nutrients

MDPI

Review

Targeting the Platelet-Activating Factor Receptor (PAF-R): Antithrombotic and Anti-Atherosclerotic Nutrients

Rajendran Harishkumar [1,2], Sakshi Hans [1,2], Janelle E. Stanton [1,2], Andreas M. Grabrucker [2,3,4], Ronan Lordan [5,6,7,*] and Ioannis Zabetakis [1,2,3,*]

1 Department of Biological Sciences, University of Limerick, V94 T9PX Limerick, Ireland
2 Bernal Institute, University of Limerick, V94 T9PX Limerick, Ireland
3 Health Research Institute, University of Limerick, V94 T9PX Limerick, Ireland
4 Cellular Neurobiology and Neuro-Nanotechnology Laboratory, Department of Biological Sciences, University of Limerick, V94 T9PX Limerick, Ireland
5 Institute for Translational Medicine and Therapeutics, Perelman School of Medicine, University of Pennsylvania, Philadelphia, PA 19104, USA
6 Department of Medicine, Perelman School of Medicine, University of Pennsylvania, Philadelphia, PA 19104, USA
7 Department of Systems Pharmacology and Therapeutics, Perelman School of Medicine, University of Pennsylvania, Philadelphia, PA 19104, USA
* Correspondence: ronan.lordan@pennmedicine.upenn.edu (R.L.); ioannis.zabetakis@ul.ie (I.Z.)

Abstract: Platelet-activating factor (PAF) is a lipid mediator that interacts with its receptor (PAF-R) to carry out cell signalling. However, under certain conditions the binding of PAF to PAF-R leads to the activation of pro-inflammatory and prothrombotic pathways that have been implicated in the onset and development of atherosclerotic cardiovascular diseases (CVD) and inflammatory diseases. Over the past four decades, research has focused on the identification and development of PAF-R antagonists that target these inflammatory diseases. Research has also shown that dietary factors such as polar lipids, polyphenols, and other nutrient constituents may affect PAF metabolism and PAF-R function through various mechanisms. In this review we focus on the inhibition of PAF-R and how this may contribute to reducing cardiovascular disease risk. We conclude that further development of PAF-R inhibitors and human studies are required to investigate how modulation of the PAF-R may prevent the development of atherosclerotic cardiovascular disease and may lead to the development of novel therapeutics.

Keywords: platelet-activating factor; platelet-activating factor receptor; polar lipids; antithrombotic activity; inflammation; atherosclerosis

Citation: Harishkumar, R.; Hans, S.; Stanton, J.E.; Grabrucker, A.M.; Lordan, R.; Zabetakis, I. Targeting the Platelet-Activating Factor Receptor (PAF-R): Antithrombotic and Anti-Atherosclerotic Nutrients. *Nutrients* **2022**, *14*, 4414. https://doi.org/10.3390/nu14204414

Academic Editor: Hayato Tada

Received: 12 September 2022
Accepted: 18 October 2022
Published: 20 October 2022

1. Introduction

Atherosclerotic cardiovascular diseases (CVD) are the leading cause of morbidity and mortality globally [1]. Various factors contribute to the development of atherosclerosis, but evidence in recent decades has demonstrated that nutrition plays a pivotal role in the prevention of atherosclerosis and other chronic inflammatory conditions including diabetes and obesity [2,3]. Hence there is a requirement to research the effects of diets and food components on cardiovascular health.

Atherosclerosis is a progressive inflammatory disease responsible for the development of atherothrombotic complications including myocardial infarction, peripheral artery disease, and ischaemic or transient stroke among other cardiac manifestations [4,5]. Atherosclerosis develops through several steps including endothelial dysfunction followed by the deposition of lipids in the intima, which accumulate in the lining of blood vessels. These lipids are then engulfed by macrophages, which eventually undergo apoptosis forming foam cells and a necrotic core that leads to the development of the characteristic lesions or fatty streaks in blood vessels. Erosion of these lesions or plaques causes microruptures that activate platelets causing fibrin netting and platelet aggregates to form on the inner walls of arteries, thus leading to the narrowing of blood vessels affecting blood supply [6]. With time, the lumen may narrow and erode further causing plaque rupture, leading to a major cardiovascular event such as myocardial infarction or stroke. The main mechanistic events that lead to these events are characterised by persistent low-grade inflammation [5].

However, inflammation is a necessary physiological response of the innate immune system, and its main role is to maintain a constant internal environment despite being subjected to constantly changing environmental pressures. These can include mechanical, physical, chemical, infectious, immunological, or reactive natural adverse events. The inflammatory response seeks to diminish and/or minimize the agents that causes tissue damage, promote adequate wound healing, and restore tissue homeostasis. However, if the inflammatory response fails to resolve owing to the persistence of the triggering factors or poor restoration of the original tissue, a prolonged underlying inflammatory process arises, leading to increased tissue dysfunction and adverse effects. At the molecular and cellular level, it has been postulated that endothelial dysfunction leading to systemic inflammation appears to be the primary underlying mechanistic factor in the onset and progression of atherosclerosis [7]. Endothelial dysfunction is often defined by an inflammatory microenvironment that acts on leukocytes and endothelial cells via interactions with other immune cells such as T lymphocytes, mast cells, dendritic cells (DC), and platelets [8].

Platelets play a key role in the onset and development of atherosclerosis [9–13]. Platelets also orchestrate the development of obstructive thrombi in the latter stages of the atherosclerotic process in response to plaque rupture through the sequential processes of haemostatic responses to vascular injury such as initiation, extension, and stabilization [14]. Each of these stages contains pro-haemostatic molecular mechanisms, in balance with anti-haemostatic processes, which restrict the reaction to the damage site and prevent inappropriate vascular occlusion. The molecular players involved in the initiation process include adhesion molecules, signalling ligands, and their associated platelet surface receptors [15]. Strong inflammatory and prothrombotic mediators such as platelet-activating factor (PAF) play pivotal roles in these processes, particularly in the activation of platelets [16]. Indeed, PAF and its receptor have previously been investigated as a pharmaceutical target for some inflammatory conditions including asthma and sepsis with limited success to date. They have also been implicated in many of the key processes that lead to the development of atherosclerosis. However, researchers over the years have postulated that dietary PAF-R antagonists may affect PAF-related signalling and inflammatory pathways [7,17,18]. This has opened several avenues of research that aim to investigate certain dietary patterns such as the Mediterranean diet, which is thought to offer protection from atherosclerotic cardiovascular disease and other inflammatory diseases due to a high concentration of these compounds in the diet [18,19]. In this review, we examine the role of various nutrients and their effects on PAF and its receptor PAF-R and how attenuating this inflammatory and thrombotic pathway may contribute to atherosclerosis prevention via altering one's diet. It is also important to recognise that while this review largely focusses on the relationship between PAF and the PAF-R, there are also ongoing developments in cardiovascular research relating to the metabolic enzymes of PAF, which have been discussed at length elsewhere [7,20].

2. Platelet-Activating Factor (PAF) and PAF-Receptor (PAF-R)

PAF (1-*O*-alkyl-2-acetyl-*sn*-glycero-3-phosphocholine) is a phospholipid mediator that functions through the PAF-receptor (PAF-R). PAF was discovered when Ig-E sensitised basophils of rabbits were challenged with antigen stimuli [21]. In physiology, PAF is an important signalling molecule in the renal, cardiovascular, immune, and reproductive systems. However, PAF is not just one molecule; there happens to be a family of PAF-like lipids (PAFLL) or PAF-like moieties, which all have varying degrees affinity with the PAF-R leading to various levels of potency [22]. The classic PAF molecule has an alkyl ether linkage at the *sn*-1 position, a characteristic acetyl group at the *sn*-2 position, and a phosphocholine group at the *sn*-3 position of the glycerol backbone [23]. The most potent PAF molecules contains a 16:0 at the *sn*-1 position, but may also have 18:0, 17:0, and 18:1 on the alkyl ether-linked side chain leading to varying degrees of affinity for the PAF-R and as a consequence, varying degrees of biological activity [7,24]. PAF is known to carry out its biological activities at concentrations as low as 10^{-12} M and almost always by 10^{-9} M as an intercellular messenger [25] and it carries out its functions in a autocrine, juxtracrine, and paracrine manner [26,27]. The history of the elucidation of the PAF structure and developments in the field has recently been reviewed [28].

The PAF-R is expressed by cells in various tissues, including the lungs, spleen, heart, kidneys, skeletal muscle, and in blood cells as shown in Figure 1 [24]. Therefore, it is also unsurprising that PAF-R signaling is implicated in many physiological processes [28]. There is an abundance of phospholipids in the brain and central nervous system (CNS) [29], where the PAF-R is expressed by various parts of the CNS including the spinal cord, substantia niagra, hypothalamus, hippocampus, frontal cortex, nucleus accumbens, cortex, cerebellum, cerebellar hemisphere, basal ganglia, and the amygdala [30]. Notably, PAF is also synthesised by neuronal tissue and its signaling is associated with neurotrophic effects [31]. Indeed, permeability of the blood-brain barrier (BBB) increases via PAF-R dependent mechanisms, consequent to calcium (Ca^{2+}) influx, increased nitric oxide levels, and alterations to proteins that regulate intercellular gaps in the BBB *in vivo* [32].

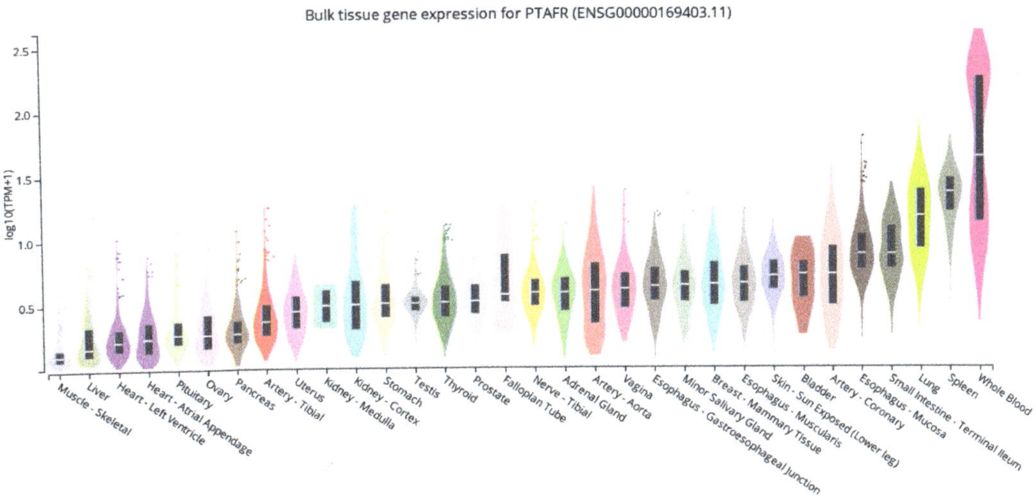

Figure 1. Bulk gene expression for the platelet-activating factor receptor (PAF-R) encoded by the gene *PTAFR* in various human tissues using data from the Genotype-Tissue Expression (GTEx) Project [30]. The expression data is shown in transcripts per million (TPM) with the plots showing the median and the 25th and 75th percentiles. Dots indicate outliers, which are above or below 1.5 times the interquartile range.

PAF-R signalling also plays a prominent role in reproductive biology, including ovulation, fertilisation, preimplantation, and parturition in women. In men, PAF is present in spermatozoa and is thought to be involved in sperm motility and in the induction of acrosome reactions [33–38]. PAF and PAF-R is also a known physiological mediator of healthy cardiovascular function via modulating inflammatory signaling, platelet function, and blood pressure [39–41]. As the name suggests, PAF is a platelet activator via binding to the PAF-R in the normal response to injury [13]. PAF-R binding by PAF induces platelet shape change and the release of platelet granules via stimulation of the phosphatidylinositol cycle and intracellular Ca^{2+} mobilization. Serotonin and platelet factor 4 are secreted, along with arachidonic acid and other bioactive lipids, including PAF, which mediate platelet aggregation [13,42,43].

While we are still learning about the roles of PAF and PAF-R in physiology, PAF is mostly known for its role as an inflammatory messenger that passes signals to cell types such as platelets, neutrophils, endothelial cells, macrophages, and lymphocytes [7]. PAF is involved in multiple communicable and non-communicable diseases through excessive binding with the PAF-R. Some studies have shown that PAF mediates metastasis in tumour cells. For example, PAF triggers human melanoma cells via stimulating the phosphorylation of cAMP-responsive element (CRE)-binding protein (CREB) and activating transcription factor-1(ATF-1). This signal transduction leads to the overexpression of major effectors involved in tumour growth, angiogenesis, and malignant progressions such as MMPs, STAT-3, and NF-κB [44]. PAF also affects other pathological processes including increased vascular permeability, hypotension, ulcerogenesis, bronchoconstriction triggering airway hyperresponsiveness, and platelet degranulation. PAF has also been implicated in septic shock, asthma, ischemia/ reperfusion injury, pancreatitis, inflammatory bowel disease, and rhinitis [45]. PAF-R activation has also been reported to manifest in communicable diseases. For example, PAF-R activation causes increased thrombocytopenia, haemoconcentration, increased systemic levels of cytokines, and lethality in wild-type mice compared with PAF-R-silenced mice in a model of dengue fever [46]. PAF is implicated in other infectious diseases characterised by inflammation including human immunodeficiency virus (HIV) [47] and severe acute respiratory syndrome coronavirus 2 (SARS-CoV-2) infection [48,49]. Considering the vast pathological functions of PAF and its receptor, many investigations have focused on preventing PAF from binding to the PAF-R with the aim of reducing prothrombotic and proinflammatory signalling.

Structurally, the PAF-R is a seven transmembrane G-protein coupled receptor encoded by the *PTAFR* gene. The gene locus has been identified in humans as chromosome 1p35-p34.5. Human and guinea pig PAF receptors are single polypeptides with 342 amino acids; rat and mouse PAF receptors omit one amino acid in the third extracellular loop. Despite various findings to the contrary, it is presently believed that a single receptor subtype mediates all PAF's actions in humans and is generally located on the plasma membrane, endomembrane, nucleus, and nuclear envelope [18,50]. The gene, *PTAFR*, is tightly regulated by two distinct promotors that are involved in the transcriptional regulation, consequently there are two alternatively spliced transcripts that differ in their untranslated regions. The first, transcript 1, is widely expressed in tissues regulated by inflammation, predominantly in leukocytes, macrophages, eosinophils, and monocytes. The second, transcript 2, is found in organs such as the heart, kidney, lung, and spleen and its expression can be influenced by oestrogen, thyroid hormone T3, retinoic acid, transforming growth factor-β (TGF-β), tumour necrosis factor-α (TNF-α), interferon-γ and others. The second transcript is not thought to be expressed by hematopoietic cells or in the brain [24,51,52]. In a positive feedback manner, PAF may upregulate the expression of its own receptor via transcript 1 through NF-κB signalling [53]. There is also evidence that PAF-R transcription is dependent on activation of the Jak/STAT pathway [51]. The upregulation of PAF synthesis and its degradation is also tightly regulated and has been extensively reviewed [16]. However, there are many aspects of PAF-R (*Ptafr*) expression that have been underexplored including whether it exhibits circadian rhythmicity as some data indicates that it might, along with genes associated with PAF metabolism, including *Pla2g7* (Circadb: Circadian Expression Profiles Data Base. Available online: http://circadb.hogeneschlab.org (accessed 12 October 2022)). Considering the PAF-R is expressed in numerous cells and tissue types there is a lot to left to be explored regarding its function and modulation.

The first binding experiment of PAF was conducted on human platelets in 1982, whereby two distinct binding sites were revealed. The first site had shown higher affinity (K_d value = 37 ± 13 nm) and the other site possessed nearly low affinity toward PAF [54]. To understand the pathophysiological function of PAF, gene modifications were applied in earlier studies. For example, cDNA encoding the PAF-R was isolated from the guinea pig lung cDNA library and was cloned into *Xenopus laevis* oocytes depicted in Figure 2. In this cloned receptor, several amino acids are highly conserved when compared to other G protein receptors, including aspartic acid (Asp) in the second transmembrane segment, one cysteine (Cys) in both the second and third intracellular loops, and three proline (Pro) in the sixth and seventh segments. The PAF receptor's cytoplasmic tail comprises four serines (Ser) and five threonines (Thr). There is a total of 12 tyrosine (Tyr) residues, with two of them located in the cytoplasmic loops. Asparagine (Asn) residues are found on the receptor's exterior surface and may serve as sites for glycosylated residue attachment [55]. Some other reports stated that cloning of human PAF receptors can be achieved by isolating cDNA from peripheral leukocytes, heart, and EoL-1 eosinophilic leukaemia cells [56]. Figure 1B shows the helical 3D structure of PAF-R (Chain-A) that was obtained from the protein data bank (PDB ID: 5ZKQ) and its bound ligands were removed by UCSF Chimera [57].

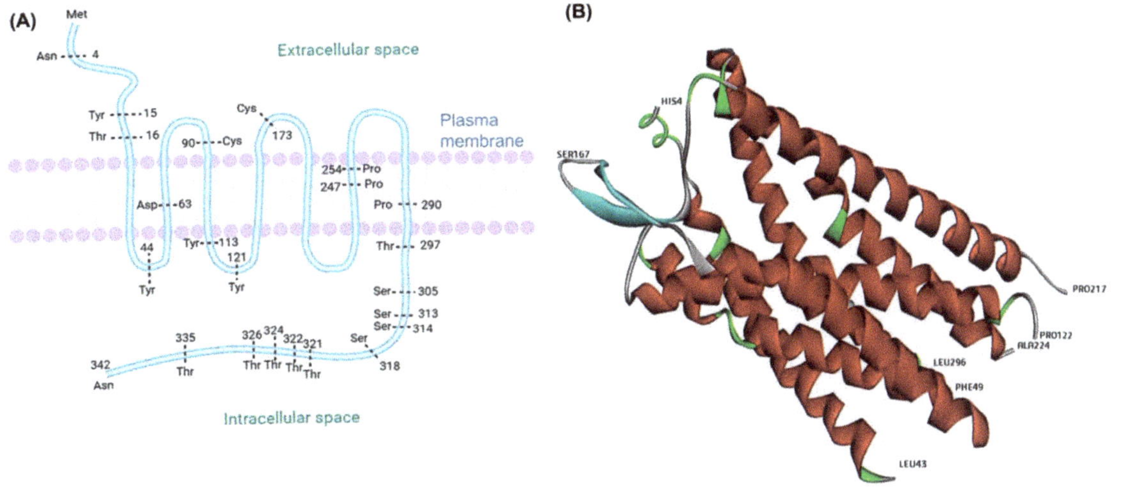

Figure 2. (**A**) Diagrammatic representation of the platelet-activating factor receptor (PAF-R) with its seven transmembrane domains within the plasma membrane bilayer [Note: PAF-R cloned from guinea pig represented with amino acid residues]; (**B**) Human PAF-R (Chain-A) with selective amino acid residues (PDB ID: 5ZKQ).

PAF and PAF-R Activation in Inflammatory Diseases

Elevated levels of PAF can be detected in tissues affected by inflammatory diseases [7]. Excessive activation of the PAF-R via PAF and PAF-like lipids (PAF-LL) in inflammatory diseases induces several biological effects including systemic pro-inflammatory, prothrombotic, and pro-proliferative signalling. Indeed, delayed immune responses have also been reported and PAF-R signalling has been implicated in cancer development. Many malignant cells have been shown to overexpress PAF-R [58]. The PAF-R receptor is related to phosphoinositide metabolism via a G-protein that is also linked to phospholipases C and A_2. PAF-R stimulation results in the brief synthesis of diacylglycerol, which activates protein kinase C, and inositol triphosphate, which triggers the release of internal calcium reserves [59]. The activation of PAF-R through PAF is represented the Figure 3.

PAF increases tyrosine phosphorylation of several proteins in neutrophils, macrophages, and platelets, as well as nuclear factor kappa B (NF-*k*B) activation and transcription of *c-fos* and c-*jun* in inflammatory cells. PAF can activate the mitogen-activated protein kinase (MAPK) kinase-3, a known activator of p38 MAPK, and the Jak/STAT pathway [59]. Following ligand activation, the PAF-R is degraded through both the proteasome and lysosomal pathways.

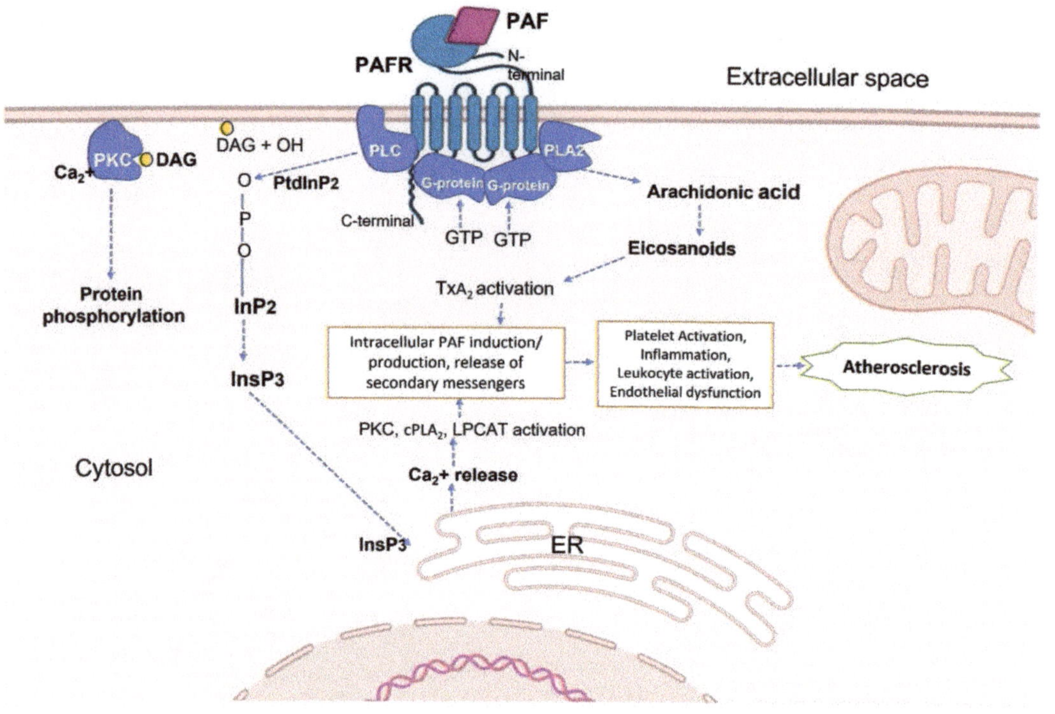

Figure 3. Mechanism of PAF-R activation and PAF-mediated signalling pathway [59,60]. Abbreviations: $cPLA_2$—cytosolic phospholipase A_2; DAG—Diacylglycerol; ER—Endoplasmic reticulum; InP_2 —inositol 4,5-bisphosphate; $InsP_3$—inositol 1,4,5-triphosphate; LPCAT—lysophosphatidylcholine acyltransferase; PKC—protein kinase C; PLA_2—Phospholipase A_2; PLC—Phospholipase C; $PtdInP_2$—phosphatidyl 4,5-bisphosphate; TxA_2—thromboxane A_2.

While platelet activation leads to aggregation as part of normal haemostatic function, under acute or systemic inflammatory conditions PAF-R activation has been shown to induce various immune and inflammatory pathways [44] that can lead to both acute and chronic conditions. For example, PAF-R activation by PAF induces histamine and prostaglandin D_2 release from mast cells [61,62] and it is involved in the chemotaxis of mast cells [63]. PAF has been shown to be a powerful chemoattractant for eosinophils [64,65] and it is responsible for the generation of chemokines and prostaglandins [65–67]. PAF along with leukotriene B4 (LTB4) and matrix metalloproteinase-9 (MMP-9) are involved in the accumulation of eosinophils in asthmatic airways via interleukin-8 (IL-8) stimulation of neutrophils [68]. Indeed, PAF has been shown to promote the recruitment of neutrophils and polymorphonuclear cells to inflammatory sites [7], and these cells can also generate PAF [7,69,70], which is thought to be one of the underlying mechanisms by which conditions such as atherosclerosis may propagate [7,71].

As a consequence of the wide-ranging inflammatory actions of PAF and the PAF-R, pharmaceutical companies and scientists have previously investigated the use of PAF inhibitors and developed pharmaceutical grade products to target these inflammatory pathways. These include products such as Lexipafant [72], Modipafant [73], and Rupatadine [74,75] among others that have previously reviewed [28] for the treatment of asthma, sepsis, and other conditions characterised by PAF-related inflammation. However, a recent study has shown that PAF and PAFLL can mediate nucleotide-binding domain, leucine-rich-repeat-containing protein 3 and never in mitosis A-related kinase 7 (NLRP3-NEK7) inflammasome induction in a PAF-R independent manner, which may explain observations of the ineffectiveness of many PAF-R antagonists [76] including those aforementioned. These findings may lead to further developments in our understanding of the role of PAF in diseases such as cancer and atherosclerosis considering the important role of the inflammasome in these diseases. Pharmaceuticals aside, research has also determined that there are a broad range of naturally occurring PAF-R antagonists present in certain edible plants and foods, which will be discussed in the ensuing sections.

3. Antiplatelet Properties of Nutrients

Diet has long been associated with the maintenance of health and the prevention of disease. It is well established that healthy dietary patterns, such as the Mediterranean diet and the dietary approaches to stop hypertension (DASH) diet, may offer protection against the development of atherosclerosis and cardiovascular diseases [77,78]. With this knowledge, the functional foods, dietary supplements, and nutraceuticals industries have grown exponentially over the last two decades offering individuals food-derived and natural product derived constituents that may confer health benefits on the consumer [79]. Historically, many cultures turn to food and natural products as a source of healing in times of ill health. These practices are particularly prevalent in areas with indigenous rural communities. The World Health Organization (WHO) defines traditional medicine as "the total of knowledge, skills, and practices based on theories, beliefs, and experiences indigenous to different cultures, whether explicable or not, used in the maintenance of health as well as the prevention, diagnosis, improvement, or treatment of physical and mental illness" [80]. Some traditional medicine systems are supported by substantial literature and recordings of theoretical notions and practical abilities; others are passed down verbally from generation to generation. Until now the turn of the 20th century, the majority of the world's population relied on their own traditional medicine to satisfy their primary health care requirements in several regions of the world [81]. Traditional medicine is commonly referred to as "complementary and alternative medicine" when practised outside of its traditional culture [82]. Traditional medicine is most popular and practiced nowadays in China, India, and many African nations among others [83].

In India, the traditional system of medicine (TSM) has been practiced before the adoption of modern medicine by traditional communities to heal any type of illness. These medical practices provide invaluable assistance in the healthcare system for current and future generations. The traditional systems of Indian medicine, presently known as the Indian System of Medicine (ISM), have a very solid conceptual foundation and have been practiced for a very long period. Ayurveda, Siddha, and Unani are three prominent traditional systems practiced in India [84]. Some Indian medicinal plants are reported to have antihyperlipidemic activity and anti-thrombolytic because of their antiplatelet aggregation activity and fibrinolytic activity [85]. Phytocompounds such as cudratricusxanthone A [86], withaferin A [87], and even some of the serine proteases were identified and tend to prevent clot formation [88].

Over the past three decades, there has been considerable research conducted investigating the potential antithrombotic and anti-inflammatory effects of dietary PAF inhibitors. In particular, there has been a focus on polar lipids, mainly phospholipids and sphingolipids, derived from natural sources such as plants and animal products, which exert antithrombotic activities due their inhibition of the PAF-R activation and other platelet agonists [28,89,90]. A recent study found that dietary supplementation with plant extract containing aloe gel, grape juice, green tea extract, etc. reduced platelet sensitivity upon stimulation with PAF [91]. In this study, it is not clear what constituent or combination of constituents are responsible for these observed effects. However, many compounds such as polyphenols, phenolipids, and polar lipids present in these capsule constituents have previously been associated with antiplatelet effects. For example, certain compounds isolated from *Spirulina* (blue-green algae) and other marine algae also possess bioactive properties beneficial to health, including antiplatelet, anti-inflammatory and antioxidant qualities.

These qualities have been traced to the glycolipid sulphoquinovosyl diacylglycerol (SQDG) present in photosynthetic plants [92,93].

However, plants are not the only food-derived antithrombotic polar lipids. Fish-derived lipids also exhibit inhibitory properties against PAF. Polar lipid fractions isolated from cod (*Gadus morhua*) and salmon (*Salmo salar*) showed platelet inhibitory capabilities, suggesting that the consumption of such lipids could protect against cardiovascular disease [90,94]. Other animal foods, such as dairy products, notably yoghurt, also exhibit inhibitory activities against PAF *in vitro* [95,96]. In humans, intake of yoghurt enriched with polar lipids from olive oil by-products resulted in lower platelet sensitivity against PAF, and reduced low-grade inflammation, assessed by monitoring serum levels of IL-10 and IL-6 [97].

Many compounds derived from traditional herbal remedies also possess potent anti-PAF activity. Curcumin is a spice derived from turmeric and commonly used in Asian cuisines. A 1999 study found that curcumin inhibits platelet aggregation induced by agonists such as PAF, epinephrine, and ADP, via the inhibition of thromboxane production and Ca^{2+} signalling [98]. Another investigation found that extracts of several species of Malaysian medicinal plants exhibited significant inhibitory activity against PAF [99]. The Korean folk medicinal plant *Alpinia officinarum* is traditionally used to treat gastrointestinal diseases. Diarylheptanoid compounds were isolated from this plant and also showed a high inhibitory effect against platelet aggregation by PAF [45]. Apart from medicinal plants, plants that are commonly found in various diets also possess bioactive compounds with antithrombotic activities with various target mechanisms as listed in a Table 1.

Table 1. Comparison of different studies investigating phytocompounds and their antithrombotic activities against PAF andother platelet agonists.

Phytocompounds	Scientific Name (Common name)	Optimum Dose Determined or Dosage Investigated	Study Outcomes
Polyphenols such as theaflavin and its gallolyl ester, geranyl gallate, farnesyl gallate and geranylgeranyl gallate.	*Camellia sinensis* (Tea)	Theaflavin and its galloyl esters (IC_{50} = 32–77 μM), geranyl gallate, farnesyl gallate and geranylgeranyl gallate (IC_{50} = 6.4–7.6 μM).	Tea polyphenol such as theaflavin and its other galloyl esters showed potential antithrombotic activity against PAF and inhibited an acetyltransferase involved in its biosynthesis [100].
Polar lipids	*Camellia sinensis* (Tea leaves)	TL (110 ± 50 μg/ μL), PL (34 ± 4 μg/μL) and NL (820 ± 460 μg/μL) from 30 min observation respectively.	Synergetic effect of the antithrombotic activity of tea polyphenol and PL were against PAF, thrombin, ADP, and collagen, due to their high unsaturated fatty acid content especially rich in omega-3 PUFA and MUFA [101].
Sulphonoglycolipid	*Polypodium decumanum* (Fern calaguala)	IC_{50} = 2 μM.	Sulphoquinovosyl diacylglycerol 1,2-di-*O*-palmitoyl-3-*O*-(6-sulpho-α-d-quinovopyranosyl)-glycerol showed inhibitory activity against PAF in an *in vitro* model using human neutrophils [102].
Curcumin	*Curcuma longa* (Turmeric)	Concentration: 0.3 mg/day in mice.	Oral administration of curcumin (0.3mg/day) in mice inhibited thromboxane levels and increased prostacyclin activity [103].
Ar-turmerone	*Curcuma longa* (Turmeric)	IC_{50} values of 14.4 μM and 43.6 μM against collagen and arachidonic acid (AA) respectively.	*In vitro* study showed that aromatic (ar-)turmerone effectively inhibits platelet aggregation induced by collagen and arachidonic acid [104].

Table 1. *Cont.*

Phytocompounds	Scientific Name (Common name)	Optimum Dose Determined or Dosage Investigated	Study Outcomes
Curcuminoids	*Curcuma longa* (Turmeric)	Concentration: 10–30 μg/mL.	The isolated PRP was exposed to various concentrations of curcuminoids (10–30 μg/mL) and showed antiplatelet activity against AA and collagen [105].
Allicin and thiosulfinates	*Allium sativum* (Garlic)	Volume: 30 μL of garlic juice.	*In vitro* studies showed that allicin and thiosulfanates are the key constituents of garlic juice resulting in antiplatelet activity against collagen-induced platelet activity [106].
Thiosulfinate	*Allium cepa* (Onion)	Volume: 220 μL of onion juice.	The study resulted that 220 μL of onion juice was enough to produce complete inhibition of platelet aggregation *in vitro* against AA [107].
AMP48 (Serine protease) of latex	*Artocarpus heterophyllus* (Jack fruit)	Amount: 1, 2, 4, 8, 16, 32 μg.	Using a thrombin and $CaCl_2$ mediated fibrin clot experiment, 4 μg of AMP48 completely hydrolyzed α-subunit of fibrinogen in 30 min. Techniques including N-terminal sequencing fibrinolysis and ATR-FTIR spectroscopy revealed this novel protein has fibrinolytic properties [88].
Eugenol, amygdalactone, cinnamic alcohol, 2-hydroxycinnamaldehyde, 2-methoxycinnamaldehyde, coniferaldehyde, acetylsalicylic acid, coumarin, cinnamaldehyde, cinnamic acid, icariside DC, dihydrocinnacasside,	*Cinnamomum cassia* (Cinnamon bark)	IC_{50} values of Eugenol and coniferaldehyde obtained as 3.8 and 0.82 μM against AA; 3.51, and 0.44 μM against U46619 (thromboxane A_2 mimic); 1.86 and 0.57 μM against epinephrine-induced aggregation.	Among the 13 compounds from the extract of cinnamon bark, eugenol, and coniferaldehyde were the two of the most active antiplatelet constituents against AA, U46619 (thromboxane A2 mimic) and epinephrine-induced platelet aggregation [108].
Aqueous extract from the bark	*Cinnamomum tamala* (Indian Bay Leaf)	Various concentrations of 100, 200, 300, 400, and 500 μg.	The aqueous extract inhibited TXB_2 formation through COX pathway (IC_{50} of 112 μg ± 16) also LP-1 by LOX pathway (IC_{50} of 120 μg ± 15), and 500 μg concentration showed complete inhibition of platelet aggregation [109].
(6S,7Z,9R)-roseoside, Eriodectyol and 2″-O-rhamnosyl vitexin	*Crataegus pinnatifida* (Chinese hawberry)	Concentration: 400 μg/mL.	The isolated compounds 7, 13 and 15 exhibited potent antithrombotic activity against ADP induced platelet aggregation *in vitro* by 87.18, 72.92 and 75.00% respectively at 400 μg/ mL, among them the 13th compound exhibited antithrombotic activity *in vivo* (zebrafish) by prolonged thrombus formation (19.04 ± 3.32 min) than heparin control (17.63 ± 2.23 min) [110].

Table 1. *Cont.*

Phytocompounds	Scientific Name (Common name)	Optimum Dose Determined or Dosage Investigated	Study Outcomes
Ethanolic extract	*Ocimum basilicum* (Basil)	Concentrations: 0.1, 1 and 10 mg/mL of Ocimum ethanolic extract.	Overall OBL and its extracts elevated 6-keto-PGF1α production while decreasing PGE_2 and TXB_2 production in a dose- and time-dependent manner. This might be due to the combined inhibition of COX-2 and activation of endothelial COX-1 [111].
Methanolic leaf extract	*Mangifera sylvatica* (Himalayan mango)	A volume of 100 µL.	Methanolic fraction showed a maximum of 46.93% clot lysis activity whereas streptokinase standard showed 80.51% [112].
Mangiferin	*Mangifera indica* L. (Mango)	Extracts from each part of the mango such as pulp, peel, seed husk and seed with various concentrations like 0.1, 0.5, and 1 mg/mL.	Mango seed showed a 72% of inhibition against adenosine 5′-diphosphate (ADP) induced by platelet aggregation. Among the identified monogalloyl compounds and benzophenones, mangiferin showed a 31% of inhibitory effect against ADP [113].
Bromelain	*Ananas comosus* (Pineapple)	Bromelain at various doses of 70, 140, and 210 µg/kg of body weight.	Antiplatelet aggregation tests from *in vivo* method exhibited that bromelain (at the dose of 210 µg/KgBW) has increased the bleeding time (515.10 ± 182.23%) on the 21st day of termination [114], indicating antiplatelet effects.
Baru almond oil	*Dipteryx alata* Vog (Baru Almond)	Ten days of Baru oil as 7.2 and 14.4 mL/kg/day.	Baru almond oil treatment has lowered about 31% of ADP-induced platelet aggregation and thrombotic processes in male Wistar rats, suggesting that it helps lower platelet activation and exert advantages in thrombotic processes [115].
Aqueous extract of strawberry fruit	*Fragaria ananassa* (Strawberry)	Extract concentrations from 0.1–1 mg/mL.	Dose-dependent reduction against AA and ADP-induced platelet aggregation was observed as 65 ± 5% and 55 ± 4% of inhibition respectively [116].
Hippuric acid	Phenol-rich fruits and plant	Concentrations: 100, 200, 500, 1 and 2 mM.	Dose-dependent inhibition against platelet surface receptor P_2Y_1/P_2Y_{12} induced by ADP [117].

Table 1. *Cont.*

Phytocompounds	Scientific Name (Common name)	Optimum Dose Determined or Dosage Investigated	Study Outcomes
Piperine, pipernonaline, piperoctadecalidine, piperlongumine	*Piper longum* L. (Black Pepper)	Concentrations: 300, 150, and 30 μM.	The most effective antiplatelet agent was piperlongumine *in vitro*. Piperlongumine inhibited collagen-induced platelet aggregation with inhibition rates of 100, 100, 49.8, and 19.9% at 300, 150, 30, and 10 μM, respectively. Piperlongumine had 100%, 76.4%, and 12% inhibitory activity in an AA test at 300, 150, and 30 μM, respectively. Furthermore, piperlongumine at doses of 300, 150, and 30 μM reduced PAF-induced platelet aggregation with inhibition rates of 100%, 100%, and 29.9%, respectively [118].
Orientin and Iso-orientin	*Vaccinium bracteatum* Thunb. (Sea bilberry or Asiatic bilberry)	*In vitro* experiment with 5 to 50 μM and *in vivo* experiment with 9, 26.9 and 44.8 μg per mouse respectively.	A dose-dependent reduction in platelet aggregation was observed *in vitro*. *In vivo* experiments showed dose-dependent inhibition against thrombin was observed in mice model. From both compounds, orientin showed potent activity in both models [119].
Oleuropein	*Olea europaea* (Olive)	IC_{50} = 0.41 mM.	The various concentrations ranging from 0.25 to 1.25 mM of oleuropein has shown dose-based inhibition against PAF *in vitro* [120].
Gomisin N and pre-gomisin	*Schisandra chinensis* (Magnolia berry)	IC_{50} values of gomisin N and pre-gomisin as 96.5 and 153.3 μM against AA and 49.3 and 122.4 μM against PAF were obtained respectively.	From the various solvents extracts of *S. chinensis* fruit, methanol and hexane have shown higher inhibitory effects as 65.7 and 94.8% respectively against AA. When compared to all agonists such as PAF, AA, collagen and thrombin, compounds gomisin N and pre-gomisin showed higher effects against AA and PAF [121].
(+)- fenchone and estragole	*Foeniculum vulgare* Gaertner (Fennel fruit)	Concentrations: (+)- fenchone (IC_{50} values 3.9μM and 27.1 μM against collagen and AA) estragole (IC_{50} values 4.7 μM against collagen).	From the *in vitro* study, (+)-fenchone's inhibitory effect against platelet aggregation caused by AA was 1.3 times greater than that of aspirin [122].

Table 1. *Cont.*

Phytocompounds	Scientific Name (Common name)	Optimum Dose Determined or Dosage Investigated	Study Outcomes
Pinocembrine, Alpinetin, Cardamonin, 2′,3′,4′,6′-Tetrahydroxychalcone, 5,6-Dehydrokawain, Flavokawain B (above all from *A. mutica*), Flavokawain A, Crotepoxide, 3-Deacetylcrotepoxide, Zerumbone (above all from *Z. zerumbet*), Xanthorrhizol (from *C. xanthorrhiza*), Curcumin, Xanthorrhizol epoxide, 1-Acetyl-2-methyl-5-(1′,5′-dimethylhex-4′enyl) benzene, 1-Methoxy-2-methyl-5-(1′,5′-dimethylhex-4′enyl) benzene (above all from *C. aromatica*)	*Alpinia mutica* Roxb. (Orchid Ginger) *Kaempferia rotunda* Linn (Blackhorm) *Curcuma xanthorhiza* Roxb (Javanese turmeric) *Curcuma aromatica* Valeton (Turmeric) *Zingiber zerumbet* Smith (Shampoo ginger)	Concentrations: 84 μM against AA and 45.7 μM against AA, collagen, and ADP.	Curcumin, cardamonin, pinocembrine, 5,6-dehydrokawain, and 3-deacetylcrotepoxide significantly inhibited platelet aggregation triggered by the AA with IC_{50} values less than 84 μM. Curcumin was the most efficient antiplatelet agent, inhibiting AA, collagen, and ADP-induced platelet aggregation with IC_{50} values of 37.5, 60.9, and 45.7 μM, respectively [123].
Vitamin C (Ascorbic acid) and total lipids (TL)	*Citrus sinensis* (Sweet orange) *Citrus sinensis* (Blood orange) *Citrus clementina* (Clementine)	IC_{50} values against PAF with various samples are as follows, Fresh juice of Navalina oranges (23.2 μg), sanguine oranges (21.4 μg), clementines (28.6 μg), TL from navalina (14.3 μg), TL from sanguine (15.3 μg), TL from clementines (17.3 μg), TL of navalina peel (1.5 μg), TL of sanguine peel (1.2 μg), TL of clementines (1.7 μg).	*In vitro* antiplatelet activity of vitamin C and TL extract of three different citrus fresh and oxidized fruit juice and peels have shown possible inhibitory effects against PAF and thrombin [124].
Aqueous extract of leaf	*Moringa oleifera* (Drumstick tree)	IC_{50} values against ADP-induced aggregation were 0.48 mg and 0.70 mg respectively.	Aqueous extract of moringa leaf (0.1 to 1mg) showed potent activity against all types of agonists used in this study such as collagen, ADP, and epinephrine. 1 mg of the extract has shown 100% inhibition against epinephrine-induced aggregation [125].
Ethanolic extract of grape pomace rich in phenolics (catechin, epicatechin and quercetin) fatty acids (linoleic acid (C18:2n6), linolenic acid (C18:3n3) and palmitic acid (C16:0))	*Vitis vinifera* (Grape tree)	IC_{50} value against PAF, ADP, and TRAP as 160.7 ± 64.2, 180.8 ± 78.8, and 158.1 ± 93.6 μg, respectively.	From the *in vitro* antiplatelet activity, the ethanolic extract of grape pomace was found to be rich in phenolics and fatty acids such as linoleic, linolenic, and palmitic acid. The IC_{50} values were calculated as 144, 176.5 and 180.5 μg of extract (healthy volunteer) and 214.2, 191.8 and 177.1 μg of extract (cardiovascular patient) against PAF, ADP and TRAP respectively [126].

Table 1. *Cont.*

Phytocompounds	Scientific Name (Common name)	Optimum Dose Determined or Dosage Investigated	Study Outcomes
Olive oil rich in glycerol−glycolipid	*Olea europaea* (Olive)	IC_{50} values of Polar lipid fractions 3 showed 437.5 μL, 4 showed 162.5 μL and 5 showed 375.0 μL against PAF.	From the various olive oil fractions, it was evident that glycerol-glucolipids, phosphatidylcholine, sphingomyelin, phosphatidylinositol, and phosphatidylserine were identified and have potent antiplatelet activity against PAF [127].

Abbreviations: ADP, adenosine diphosphate; AA, Arachidonic acid; COX_2, Cyclooxygenase-2; HUVEC, human umbilical vein endothelial cells; IC_{50}, 50% inhibitory concentration; 6-keto-PGF1α, 6-keto prostaglandin F1α; MUFA, Monounsaturated fatty acids; NL, neutral lipids; OBL, *Ocimum basilicum* L; PAF, Platelet activating factor; PGE_2- Prostaglandin E_2; PL, polar lipids; PUFA, Polyunsaturated fatty acids; TL, Total lipids; TXB_2, Thromboxane B_2; TRAP, Thrombin receptor activator peptide.

4. Antiplatelet Properties of Polar Lipids

Polar lipids are amphipathic in nature, possessing both a hydrophilic head group and a hydrophobic tail. Polar lipids are key structural components of cellular membranes, and they play a role in signaling cascades with membrane proteins [128]. Polar lipids are mostly phospholipids and sphingolipids. In contrast, neutral lipids are non-polar and hydrophobic. Neutral lipids include triacylglycerols, cholesterols, waxes, fatty acids, and esters [129]. Polar lipids have been identified as PAF inhibitors that interact and inhibit the PAF-R through various mechanisms, both direct and indirect, as previously reviewed [7]. In contrast, neutral lipids mostly do not exhibit potent antiplatelet activities [130]. In the following sections we discuss the existing evidence involving *in vitro*, *in vivo*, and *ex vivo* studies that investigate the potential anti-PAF properties of polar lipids.

4.1. In Vitro Studies of Platelet-Activating Factor Receptor (PAF-R) Antagonists

Several *in vitro* studies have been published that reported that polar lipids exhibit antiplatelet properties likely mediated by interactions between the PAF-R. These polar lipids tend to be mostly researched in foods of animal origin, particularly dairy and marine sources. In dairy, it has been reported that the beneficial properties of polar lipids may be altered or enhanced by fermentation of the dairy product. Fermented dairy products, such as yoghurt and cheeses have also been noted for their high inhibitory activity against PAF and other agonists. Many fermented foods that are traditionally part of the Mediterranean diet are rich in omega-3 polyunsaturated fatty acids that support cardiovascular health [131]. Cheeses made from goat's or sheep's milk are an important part of the Greek diet. For example, the traditional Greek cheeses Kefalotyri and Ladotyri have strong inhibitory activity against PAF-induced platelet aggregation [96]. Certain bacterial cultures, such as *Lactobacillus acidophilus* and *Streptococcus thermophilus* can increase the bioactivity of ovine yoghurt milk and alters its anti-thrombotic activity in presence of PAF [132]. These starter cultures are capable of producing and altering bioactive polar lipids by some mechanism, possibly by producing antimicrobial peptides known as bacteriocins which can alter the fatty acid composition. The bacterium *L. acidophilus* has been shown to reduce PAF-induced inflammatory response in human intestinal cells [133]. A similar investigation [134] found that fermentation increases the antithrombotic properties of bovine dairy and plant-based dairy alternative drinks. Homemade dairy alternatives prepared from almond, coconut and rice and bovine dairy milk showed significantly higher antiplatelet activity against PAF, in comparison to their non-fermented counterparts, with the rice-based drink displaying the strongest inhibitory activity.

Other sources of polar lipids include marine sources such as fish and algae [135]. Marine omega-3 PUFA are derived from fish, krill, and roe (fish eggs) and possesses significant antiplatelet activity [136], which may be more bioavailable in polar lipid forms. Polar lipid fractions isolated from codfish (*Gadus morhua*) showed platelet inhibitory capabilities, suggesting that consumption of such lipids could protect against cardiovascular disease [94]. Significant quantities of unused fish by-products by-catch and are generated from the fishing industry, including salmon heads, herring heads and off cuts, and boarfish. While these by-products and by-catch are conventionally regarded as undesirable, valorisation of their antithrombotic and cardioprotective properties could establish these products as important bioactive functional foods [137]. In a 2019 study, polar lipids derived from bycatch and by-products of these fish were assessed for their antiplatelet activity against various platelet agonists, and they exhibited strong inhibitory activities against PAF, thrombin, collagen, and ADP [89]. Another study focusing on salmon [90] demonstrated the potent *in vitro* antithrombotic effects of a food-grade polar lipid extract (FGE) prepared from salmon (*Salmo salar*) fillets in human platelets, in the presence of the platelet agonists PAF and thrombin. Among the lipid subfractions, phosphatidylcholines (PC) and phosphatidylethanolamines (PE) showed the strongest inhibitory capacity against PAF in human platelets. A later investigation found that salmon cooked *sous vide* at higher temperatures (80 °C and above) significantly reduced these antithrombotic properties, along with decreased PUFA content in salmon prepared without brining [138].

Another rich animal source of polar lipids is eggs. Egg yolks are a rich source of sphingomyelin, lysophosphatidylcholine (L-PC), and lyso-phosphatidylethanolamine (L-PE), along with other nutrients including protein, vitamins, and minerals [139,140]. Cage-free, organic, and daily fresh eggs were assessed to determine if their polar lipids exhibited antiplatelet properties. Out of the three varieties, lipid fractions from cage-free eggs showed the highest inhibition against PAF, owing mainly to the polar lipid component of the total lipid fraction [140]. Significant advances in poultry science have led to the natural fortification of eggs to contain higher levels of PUFA. It would be interesting to assess whether PUFA-rich eggs have different polar lipid compositions with even more effective antiplatelet properties considering the other potential cardioprotective effects that have been documented [141].

Overall, it appears that animal sources of polar lipids including dairy, meat, and egg products exhibit antithrombotic effects (Table 2). However, it should be noted that lipids sourced from non-animal sources such as vegetable oils are also known for their cardioprotective and antithrombotic properties, especially olive oil. A 2002 investigation [127] compared the *in vitro* antiplatelet properties of olive oil and other seed oils (sunflower, corn, sesame, and soybean) against PAF. Out of all the polar lipid samples, olive oil was the most bioactive and inhibited both PAF and thrombin in washed rabbit platelets [127]. Indeed, olive oil and related by-products have also been shown to affect PAF metabolism [142].

Table 2. Comparison of in vitro studies investigating dairy and marine lipids possessing antithrombotic activity against PAF and other platelet agonists.

Lipid Source	Study Aim	Result
Fermented Irish ovine yoghurt milk	Comparison of *in vitro* inhibition against PAF-induced aggregation, among different yoghurts and unfermented ovine milk.	Fermentation enhances the antiplatelet nature of ovine milk, due to specific starter cultures, e.g., *Lactobacillus* (demonstrated by decreased IC_{50} values) [132].
Fermented bovine yoghurts and coconut, almond and rice-based dairy alternative drinks	Comparison of *in vitro* inhibition by PL of platelet aggregation.	Fermented plant-based dairy alternatives show much higher antiplatelet activity compared to non-fermented counterparts. The PL from rice-based fermented products shows the highest platelet inhibition of all products, against aggregation induced by PAF and ADP [134].

Table 2. *Cont.*

Lipid Source	Study Aim	Result
Kefalotyri and Ladotyri Greek cheeses	Investigate the *in vitro* inhibition of cheese PL against PAF-induced aggregation.	Lipid fractions of both kinds of cheese inhibit platelet activation, Ladotyri has stronger inhibition [96].
Greek yogurts derived from cow, ewe, and goat milk	Evaluate the *in vitro* anti-thrombotic properties of yogurts in presence of PAF.	TPL and TL of all yogurts showed platelet inhibition, with TPL of goat and ewe yogurt demonstrated highest inhibition against PAF in WRP [143].
Irish organic farmed salmon filet	Investigate the *in vitro* inhibition by salmon PL extract against PAF and thrombin-induced platelet aggregation.	Salmon PL, TNL and TL fractions from PE and PC showed higher inhibitory activity [90].
Fresh and fried cod (*Gadus morhua*)	Test the PAF-like and anti-PAF properties of lipid fractions of fresh and fried cod, against PAF-induced platelet aggregation.	Lipid fractions (TPL and TNL) from fried and fresh cod showed inhibitory activity as well as slight platelet aggregation, indicating presence of both PAF agonists and inhibitors [94].
Hen's egg yolk	Comparison of the antiplatelet activity of TL, TPL and TNL of different types of hen's egg yolk (daily fresh, organic, and cage-free hen's eggs).	All 3 types of hen's egg yolks displayed potent inhibition against PAF-induced aggregation, with cage-free egg yolk having the highest bioactivity of all, in washed rabbit platelets (WRP) [140].
Red and white wines and musts	Assess the biological activity of lipid fraction from wines/must *in vitro*.	All lipid fractions of all samples exhibited inhibition against PAF-induced aggregation in washed rabbit platelets, with TPL of Ambelon (white wine) and Cabernet Sauvignon (red wine) having the most potent antiplatelet activity of all [144].

Abbreviations: ADP, adenosine diphosphate; TNL, total neutral lipids; PAF, platelet-activating factor; PC, phosphatidylcholine; PE, phosphatidylethanolamine; PL, polar lipids; TL, total lipids; TPL, total polar lipids; WRP, washed rabbit platelets.

4.2. *Ex Vivo and Human Studies*

Ex vivo and human studies are important to conduct to gain an understanding of how polar lipids affect platelet and cardiometabolic homeostasis. Certain populations in which the local diet is rich in omega-3 PUFA, such as the Greenland Eskimos [145] and Mediterranean people [146] exhibit a lower rate of cardiovascular diseases. It has been speculated that dietary components such as polar lipids or PUFA may contribute to the observed benefits of these diets. As aforementioned, marine lipid sources, notably polar lipids and potentially PUFA sourced from oily fish species, exhibit antiplatelet activity. A 2019 crossover study involving healthy human volunteers found that intake of enriched marine oil supplements resulted in reduced platelet and leukocyte activation, among other beneficial effects on immune cell functioning [147]. However, similarly to the *in vitro* studies presented, foods and food derivatives other than marine sources exert antithrombotic effects.

A recent investigation found that intake of yoghurt enriched with polar lipids from olive oil by-products resulted in lower platelet sensitivity against PAF and reduced low-grade inflammation, which was assessed by monitoring serum levels of IL-10 and IL-6 [97]. Alcoholic beverages are also known to contain anti-inflammatory and antithrombotic properties against PAF and other platelet agonists [148,149]. A crossover study found that the intake of Cabernet Sauvignon red wine and Robola white wine results in decreased postprandial platelet activity against PAF in human platelet-rich plasma (PRP) [150]. In this study, healthy male volunteers were provided with a standardized meal along with portions of either wine, ethanol solution or water, following which plasma samples were obtained at multiple time points. Platelet sensitivity against PAF was significantly affected following the intake of either red or white wine, compared to samples after intake of water in place of wines. Indeed, a related study investigated the consumption of wine and its effects on PAF metabolism and found that wine beneficially decreases the biosynthesis of PAF [151]. Collectively, these finding contribute to a growing body of literature that indicates there are bioactive constituents including polar lipids in alcoholic beverages such as wine [152] and beer [153]. Results from examples of these *ex vivo* studies are presented in Table 3.

Table 3. Studies investigating the *ex vivo* antiplatelet properties of animal lipids and alcoholic beverages.

Lipid Source	Study Aim	Study Type	Number of Volunteers	Control	Result
Marine oil omega-3 supplement	Establish the relationship between marine oil supplementation and specialized pro-resolving mediators (SPM)	A double-blinded, placebo-controlled crossover	22	Placebo	Platelet aggregates induced by PAF stimulation are reduced after consumption of marine oil supplement [147].
Yoghurt enriched with olive oil pomace polar lipids	To determine the effect of the incorporation of olive oil pomace polar lipids in yoghurt and their effects on platelet function	Randomised double-blinded, placebo-controlled	30	Plain yoghurt	Consumption of yoghurt enriched with olive oil PL resulted in lower platelet sensitivity to PAF [97].
Cabernet sauvignon red wine or Robola white wine	Assess the beneficial effects of wine intake in the postprandial state in human volunteers	Crossover study	10	Water and ethanol	Consumption of red or white wine along with a standardized meal resulted in reduced postprandial PAF-induced platelet aggregation in healthy male volunteers [150].

4.3. PAF Modulation by Micronutrients

Several dietary micronutrients such as vitamins, trace minerals and elements have exhibited anti-inflammatory, antithrombotic [154,155], and antioxidant functions [156] (Table 4). Among those, carotenoids, one of the main sources of vitamin A, are highly bioactive, with antioxidant, anti-inflammatory, and immunoregulatory properties [156,157]. The other form of vitamin A, retinol is known to affect PAF-R expression [158]. Vitamin E has also been linked to the metabolism of PAF and is capable of regulating platelet function [159]. A deficiency of vitamin E (alpha-tocopherol) was shown to stimulate the biosynthesis of PAF in rat polymorphonuclear leukocytes [160]. A study involving pregnant women found that oral supplementation with alpha-tocopherol inhibits platelet aggregation induced by ADP and PAF, using a range of concentrations from 6.55–500 mg/mL [161]. However, yet another *ex vivo* study in male volunteers found that short-term vitamin E supplementation does not significantly affect platelet function or phospholipase A_2 (PLA_2) and lyso-PAF activity [162], enzymes involved in PAF metabolism.

Vitamin D is a fat-soluble vitamin that exists in two major forms, namely cholecalciferol (D_3) and ergocalciferol (D_2). It is typically associated with bone and calcium homeostasis, and the risk of developing diseases such as osteoporosis and rickets [163]. However, vitamin D has diverse physiological functions and is involved in inflammatory and procoagulatory pathways in the body due to its important role in immune function [164]. A randomized study found that vitamin D supplementation can reduce platelet-mediated inflammation and oxidative stress in diabetic patients [165]. Vitamin D can also regulate haemostasis, and its deficiency is associated with increased platelet aggregation in the presence of the agonist ADP [166]. An *in vitro* experiment demonstrated that 25-hydroxyvitamin D, a metabolite of vitamin D, attenuated increased expression of *PTAFR* in a human respiratory epithelial cancer cell line in response to rhinovirus infection [167], indicating that vitamin D might regulate PAF-R expression. It has also been hypothesized that vitamin D may attenuate PAF signalling in other viral infections via the PAF-R such as in SARS-CoV-2 infection and coronavirus disease 2019 (COVID-19) [168]. Indeed, paricalcitol, a vitamin D analogue, is a known PAF-inhibitor as demonstrated *in vitro* and *in vivo* [169].

Vitamin C is a water-soluble vitamin abundantly found in plant sources such as citrus fruits and leafy vegetables. In addition to its well-documented roles in immune function and wound healing, vitamin C possesses antioxidant and antiplatelet functions [170]. In an *ex vivo* study, the addition of vitamin C effectively halted platelet aggregation and scavenged reactive oxygen species (ROS) in human platelets [171]. Another study found that dietary supplementation with vitamin C prevented the accumulation of PAF-LL agonists and cigarette-smoke-induced platelet adhesion and aggregation [172]. This also has important implications for vitamin C supplementation as a dietary intervention to reduce the risk of cardiovascular disease linked to smoking. These findings are in accordance with studies in rabbits that have shown that vitamin C downregulates PAF and PAF-LL and improves postischemic oxidative and inflammatory responses [173].

Table 4. Studies investigating the *in vitro*, *in vivo*, and *ex vivo* antiplatelet properties of micronutrients.

Micronutrient	Study Aim	Study Type	Result
Vitamin C	Effect of vitamin C on the release of PAF and PAF-like phospholipids during reperfusion injury.	*In vivo*	Vitamin C attenuated oxidative stress and reduced PAF and PAF-like lipid levels in rabbits [173].
Vitamin D	Study the effect of vitamin D supplementation in volunteers with Type 2 diabetes in a placebo-controlled trial.	*Ex vivo*	Six months of vitamin D supplementation decreased platelet activation and inflammatory markers such as IL-18, TNF-α and IFN-γ [165].
Vitamin D	Study the inhibitory effect of paricalcitol against PAF and thrombin-induced platelet aggregation	*In vitro*	Addition of paricalcitol effectively inhibited platelet aggregation as well as modulating the activity of metabolic enzymes PAF-CPT and PAF-AH in platelets and leukocytes [169].
Vitamin E	Establish the role of vitamin E (alpha-tocopherol) during pregnancy in platelet function	*In vivo*	Vitamin E supplementation almost completely inhibited platelet aggregation in presence of PAF and ADP, with very high inhibition observed in the brush border membrane vesicles [161].
Selenium (Se)	Investigate the mechanism by which selenium modulates PAF production in endothelial cells	*In vitro*	Selenium deficiency reduces PAF biosynthesis in bovine endothelial cells by downregulating the activity of anabolic enzymes [174].
Zinc (Zn)	Consequences of abnormal Zn storage and release in mouse platelets	*In vivo*	Ionic Zn^{2+} accumulated in secretory granules is released upon platelet activation and has a procoagulant effect [175].
Copper (Cu)	Role of dietary copper in platelet activation using rat models	*In vivo*	Platelet aggregation induced by ADP is significantly higher in copper-deficient rats compared to rats with an adequate amount of copper in their diet [176].

5. Importance of Essential Trace Metals on PAF-R Targets

Dietary trace metals are principal components and regulators of various metabolic processes in the body. These elements form only 5% of the average human diet and are typically required in doses of 1–100 mg daily in adults [177]. Trace elements such as zinc (Zn), and copper (Cu) have been shown to affect platelet function in health and disease, but these elements may also affect the PAF pathways. Deficiencies in the trace element Se have been shown to upregulate PAF production in human [178] and bovine endothelial cells [174], by enhancing the activity of two important enzymes involved in the remodelling pathway of PAF biosynthesis, PLA_2 and lyso-PAF-AT.

Zinc (Zn^{2+}) is a known antioxidant and anti-inflammatory agent [179]. In rat models, zinc deficiency studies have shown a decrease in platelet aggregation and impaired reactivity to agonists, including ADP and thrombin [94,137,138]. Furthermore, recent studies have shown that altered levels of zinc impact platelet reactivity in zinc deficient conditions [180]. Chelation of intracellular zinc can also inhibit the tyrosine phosphorylation cascade, which reduces platelet reactivity and aggregation *in vitro* [181]. In turn, increased dietary zinc increases platelet responses to ADP and thrombin in human plasma [180]. In line with this, zinc supplementation of 50 mg Zn/day demonstrated increased platelet reactivity and serum zinc levels in humans [182]. Zinc supplements have also been shown to decrease oxidative stress and the production of inflammatory cytokines in elderly individuals [179]. The role of zinc in platelet aggregation has, however, not been fully elucidated and some studies also suggest a direct inhibitory role of zinc. It has been suggested that zinc interacts with PAF at the functional receptor site or contiguous site due to its specific inhibition of PAF-induced platelet activation [183]. A further study has shown that zinc levels must be inversely proportional to PAF levels to carry out these inhibitory effects [184]. Additionally, zinc must be present before PAF exposure. This suggests that PAF and receptor binding may be limited by zinc and phospholipid (PAF) interaction [143,144]. This model is supported by zinc's ability to bind to phospholipids in a 2:1–1:1 complex, particularly to the negatively charged phosphate groups [185].

Like zinc, copper is an essential trace metal for the human body. The delicate balance of copper levels in the body is crucial to maintaining terminal oxidation, elimination of free radicals, and iron metabolism [186]. Several studies have shown the effects of altered copper levels on platelet aggregation and thrombin activity. For example, a study using mice subjected to copper deficient diets demonstrated a significant increase in prothrombin time, a parameter used to evaluate blood clotting [187]. This was followed by another study in rats fed a copper-deficient diet (0.3 μg copper/g of diet), which demonstrated impaired platelet adhesion to endothelial cells with an increase in ADP-induced platelet aggregation [176]. However, an *ex vivo* study using blood samples obtained from males found that copper alone, as well as combined with manganese accelerated platelet activation and led to the deformation of erythrocytes [188]. Thus, balanced levels of copper are necessary for healthy platelet activation and aggregation. The relationship between PAF and copper has also been shown to be similar to that of iron in terms of oxidation of lipids and PAF-associated enzymes, whereby the iron-catalysed production of hydroxyl radicals can promptly and conclusively inactivate PAF acetyl hydrolase, which can lead to the prolonged inflammatory effect of PAF. Furthermore, metal-induced oxidative stress and superoxide can activate PAF acetyl hydrolase, increasing PAF levels [149,150]. Trace metals such as copper and iron may indirectly affect PAF signalling through increasing reactive oxygen species and lipid oxidation.

The interplay between trace metals and the PAF/PAF-R pathway has clinical implications. For example, pre-eclampsia is one of many conditions characterized by increased platelet aggregation and superoxide production and has been linked to alterations in trace metal levels, such as a decrease in manganese, copper, and zinc. As such, precautions during pregnancy to ensure balanced levels of essential trace elements are necessary to avoid conditions such as pre-eclampsia [189–194]. Indeed, elevated magnesium (mg) appears to exert protective effects against lesion formation as well as antiarrhythmic and antihypertensive effects [195]. Collectively, these studies show the important of trace metals in PAF biology, but little is known about whether trace metals affect PAF-R expression or function.

6. Conclusions and Future Perspectives

Although pharmaceutical options exist for PAF-R antagonists, they are sparse, and they are not currently utilized against CVD. However, targeting the inhibition of PAF via the PAF-R through dietary means may be a strategy to reduce the risk of atherosclerosis and CVD by reducing the activities of PAF. In this review, we have presented the *in vitro*, *in vivo*, and human studies that have examined the dietary inhibition of PAF. It appears that dietary PAF inhibitors exert their beneficial effects is through their anti-inflammatory and antithrombotic properties. Indeed, many authors have suggested that the longstanding beneficial effects of the Mediterranean diet may be due to the abundance of PAF inhibitors present in the diet. However, there is still a paucity of research investigating polar lipid consumption in humans. Although outside the scope of this review, there is also significant research in animals and humans demonstrating that polar lipids may be cardioprotective via modulating lipid metabolism. Collectively, these advances in research may lead to the development of dietary interventions or nutraceuticals with the aim to deliver dietary PAF inhibitors. However, there are vast gaps in our knowledge regarding the modulation of PAF-R expression directly in health and disease that requires further investigation.

Author Contributions: Conceptualization, R.H., S.H., R.L. and I.Z.; writing—original draft preparation, R.H., S.H., and J.E.S.; writing—review and editing, R.L., R.H., A.M.G. and I.Z.; visualization, R.H., S.H. and R.L.; supervision, I.Z., A.M.G. and R.L.; funding acquisition, I.Z. and R.L. All authors have read and agreed to the published version of the manuscript.

Funding: The financial support from Enterprise Ireland (IP20210972) is acknowledged.

Acknowledgments: The authors would like to acknowledge the support of the Department of Biological Sciences at the University of Limerick, Enterprise Ireland, and the Perelman School of Medicine at the University of Pennsylvania. Furthermore, we acknowledge the Genotype-Tissue Expression (GTEx) Project and their funders. The data used for the analyses presented in Figure 1 were obtained from the GTEx Portal on 15 October 2022 (https://www.gtexportal.org/home/aboutGTEx). Other figures were created using Biorender.

Conflicts of Interest: The authors declare no conflict of interest.

References

1. Barquera, S.; Pedroza-Tobías, A.; Medina, C.; Hernández-Barrera, L.; Bibbins-Domingo, K.; Lozano, R.; Moran, A.E. Global Overview of the Epidemiology of Atherosclerotic Cardiovascular Disease. *Arch. Med. Res.* **2015**, *46*, 328–338. [CrossRef] [PubMed]
2. Yu, E.; Malik, V.S.; Hu, F.B. Cardiovascular disease prevention by diet modification: JACC Health Promotion Series. *J. Am. Coll. Cardiol.* **2018**, *72*, 914–926. [CrossRef] [PubMed]
3. Mozaffarian, D.; Appel, L.J.; Van Horn, L. Components of a cardioprotective diet new insights. *Circulation* **2011**, *123*, 2870–2891. [CrossRef] [PubMed]
4. Libby, P. The changing landscape of atherosclerosis. *Nature* **2021**, *592*, 524–533. [CrossRef]
5. Libby, P.; Ridker, P.M.; Maseri, A. Inflammation and atherosclerosis. *Circulation* **2002**, *105*, 1135–1143. [CrossRef]
6. Jebari-Benslaiman, S.; Galicia-García, U.; Larrea-Sebal, A.; Olaetxea, J.R.; Alloza, I.; Vandenbroeck, K.; Benito-Vicente, A.; Martín, C. Pathophysiology of Atherosclerosis. *Int. J. Mol. Sci.* **2022**, *23*, 3346. [CrossRef]
7. Tsoupras, A.; Lordan, R.; Zabetakis, I. Inflammation, not Cholesterol, Is a Cause of Chronic Disease. *Nutrients* **2018**, *10*, 604. [CrossRef]
8. Davignon, J.; Ganz, P. Role of endothelial dysfunction in atherosclerosis. *Circulation* **2004**, *109*, III-27–III-32. [CrossRef]
9. Nording, H.M.; Seizer, P.; Langer, H.F. Platelets in inflammation and atherogenesis. *Front. Immunol.* **2015**, *6*. [CrossRef]
10. Lindemann, S.; KrÄMer, B.; Seizer, P.; Gawaz, M. Platelets, inflammation and atherosclerosis. *J. Thromb. Haemost.* **2007**, *5*, 203–211. [CrossRef]
11. Massberg, S.; Brand, K.; Grüner, S.; Page, S.; Müller, E.; Müller, I.; Bergmeier, W.; Richter, T.; Lorenz, M.; Konrad, I.; et al. A critical role of platelet adhesion in the initiation of atherosclerotic lesion formation. *J. Exp. Med.* **2002**, *196*, 887–896. [CrossRef]
12. Huo, Y.; Schober, A.; Forlow, S.B.; Smith, D.F.; Hyman, M.C.; Jung, S.; Littman, D.R.; Weber, C.; Ley, K. Circulating activated platelets exacerbate atherosclerosis in mice deficient in apolipoprotein E. *Nat. Med.* **2002**, *9*, 61. [CrossRef]
13. Lordan, R.; Tsoupras, A.; Zabetakis, I. Platelet activation and prothrombotic mediators at the nexus of inflammation and atherosclerosis: Potential role of antiplatelet agents. *Blood Rev.* **2020**, *45*, 100694. [CrossRef]
14. Huilcaman, R.; Venturini, W.; Fuenzalida, L.; Cayo, A.; Segovia, R.; Valenzuela, C.; Brown, N.; Moore-Carrasco, R. Platelets, a Key Cell in Inflammation and Atherosclerosis Progression. *Cells* **2022**, *11*, 1014. [CrossRef]
15. Tomaiuolo, M.; Brass, L.F.; Stalker, T.J. Regulation of Platelet Activation and Coagulation and Its Role in Vascular Injury and Arterial Thrombosis. *Interv. Cardiol. Clin.* **2017**, *6*, 1–12. [CrossRef]
16. Lordan, R.; Tsoupras, A.; Zabetakis, I. Investigation of Platelet Aggregation in Atherosclerosis. In *Atherosclerosis: Methods and Protocols*; Ramji, D., Ed.; Springer: New York, NY, USA, 2022; pp. 333–347. [CrossRef]
17. Lordan, R.; Tsoupras, A.; Zabetakis, I. The Potential Role of Dietary Platelet-Activating Factor Inhibitors in Cancer Prevention and Treatment. *Adv. Nutr.* **2019**, *10*, 148–164. [CrossRef]

18. Papakonstantinou, V.D.; Lagopati, N.; Tsilibary, E.C.; Demopoulos, C.A.; Philippopoulos, A.I. A Review on Platelet Activating Factor Inhibitors: Could a New Class of Potent Metal-Based Anti-Inflammatory Drugs Induce Anticancer Properties? *Bioinorg. Chem. Appl.* **2017**, *2017*, 6947034. [CrossRef]
19. Nomikos, T.; Fragopoulou, E.; Antonopoulou, S.; Panagiotakos, D.B. Mediterranean diet and platelet-activating factor; a systematic review. *Clin. Biochem.* **2018**, *60*, 1–10. [CrossRef]
20. English, C.J.; Mayr, H.L.; Lohning, A.E.; Reidlinger, D.P. The association between dietary patterns and the novel inflammatory markers platelet-activating factor and lipoprotein-associated phospholipase A2: A systematic review. *Nutr. Rev.* **2022**, *80*, 1371–1391. [CrossRef]
21. Benveniste, J.; Henson, P.M.; Cochrane, C.G. Leukocyte-dependent histamine release from rabbit platelets: The role of IgE, basophils, and a platelet-activating factor. *J. Exp. Med.* **1972**, *136*, 1356–1377. [CrossRef]
22. Demopoulos, C.A.; Antonopoulou, S. A discovery trip to compounds with PAF-like activity. In *Platelet-Activating Factor and Related Lipid Mediators 2*; Springer: Berlin/Heidelberg, Germany, 1996; pp. 59–63.
23. Demopoulos, C.; Pinckard, R.; Hanahan, D.J. Platelet-activating factor. Evidence for 1-*O*-alkyl-2-acetyl-*sn*-glyceryl-3-phosphorylcholine as the active component (a new class of lipid chemical mediators). *J. Biol. Chem.* **1979**, *254*, 9355–9358. [CrossRef]
24. Upton, J.E.M.; Grunebaum, E.; Sussman, G.; Vadas, P. Platelet Activating Factor (PAF): A Mediator of Inflammation. *Biofactors* **2022**. [CrossRef]
25. Antonopoulou, S.; Nomikos, T.; Karantonis, H.; Fragopoulou, E.; Demopoulos, C.A. PAF, a potent lipid mediator. In *Bioactive Phospholipids: Role in Inflammation and Atherosclerosis*; Tselepis, A.D., Ed.; Transworld Research Network: Kerala, India, 2008; pp. 85–134.
26. Zimmerman, G.A.; McIntyre, T.M.; Prescott, S.M.; Stafforini, D.M. The platelet-activating factor signaling system and its regulators in syndromes of inflammation and thrombosis. *Crit. Care Med.* **2002**, *30*, S294–S301. [CrossRef]
27. Prescott, S.M.; Zimmerman, G.A.; Stafforini, D.M.; McIntyre, T.M. Platelet-activating factor and related lipid mediators. *Annu. Rev. Biochem.* **2000**, *69*, 419–445. [CrossRef]
28. Lordan, R.; Tsoupras, A.; Zabetakis, I.; Demopoulos, A.C. Forty years since the structural elucidation of platelet-activating factor (PAF): Historical, current, and future research perspectives. *Molecules* **2019**, *24*, 4414. [CrossRef]
29. Senanayake, V.; Goodenowe, D.B. Plasmalogen deficiency and neuropathology in Alzheimer's disease: Causation or coincidence? *Alzheimer's Dement. Transl. Res. Clin. Interv.* **2019**, *5*, 524–532. [CrossRef]
30. GTExPortal. Bulk Tissue Gene Expression for PTAFR (ENSG00000169403.11). Available online: https://www.gtexportal.org/home/gene/PTAFR (accessed on 15 October 2022).
31. Farooqui, F.A.F.T.; Horrocks, L.A. Roles of Platelet-Activating Factor in Brain. In *Metabolism and Functions of Bioactive Ether Lipids in the Brain*; Springer: New York, NY, USA, 2008; pp. 171–195. [CrossRef]
32. Brailoiu, E.; Barlow, C.L.; Ramirez, S.H.; Abood, M.E.; Brailoiu, G.C. Effects of Platelet-Activating Factor on Brain Microvascular Endothelial Cells. *Neuroscience* **2018**, *377*, 105–113. [CrossRef]
33. Angle, M.J.; Tom, R.; Jarvi, K.; McClure, R.D. Effect of platelet-activating factor (PAF) on human spermatozoa–oocyte interactions. *Reproduction* **1993**, *98*, 541–548. [CrossRef]
34. Minhas, B.S.; Kumar, R.; Ricker, D.D.; Roudebush, W.E.; Dodson, M.G.; Fortunato, S.J. Effects of platelet activating factor on mouse oocyte fertilization in vitro. *Am. J. Obstet. Gynecol.* **1989**, *161*, 1714–1717. [CrossRef]
35. Sakellariou, M.; Drakakis, P.; Antonopoulou, S.; Anagnostou, E.; Loutradis, D.; Patargias, T. Intravenous infusion of PAF affects ovulation, fertilization and preimplantation embryonic development in NZB x NZW F1 hybrid mice. *Prostaglandins Other Lipid Mediat.* **2008**, *85*, 125–133. [CrossRef]
36. Lecewicz, M.; Kordan, W.; Majewska, A.; Kamiński, S.; Dziekońska, A.; Mietelska, K. Effects of the platelet-activating factor (PAF) on selected quality parameters of cryopreserved bull semen (AI) with reduced sperm motility. *Pol. J. Vet. Sci.* **2016**, *19*, 147–158. [CrossRef] [PubMed]
37. Tiemann, U. The Role of Platelet-activating Factor in the Mammalian Female Reproductive Tract. *Reprod. Domest. Anim.* **2008**, *43*, 647–655. [CrossRef] [PubMed]
38. Wu, H.; Gao, J.; Wang, X.; Leung, T.Y.; Duan, Y.-G.; Chiu, P.C.N. Platelet-activating factor induces acrosome reaction via the activation of extracellular signal-regulated kinase in human spermatozoa. *Andrologia* **2020**, *52*, e13565. [CrossRef] [PubMed]
39. Evangelou, A.M. Platelet-activating factor (PAF): Implications for coronary heart and vascular diseases. *Prostaglandins Leukot. Essent. Fat. Acids* **1994**, *50*, 1–28. [CrossRef]
40. Montrucchio, G.; Alloatti, G.; Camussi, G. Role of Platelet-Activating Factor in Cardiovascular Pathophysiology. *Physiol. Rev.* **2000**, *80*, 1669–1699. [CrossRef]
41. Tselepis, A.D.; Evangelou, A.; Tsoukatos, D.; Demopoulos, C.A.; Kapoulas, V.M. Electrocardiographic alterations induced by AGEPC in Wistar rats in relation to its hypotensive and hematologic effects. *Comp. Biochem. Physiol. Part C Comp. Pharmacol.* **1987**, *87*, 41–46. [CrossRef]
42. Chignard, M.; Le Couedic, J.; Tence, M.; Vargaftig, B.; Benveniste, J. The role of platelet-activating factor in platelet aggregation. *Nature* **1979**, *279*, 799–800. [CrossRef]
43. Chesney, C.; Pifer, D.; Byers, L.; Muirhead, E. Effect of platelet-activating factor (PAF) on human platelets. *Blood* **1982**, *59*, 582–585. [CrossRef]

44. Melnikova, V.; Bar-Eli, M. Inflammation and melanoma growth and metastasis: The role of platelet-activating factor (PAF) and its receptor. *Cancer Metastasis Rev.* **2007**, *26*, 359–371. [CrossRef]
45. Fan, G.J.; Kang, Y.H.; Han, Y.N.; Han, B.H. Platelet-activating factor (PAF) receptor binding antagonists from Alpinia officinarum. *Bioorg. Med. Chem. Lett.* **2007**, *17*, 6720–6722. [CrossRef]
46. Souza Danielle, G.; Fagundes Caio, T.; Sousa Lirlandia, P.; Amaral Flavio, A.; Souza Rafael, S.; Souza Adriano, L.; Kroon Erna, G.; Sachs, D.; Cunha Fernando, Q.; Bukin, E.; et al. Essential role of platelet-activating factor receptor in the pathogenesis of Dengue virus infection. *Proc. Natl. Acad. Sci. USA* **2009**, *106*, 14138–14143. [CrossRef]
47. Kelesidis, T.; Papakonstantinou, V.; Detopoulou, P.; Fragopoulou, E.; Chini, M.; Lazanas, M.C.; Antonopoulou, S. The role of platelet-activating factor in chronic inflammation, immune activation, and comorbidities associated with HIV infection. *AIDS Rev.* **2015**, *17*, 191.
48. Theoharides, T.C.; Antonopoulou, S.; Demopoulos, C.A. Coronavirus 2019, Microthromboses, and Platelet Activating Factor. *Clin. Ther.* **2020**, *42*, 1850–1852. [CrossRef]
49. Tsoupras, A.; Lordan, R.; Zabetakis, I. Thrombosis and COVID-19: The Potential role of nutrition. *Front. Nutr.* **2020**, *7*, 583080. [CrossRef]
50. Honda, Z.I.; Ishii, S.; Shimizu, T. Platelet-Activating Factor Receptor. *J. Biochem.* **2002**, *131*, 773–779. [CrossRef]
51. Stafforini, D.M.; McIntyre, T.M.; Zimmerman, G.A.; Prescott, S.M. Platelet-Activating Factor, a Pleiotrophic Mediator of Physiological and Pathological Processes. *Crit. Rev. Clin. Lab. Sci.* **2003**, *40*, 643–672. [CrossRef]
52. Shimizu, T.; Mutoh, H.; Kato, S. Platelet-Activating Factor Receptor. In *Platelet-Activating Factor and Related Lipid Mediators 2: Roles in Health and Disease*; Nigam, S., Kunkel, G., Prescott, S.M., Eds.; Springer: Boston, MA, USA, 1996; pp. 79–84. [CrossRef]
53. Mutoh, H.; Ishii, S.; Izumi, T.; Kato, S.; Shimizu, T. Platelet-Activating Factor (PAF) Positively Auto-Regulates the Expression of Human PAF Receptor Transcript 1 (Leukocyte-Type) Through NF-κB. *Biochem. Biophys. Res. Commun.* **1994**, *205*, 1137–1142. [CrossRef]
54. Chao, W.; Olson, M.S. Platelet-activating factor: Receptors and signal transduction. *Biochem. J.* **1993**, *292*, 617–629. [CrossRef]
55. Shukla, S.D. Platelet-activating factor receptor and signal transduction mechanisms 1. *FASEB J.* **1992**, *6*, 2296–2301. [CrossRef]
56. Ishii, S.; Shimizu, T. Platelet-activating factor (PAF) receptor and genetically engineered PAF receptor mutant mice. *Prog. Lipid Res.* **2000**, *39*, 41–82. [CrossRef]
57. Pettersen, E.F.; Goddard, T.D.; Huang, C.C.; Couch, G.S.; Greenblatt, D.M.; Meng, E.C.; Ferrin, T.E. UCSF Chimera?A visualization system for exploratory research and analysis. *J. Comput. Chem.* **2004**, *25*, 1605–1612. [CrossRef]
58. Yost, C.C.; Weyrich, A.S.; Zimmerman, G.A. The platelet activating factor (PAF) signaling cascade in systemic inflammatory responses. *Biochimie* **2010**, *92*, 692–697. [CrossRef]
59. Chung, K.F. LIPID MEDIATORS | Platelet-activating factors. In *Encyclopedia of Respiratory Medicine*; Laurent, G.J., Shapiro, S.D., Eds.; Academic Press: Oxford, UK, 2006; pp. 589–594. [CrossRef]
60. Lordan, R.; Vidal, N.P.; Huong Pham, T.; Tsoupras, A.; Thomas, R.H.; Zabetakis, I. Yoghurt fermentation alters the composition and antiplatelet properties of milk polar lipids. *Food Chem.* **2020**, *332*, 127384. [CrossRef]
61. Kajiwara, N.; Sasaki, T.; Bradding, P.; Cruse, G.; Sagara, H.; Ohmori, K.; Saito, H.; Ra, C.; Okayama, Y. Activation of human mast cells through the platelet-activating factor receptor. *J. Allergy Clin. Immunol.* **2010**, *125*, 1137–1145.e6. [CrossRef]
62. Petersen, L.J.; Church, M.K.; Skov, P.S. Platelet-activating factor induces histamine release from human skin mast cells in vivo, which is reduced by local nerve blockade. *J. Allergy Clin. Immunol.* **1997**, *99*, 640–647. [CrossRef]
63. Nilsson, G.; Metcalfe, D.D.; Taub, D.D. Demonstration that platelet-activating factor is capable of activating mast cells and inducing a chemotactic response. *Immunology* **2000**, *99*, 314–319. [CrossRef]
64. Nakagome, K.; Nagata, M. Involvement and Possible Role of Eosinophils in Asthma Exacerbation. *Front. Immunol.* **2018**, *9*. [CrossRef]
65. Kato, M.; Kita, H.; Tachibana, A.; Hayashi, Y.; Tsuchida, Y.; Kimura, H. Dual signaling and effector pathways mediate human eosinophil activation by platelet-activating factor. *Int. Arch. Allergy Immunol.* **2004**, *134*, 37–43. [CrossRef]
66. Pałgan, K.; Bartuzi, Z. Platelet activating factor in allergies. *Int. J. Immunopathol. Pharmacol.* **2015**, *28*, 584–589. [CrossRef]
67. Muñoz-Cano, R.M.; Casas-Saucedo, R.; Valero Santiago, A.; Bobolea, I.; Ribó, P.; Mullol, J. Platelet-Activating Factor (PAF) in Allergic Rhinitis: Clinical and Therapeutic Implications. *J. Clin. Med.* **2019**, *8*, 1338. [CrossRef]
68. Nakagome, K.; Matsushita, S.; Nagata, M. Neutrophilic inflammation in severe asthma. *Int. Arch. Allergy Immunol.* **2012**, *158*, 96–102. [CrossRef] [PubMed]
69. Bélanger, C.; Elimam, H.; Lefebvre, J.; Borgeat, P.; Marleau, S. Involvement of endogenous leukotriene B4 and platelet-activating factor in polymorphonuclear leucocyte recruitment to dermal inflammatory sites in rats. *Immunology* **2008**, *124*, 295–303. [CrossRef] [PubMed]
70. Lotner, G.; Lynch, J.; Betz, S.; Henson, P. Human neutrophil-derived platelet activating factor. *J. Immunol.* **1980**, *124*, 676–684. [PubMed]
71. Demopoulos, C.A.; Karantonis, H.C.; Antonopoulou, S. Platelet-activating factor—A molecular link between atherosclerosis theories. *Eur. J. Lipid Sci. Technol.* **2003**, *105*, 705–716. [CrossRef]
72. Taggart, D.P. Neuroprotection during cardiac surgery: A randomised trial of a platelet activating factor antagonist. *Heart* **2003**, *89*, 897–900. [CrossRef]

73. Kuitert, L.M.; Angus, R.M.; Barnes, N.C.; Barnes, P.J.; Bone, M.F.; Chung, K.F.; Fairfax, A.J.; Higenbotham, T.W.; O'Connor, B.J.; Piotrowska, B. Effect of a novel potent platelet-activating factor antagonist, modipafant, in clinical asthma. *Am. J. Respir. Crit. Care Med.* **1995**, *151*, 1331–1335. [CrossRef]
74. Mullol, J.; Bousquet, J.; Bachert, C.; Canonica, W.G.; Gimenez-Arnau, A.; Kowalski, M.L.; Martí-Guadaño, E.; Maurer, M.; Picado, C.; Scadding, G.; et al. Rupatadine in allergic rhinitis and chronic urticaria. *Allergy* **2008**, *63*, 5–28. [CrossRef]
75. Mullol, J.; Bousquet, J.; Bachert, C.; Canonica, G.W.; Giménez-Arnau, A.; Kowalski, M.L.; Simons, F.E.R.; Maurer, M.; Ryan, D.; Scadding, G. Update on rupatadine in the management of allergic disorders. *Allergy* **2015**, *70*, 1–24. [CrossRef]
76. Deng, M.; Guo, H.; Tam, J.W.; Johnson, B.M.; Brickey, W.J.; New, J.S.; Lenox, A.; Shi, H.; Golenbock, D.T.; Koller, B.H.; et al. Platelet-activating factor (PAF) mediates NLRP3-NEK7 inflammasome induction independently of PAFR. *J. Exp. Med.* **2019**, *216*, 2838–2853. [CrossRef]
77. Siervo, M.; Lara, J.; Chowdhury, S.; Ashor, A.; Oggioni, C.; Mathers, J.C. Effects of the Dietary Approach to Stop Hypertension (DASH) diet on cardiovascular risk factors: A systematic review and meta-analysis. *Br. J. Nutr.* **2015**, *113*, 1–15. [CrossRef]
78. Widmer, R.J.; Flammer, A.J.; Lerman, L.O.; Lerman, A. The Mediterranean Diet, its Components, and Cardiovascular Disease. *Am. J. Med.* **2015**, *128*, 229–238. [CrossRef]
79. Chopra, A.S.; Lordan, R.; Horbańczuk, O.K.; Atanasov, A.G.; Chopra, I.; Horbańczuk, J.O.; Jóźwik, A.; Huang, L.; Pirgozliev, V.; Banach, M.; et al. The current use and evolving landscape of nutraceuticals. *Pharmacol. Res.* **2022**, *175*, 106001. [CrossRef]
80. Rabassa, M.; Hernandez Ponce, Y.; Garcia-Ribera, S.; Johnston, B.C.; Salvador Castell, G.; Manera, M.; Perez Rodrigo, C.; Aranceta-Bartrina, J.; Martínez-González, M.Á.; Alonso-Coello, P. Food-based dietary guidelines in Spain: An assessment of their methodological quality. *Eur. J. Clin. Nutr.* **2022**, *76*, 350–359. [CrossRef]
81. Hassan, M.; Haq, S.M.; Majeed, M.; Umair, M.; Sahito, H.A.; Shirani, M.; Waheed, M.; Aziz, R.; Ahmad, R.; Bussmann, R.W. Traditional Food and Medicine: Ethno-Traditional Usage of Fish Fauna across the Valley of Kashmir: A Western Himalayan Region. *Diversity* **2022**, *14*, 455. [CrossRef]
82. Ernst, E. The role of complementary and alternative medicine. *BMJ* **2000**, *321*, 1133. [CrossRef]
83. Che, C.-T.; George, V.; Ijinu, T.; Pushpangadan, P.; Andrae-Marobela, K. Traditional medicine. In *Pharmacognosy*; Elsevier: Amsterdam, The Netherlands, 2017; pp. 15–30.
84. Rajendran, H.; Deepika, S.; Immanuel, S.C. An Overview of Medicinal plants for Potential Cardio-Protective Activity. *Res. J. Biotechnol.* **2017**, *4*, 104–113.
85. Kumar, S.; Joseph, L.; George, M.; Sharma, A. A review on anticoagulant/antithrombotic activity of natural plants used in traditional medicine. *Int. J. Pharm. Sci. Rev. Res.* **2011**, *8*, 70–74.
86. Yoo, H.; Ku, S.-K.; Lee, W.; Kwak, S.; Baek, Y.-D.; Min, B.-W.; Jeong, G.-S.; Bae, J.-S. Antiplatelet, anticoagulant, and profibrinolytic activities of cudratricusxanthone A. *Arch. Pharmacal Res.* **2014**, *37*, 1069–1078. [CrossRef]
87. Ku, S.-K.; Bae, J.-S. Antiplatelet, anticoagulant, and profibrinolytic activities of withaferin A. *Vasc. Pharmacol.* **2014**, *60*, 120–126. [CrossRef]
88. Siritapetawee, J.; Thumanu, K.; Sojikul, P.; Thammasirirak, S. A novel serine protease with human fibrino(geno)lytic activities from Artocarpus heterophyllus latex. *Biochim. Biophys. Acta (BBA)-Proteins Proteom.* **2012**, *1824*, 907–912. [CrossRef]
89. Tsoupras, A.; O'Keeffe, E.; Lordan, R.; Redfern, S.; Zabetakis, I. Bioprospecting for Antithrombotic Polar Lipids from Salmon, Herring, and Boarfish By-Products. *Foods* **2019**, *8*, 416. [CrossRef]
90. Tsoupras, A.; Lordan, R.; Shiels, K.; Saha, S.K.; Nasopoulou, C.; Zabetakis, I. In Vitro Antithrombotic Properties of Salmon (Salmo salar) Phospholipids in a Novel Food-Grade Extract. *Mar. Drugs* **2019**, *17*, 62. [CrossRef]
91. Gavriil, L.; Detopoulou, M.; Petsini, F.; Antonopoulou, S.; Fragopoulou, E. Consumption of plant extract supplement reduces platelet activating factor-induced platelet aggregation and increases platelet activating factor catabolism: A randomised, double-blind and placebo-controlled trial. *Br. J. Nutr.* **2019**, *121*, 982–991. [CrossRef]
92. Berge, J.P.; Debiton, E.; Dumay, J.; Durand, P.; Barthomeuf, C. In vitro anti-inflammatory and anti-proliferative activity of sulfolipids from the red alga Porphyridium cruentum. *J. Agric Food Chem.* **2002**, *50*, 6227–6232. [CrossRef]
93. Shiels, K.; Tsoupras, A.; Lordan, R.; Zabetakis, I.; Murray, P.; Kumar Saha, S. Anti-inflammatory and antithrombotic properties of polar lipid extracts, rich in unsaturated fatty acids, from the Irish marine cyanobacterium Spirulina subsalsa. *J. Funct. Foods* **2022**, *94*, 105124. [CrossRef]
94. Panayiotou, A.; Samartzis, D.; Nomikos, T.; Fragopoulou, E.; Karantonis, H.C.; Demopoulos, C.A.; Zabetakis, I. Lipid Fractions with Aggregatory and Antiaggregatory Activity toward Platelets in Fresh and Fried Cod (*Gadus morhua*): Correlation with Platelet-Activating Factor and Atherogenesis. *J. Agric. Food Chem.* **2000**, *48*, 6372–6379. [CrossRef]
95. Lordan, R.; Walsh, A.; Crispie, F.; Finnegan, L.; Demuru, M.; Tsoupras, A.; Cotter, P.D.; Zabetakis, I. Caprine milk fermentation enhances the antithrombotic properties of cheese polar lipids. *J. Funct. Foods* **2019**, *61*, 103507. [CrossRef]
96. Tsorotioti, S.E.; Nasopoulou, C.; Detopoulou, M.; Sioriki, E.; Demopoulos, C.A.; Zabetakis, I. In vitro anti-atherogenic properties of traditional Greek cheese lipid fractions. *Dairy Sci. Technol.* **2014**, *94*, 269–281. [CrossRef]
97. Antonopoulou, S.; Detopoulou, M.; Fragopoulou, E.; Nomikos, T.; Mikellidi, A.; Yannakoulia, M.; Kyriacou, A.; Mitsou, E.; Panagiotakos, D.; Anastasiou, C. Consumption of yogurt enriched with polar lipids from olive oil by-products reduces platelet sensitivity against platelet activating factor and inflammatory indices: A randomized, double-blind clinical trial. *Hum. Nutr. Metab.* **2022**, *28*, 200145. [CrossRef]

98. Shah, B.H.; Nawaz, Z.; Pertani, S.A.; Roomi, A.; Mahmood, H.; Saeed, S.A.; Gilani, A.H. Inhibitory effect of curcumin, a food spice from turmeric, on platelet-activating factor- and arachidonic acid-mediated platelet aggregation through inhibition of thromboxane formation and Ca^{2+} signaling. *Biochem. Pharmacol.* **1999**, *58*, 1167–1172. [CrossRef]
99. Jantan, I.; Rafi, I.A.; Jalil, J. Platelet-activating factor (PAF) receptor-binding antagonist activity of Malaysian medicinal plants. *Phytomedicine* **2005**, *12*, 88–92. [CrossRef] [PubMed]
100. Sugatani, J.; Fukazawa, N.; Ujihara, K.; Yoshinari, K.; Abe, I.; Noguchi, H.; Miwa, M. Tea polyphenols inhibit acetyl-CoA:1-alkyl-sn-glycero-3-phosphocholine acetyltransferase (a key enzyme in platelet-activating factor biosynthesis) and platelet-activating factor-induced platelet aggregation. *Int. Arch. Allergy Immunol.* **2004**, *134*, 17–28. [CrossRef] [PubMed]
101. Tsoupras, A.; Lordan, R.; Harrington, J.; Pienaar, R.; Devaney, K.; Heaney, S.; Koidis, A.; Zabetakis, I. The Effects of Oxidation on the Antithrombotic Properties of Tea Lipids Against PAF, Thrombin, Collagen, and ADP. *Foods* **2020**, *9*, 385. [CrossRef] [PubMed]
102. Vasange, M.; Rolfsen, W.; Bohlin, L. A sulphonoglycolipid from the fern Polypodium decumanum and its effect on the platelet activating-factor receptor in human neutrophils. *J. Pharm Pharm.* **1997**, *49*, 562–566. [CrossRef]
103. Olszanecki, R.; Jawień, J.; Gajda, M.; Mateuszuk, L.; Gebska, A.; Korabiowska, M.; Chłopicki, S.; Korbut, R. Effect of curcumin on atherosclerosis in apoE/LDLR-double knockout mice. *J. Physiol. Pharmacol. Off. J. Pol. Physiol. Soc.* **2005**, *56*, 627–635.
104. Lee, H. Antiplatelet property of *Curcuma longa* L. rhizome-derived ar-turmerone. *Bioresour. Technol.* **2006**, *97*, 1372–1376. [CrossRef]
105. Maheswaraiah, A.; Jaganmohan Rao, L.; Naidu, K.A. Anti-platelet activity of water dispersible curcuminoids in rat platelets. *Phytother. Res.* **2015**, *29*, 450–458. [CrossRef]
106. Cavagnaro, P.F.; Camargo, A.; Galmarini, C.R.; Simon, P.W. Effect of Cooking on Garlic (*Allium sativum* L.) Antiplatelet Activity and Thiosulfinates Content. *J. Agric. Food Chem.* **2007**, *55*, 1280–1288. [CrossRef]
107. Cavagnaro, P.F.; Galmarini, C.R. Effect of Processing and Cooking Conditions on Onion (*Allium cepa* L.) Induced Antiplatelet Activity and Thiosulfinate Content. *J. Agric. Food Chem.* **2012**, *60*, 8731–8737. [CrossRef]
108. Kim, S.Y.; Koo, Y.K.; Koo, J.Y.; Ngoc, T.M.; Kang, S.S.; Bae, K.; Kim, Y.S.; Yun-Choi, H.S. Platelet Anti-Aggregation Activities of Compounds from Cinnamomum cassia. *J. Med. Food* **2010**, *13*, 1069–1074. [CrossRef]
109. Ahmed, S.; Gul, S.; Gul, H.; Zia-Ul-Haq, M.; Iram, S.; Jaafar, H.; Moga, M. Scientific basis for the use of Cinnamonum tamala in cardiovascular and inflammatory diseases. *Exp. Clin. Card* **2014**, *20*, 784–800.
110. Li, L.-Z.; Gao, P.-Y.; Song, S.-J.; Yuan, Y.-Q.; Liu, C.-T.; Huang, X.-X.; Liu, Q.-B. Monoterpenes and flavones from the leaves of Crataegus pinnatifida with anticoagulant activities. *J. Funct. Foods* **2015**, *12*, 237–245. [CrossRef]
111. Umar, A.; Zhou, W.; Abdusalam, E.; Tursun, A.; Reyim, N.; Tohti, I.; Moore, N. Effect of *Ocimum basilicum* L. on cyclo-oxygenase isoforms and prostaglandins involved in thrombosis. *J. Ethnopharmacol.* **2014**, *152*, 151–155. [CrossRef]
112. Zaman, R.; Parvez, M.; Jakaria, M.; Sayeed, M.A.; Islam, M. In vitro clot lysis activity of different extracts of mangifera sylvatica roxb. Leaves. *Res. J. Med. Plant* **2015**, *9*, 135–140.
113. Alañón, M.E.; Palomo, I.; Rodríguez, L.; Fuentes, E.; Arráez-Román, D.; Segura-Carretero, A. Antiplatelet Activity of Natural Bioactive Extracts from Mango (*Mangifera Indica* L.) and its By-Products. *Antioxidants* **2019**, *8*, 517. [CrossRef]
114. Musfiroh, F.; Setiasih, S.; Handayani, S.; Hudiyono, S.; Ilyas, N. In Vivo antiplatelet activity aggregation assay of bromelain fractionate by ethanol from extract pineapple core (Ananas comosus [l.] merr.). In *IOP Conference Series: Materials Science and Engineering*; IOP Publishing: Bristol, England, 2018; p. 012017.
115. Silva-Luis, C.C.; de Brito Alves, J.L.; de Oliveira, J.C.P.L.; de Sousa Luis, J.A.; Araújo, I.G.A.; Tavares, J.F.; do Nascimento, Y.M.; Bezerra, L.S.; Araújo de Azevedo, F.d.L.A.; Sobral, M.V.; et al. Effects of Baru Almond Oil (*Dipteryx alata* Vog.) Treatment on Thrombotic Processes, Platelet Aggregation, and Vascular Function in Aorta Arteries. *Nutrients* **2022**, *14*, 2098. [CrossRef]
116. Alarcón, M.; Fuentes, E.; Olate, N.; Navarrete, S.; Carrasco, G.; Palomo, I. Strawberry extract presents antiplatelet activity by inhibition of inflammatory mediator of atherosclerosis (sP-selectin, sCD40L, RANTES, and IL-1β) and thrombus formation. *Platelets* **2015**, *26*, 224–229. [CrossRef]
117. Santhakumar, A.B.; Stanley, R.; Singh, I. The ex vivo antiplatelet activation potential of fruit phenolic metabolite hippuric acid. *Food Funct.* **2015**, *6*, 2679–2683. [CrossRef]
118. Park, B.S.; Son, D.J.; Park, Y.H.; Kim, T.W.; Lee, S.E. Antiplatelet effects of acidamides isolated from the fruits of *Piper longum* L. *Phytomedicine* **2007**, *14*, 853–855. [CrossRef]
119. Lee, W.; Bae, J.-S. Antithrombotic and antiplatelet activities of orientin in vitro and in vivo. *J. Funct. Foods* **2015**, *17*, 388–398. [CrossRef]
120. Andrikopoulos, N.K.; Antonopoulou, S.; Kaliora, A.C. Oleuropein Inhibits LDL Oxidation Induced by Cooking Oil Frying By-products and Platelet Aggregation Induced by Platelet-Activating Factor. *LWT-Food Sci. Technol.* **2002**, *35*, 479–484. [CrossRef]
121. Kim, M.-G.; Lee, C.-H.; Lee, H.-S. Anti-platelet aggregation activity of lignans isolated from Schisandra chinensis fruits. *J. Korean Soc. Appl. Biol. Chem.* **2010**, *53*, 740–745. [CrossRef]
122. Lee, H.-S. Anticoagulant properties of compounds derived from Fennel (*Foeniculum vulgare* Gaertner) fruits. *Food Sci. Biotechnol.* **2006**, *15*, 763–767.
123. Jantan, I.; Raweh, S.M.; Sirat, H.M.; Jamil, S.; Mohd Yasin, Y.H.; Jalil, J.; Jamal, J.A. Inhibitory effect of compounds from Zingiberaceae species on human platelet aggregation. *Phytomedicine* **2008**, *15*, 306–309. [CrossRef]
124. Tsoupras, A. The Anti-Inflammatory and Antithrombotic Properties of Bioactives from Orange, Sanguine and Clementine Juices and from Their Remaining By-Products. *Beverages* **2022**, *8*, 39. [CrossRef]

125. Arabshahi-Delouee, S.; Aalami, M.; Urooj, A.; Krishnakantha, T.P. *Moringa oleifera* leaves as an inhibitor of human platelet aggregation. *Pharm. Biol.* **2009**, *47*, 734–739. [CrossRef]
126. Choleva, M.; Boulougouri, V.; Panara, A.; Panagopoulou, E.; Chiou, A.; Thomaidis, N.S.; Antonopoulou, S.; Fragopoulou, E. Evaluation of anti-platelet activity of grape pomace extracts. *Food Funct.* **2019**, *10*, 8069–8080. [CrossRef]
127. Karantonis, H.C.; Antonopoulou, S.; Demopoulos, C.A. Antithrombotic Lipid Minor Constituents from Vegetable Oils. Comparison between Olive Oils and Others. *J. Agric. Food Chem.* **2002**, *50*, 1150–1160. [CrossRef]
128. Spector, A.A.; Yorek, M.A. Membrane lipid composition and cellular function. *J. Lipid Res.* **1985**, *26*, 1015–1035. [CrossRef]
129. Turkish, A.R.; Sturley, S.L. The genetics of neutral lipid biosynthesis: An evolutionary perspective. *Am. J. Physiol.-Endocrinol. Metab.* **2009**, *297*, E19–E27. [CrossRef]
130. Lordan, R.; Redfern, S.; Tsoupras, A.; Zabetakis, I. Inflammation and cardiovascular disease: Are marine phospholipids the answer? *Food Funct.* **2020**, *11*, 2861–2885. [CrossRef] [PubMed]
131. De Lorgeril, M.; Salen, P. The Mediterranean-style diet for the prevention of cardiovascular diseases. *Public Health Nutr.* **2006**, *9*, 118–123. [CrossRef] [PubMed]
132. Lordan, R.; Walsh, A.M.; Crispie, F.; Finnegan, L.; Cotter, P.D.; Zabetakis, I. The effect of ovine milk fermentation on the antithrombotic properties of polar lipids. *J. Funct. Foods* **2019**, *54*, 289–300. [CrossRef]
133. Borthakur, A.; Bhattacharyya, S.; Kumar, A.; Anbazhagan, A.N.; Tobacman, J.K.; Dudeja, P.K. Lactobacillus acidophilus alleviates platelet-activating factor-induced inflammatory responses in human intestinal epithelial cells. *PLoS ONE* **2013**, *8*, e75664. [CrossRef]
134. Glenn-Davi, K.; Hurley, A.; Brennan, E.; Coughlan, J.; Shiels, K.; Moran, D.; Saha, S.K.; Zabetakis, I.; Tsoupras, A. Fermentation Enhances the Anti-Inflammatory and Anti-Platelet Properties of Both Bovine Dairy and Plant-Derived Dairy Alternatives. *Fermentation* **2022**, *8*, 292. [CrossRef]
135. Shiels, K.; Tsoupras, A.; Lordan, R.; Nasopoulou, C.; Zabetakis, I.; Murray, P.; Saha, S.K. Bioactive Lipids of Marine Microalga Chlorococcum sp. SABC 012504 with Anti-Inflammatory and Anti-Thrombotic Activities. *Mar. Drugs* **2021**, *19*, 28. [CrossRef]
136. Burri, L.; Hoem, N.; Banni, S.; Berge, K. Marine omega-3 phospholipids: Metabolism and biological activities. *Int. J. Mol. Sci* **2012**, *13*, 15401–15419. [CrossRef]
137. Shavandi, A.; Hou, Y.; Carne, A.; McConnell, M.; Bekhit, A.E.A. Marine Waste Utilization as a Source of Functional and Health Compounds. *Adv. Food Nutr. Res.* **2019**, *87*, 187–254. [CrossRef]
138. Redfern, S.; Dermiki, M.; Fox, S.; Lordan, R.; Shiels, K.; Kumar Saha, S.; Tsoupras, A.; Zabetakis, I. The effects of cooking salmon sous-vide on its antithrombotic properties, lipid profile and sensory characteristics. *Food Res. Int.* **2021**, *139*, 109976. [CrossRef]
139. Lordan, R.; Tsoupras, A.; Zabetakis, I. Phospholipids of Animal and Marine Origin: Structure, Function, and Anti-Inflammatory Properties. *Molecules* **2017**, *22*, 1964. [CrossRef]
140. Nasopoulou, C.; Gogaki, V.; Panagopoulou, E.; Demopoulos, C.; Zabetakis, I. Hen egg yolk lipid fractions with antiatherogenic properties. *Anim Sci J.* **2013**, *84*, 264–271. [CrossRef]
141. Nowacki, D.; Martynowicz, H.; Skoczyńska, A.; Wojakowska, A.; Turczyn, B.; Bobak, Ł.; Trziszka, T.; Szuba, A. Lecithin derived from ω-3 PUFA fortified eggs decreases blood pressure in spontaneously hypertensive rats. *Sci. Rep.* **2017**, *7*, 12373. [CrossRef]
142. Karantonis, H.C.; Antonopoulou, S.; Perrea, D.N.; Sokolis, D.P.; Theocharis, S.E.; Kavantzas, N.; Iliopoulos, D.G.; Demopoulos, C.A. In vivo antiatherogenic properties of olive oil and its constituent lipid classes in hyperlipidemic rabbits. *Nutr. Metab. Cardiovasc. Dis.* **2006**, *16*, 174–185. [CrossRef]
143. Megalemou, K.; Sioriki, E.; Lordan, R.; Dermiki, M.; Nasopoulou, C.; Zabetakis, I. Evaluation of sensory and in vitro antithrombotic properties of traditional Greek yogurts derived from different types of milk. *Heliyon* **2017**, *3*, e00227. [CrossRef]
144. Fragopoulou, E.; Nomikos, T.; Tsantila, N.; Mitropoulou, A.; Zabetakis, I.; Demopoulos, C.A. Biological activity of total lipids from red and white wine/must. *J. Agric. Food Chem.* **2001**, *49*, 5186–5193. [CrossRef]
145. Bang, H.O.; Dyerberg, J. Plasma lipids and lipoproteins in Greenlandic west coast Eskimos. *Acta Med. Scand.* **1972**, *192*, 85–94. [CrossRef]
146. Dontas, A.S.; Zerefos, N.S.; Panagiotakos, D.B.; Vlachou, C.; Valis, D.A. Mediterranean diet and prevention of coronary heart disease in the elderly. *Clin. Interv. Aging* **2007**, *2*, 109–115. [CrossRef]
147. Souza, P.R.; Marques, R.M.; Gomez, E.A.; Colas, R.A.; De Matteis, R.; Zak, A.; Patel, M.; Collier, D.J.; Dalli, J. Enriched Marine Oil Supplements Increase Peripheral Blood Specialized Pro-Resolving Mediators Concentrations and Reprogram Host Immune Responses: A Randomized Double-Blind Placebo-Controlled Study. *Circ. Res.* **2020**, *126*, 75–90. [CrossRef]
148. Rimm, E.B.; Williams, P.; Fosher, K.; Criqui, M.; Stampfer, M.J. Moderate alcohol intake and lower risk of coronary heart disease: Meta-analysis of effects on lipids and haemostatic factors. *BMJ* **1999**, *319*, 1523–1528. [CrossRef]
149. Renaud, S.; de Lorgeril, M. Wine, alcohol, platelets, and the French paradox for coronary heart disease. *Lancet* **1992**, *339*, 1523–1526. [CrossRef]
150. Xanthopoulou, M.N.; Kalathara, K.; Melachroinou, S.; Arampatzi-Menenakou, K.; Antonopoulou, S.; Yannakoulia, M.; Fragopoulou, E. Wine consumption reduced postprandial platelet sensitivity against platelet activating factor in healthy men. *Eur J. Nutr.* **2017**, *56*, 1485–1492. [CrossRef]
151. Argyrou, C.; Vlachogianni, I.; Stamatakis, G.; Demopoulos, C.A.; Antonopoulou, S.; Fragopoulou, E. Postprandial effects of wine consumption on Platelet Activating Factor metabolic enzymes. *Prostaglandins Other Lipid Mediat.* **2017**, *130*, 23–29. [CrossRef]

152. Fragopoulou, E.; Choleva, M.; Antonopoulou, S.; Demopoulos, C.A. Wine and its metabolic effects. A comprehensive review of clinical trials. *Metabolism* **2018**, *83*, 102–119. [CrossRef]
153. Lordan, R.; O'Keeffe, E.; Dowling, D.; Mullally, M.; Heffernan, H.; Tsoupras, A.; Zabetakis, I. The in vitro antithrombotic properties of ale, lager, and stout beers. *Food Biosci.* **2019**, *28*, 83–88. [CrossRef]
154. Garcia-Carrasco, M.; Jimenez-Herrera, E.A.; Galvez-Romero, J.L.; Mendoza-Pinto, C.; Mendez-Martinez, S.; Etchegaray-Morales, I.; Munguia-Realpozo, P.; Vazquez de Lara-Cisneros, L.; Santa Cruz, F.J.; Cervera, R. The anti-thrombotic effects of vitamin D and their possible relationship with antiphospholipid syndrome. *Lupus* **2018**, *27*, 2181–2189. [CrossRef]
155. Singh, U.; Devaraj, S.; Jialal, I. Vitamin E, oxidative stress, and inflammation. *Annu. Rev. Nutr.* **2005**, *25*, 151–174. [CrossRef]
156. Detopoulou, P.; Demopoulos, C.A.; Antonopoulou, S. Micronutrients, Phytochemicals and Mediterranean Diet: A Potential Protective Role against COVID-19 through Modulation of PAF Actions and Metabolism. *Nutrients* **2021**, *13*, 462. [CrossRef]
157. Chew, B.P.; Park, J.S. Carotenoid action on the immune response. *J. Nutr.* **2004**, *134*, 257S–261S. [CrossRef]
158. Mutoh, H.; Fukuda, T.; Kitamaoto, T.; Masushige, S.; Sasaki, H.; Shimizu, T.; Kato, S. Tissue-specific response of the human platelet-activating factor receptor gene to retinoic acid and thyroid hormone by alternative promoter usage. *Proc. Natl. Acad. Sci. USA* **1996**, *93*, 774–779. [CrossRef]
159. Freedman, J.E.; Farhat, J.H.; Loscalzo, J.; Keaney, J.F., Jr. Alpha-tocopherol inhibits aggregation of human platelets by a protein kinase C-dependent mechanism. *Circulation* **1996**, *94*, 2434–2440. [CrossRef]
160. Fukuzawa, K.; Kurotori, Y.; Tokumura, A.; Tsukatani, H.; Vitamin, E. deficiency increases the synthesis of platelet-activating factor (PAF) in rat polymorphonuclear leucocytes. *Lipids* **1989**, *24*, 236–239. [CrossRef] [PubMed]
161. Akada, S.; Iioka, H.; Moriyama, I.; Hisanaga, H.; Morimoto, K.; Ichijo, M. The role of vitamin E during pregnancy–anti-platelet aggregation activity of alpha-tocopherol. *Nihon Sanka Fujinka Gakkai Zasshi* **1991**, *43*, 523–528. [PubMed]
162. Silbert, P.L.; Leong, L.L.; Sturm, M.J.; Strophair, J.; Taylor, R.R. Short term vitamin E supplementation has no effect on platelet function, plasma phospholipase A2 and lyso-PAF in male volunteers. *Clin. Exp. Pharm. Physiol* **1990**, *17*, 645–651. [CrossRef] [PubMed]
163. Charoenngam, N.; Holick, M.F. Immunologic Effects of Vitamin D on Human Health and Disease. *Nutrients* **2020**, *12*, 2097. [CrossRef] [PubMed]
164. Mohammad, S.; Mishra, A.; Ashraf, M.Z. Emerging Role of Vitamin D and its Associated Molecules in Pathways Related to Pathogenesis of Thrombosis. *Biomolecules* **2019**, *9*, 649. [CrossRef] [PubMed]
165. Johny, E.; Jala, A.; Nath, B.; Alam, M.J.; Kuladhipati, I.; Das, R.; Borkar, R.M.; Adela, R. Vitamin D Supplementation Modulates Platelet-Mediated Inflammation in Subjects With Type 2 Diabetes: A Randomized, Double-Blind, Placebo-Controlled Trial. *Front. Immunol.* **2022**, *13*, 869591. [CrossRef] [PubMed]
166. Verdoia, M.; Pergolini, P.; Nardin, M.; Rolla, R.; Negro, F.; Kedhi, E.; Suryapranata, H.; Marcolongo, M.; Carriero, A.; De Luca, G.; et al. Vitamin D levels and platelet reactivity in diabetic patients receiving dual antiplatelet therapy. *Vasc. Pharm.* **2019**, *120*, 106564. [CrossRef] [PubMed]
167. Greiller, C.L.; Suri, R.; Jolliffe, D.A.; Kebadze, T.; Hirsman, A.G.; Griffiths, C.J.; Johnston, S.L.; Martineau, A.R. Vitamin D attenuates rhinovirus-induced expression of intercellular adhesion molecule-1 (ICAM-1) and platelet-activating factor receptor (PAFR) in respiratory epithelial cells. *J. Steroid Biochem. Mol. Biol.* **2019**, *187*, 152–159. [CrossRef]
168. Lordan, R. Notable Developments for Vitamin D Amid the COVID-19 Pandemic, but Caution Warranted Overall: A Narrative Review. *Nutrients* **2021**, *13*, 740. [CrossRef]
169. Verouti, S.N.; Tsoupras, A.B.; Alevizopoulou, F.; Demopoulos, C.A.; Iatrou, C. Paricalcitol effects on activities and metabolism of platelet activating factor and on inflammatory cytokines in hemodialysis patients. *Int. J. Artif. Organs* **2013**, *36*, 87–96. [CrossRef]
170. Chambial, S.; Dwivedi, S.; Shukla, K.K.; John, P.J.; Sharma, P. Vitamin C in disease prevention and cure: An overview. *Indian J. Clin. Biochem.* **2013**, *28*, 314–328. [CrossRef]
171. Liu, D.; Pei, D.; Hu, H.; Gu, G.; Cui, W. Effects and Mechanisms of Vitamin C Post-Conditioning on Platelet Activation after Hypoxia/Reoxygenation. *Transfus Med. Hemother* **2020**, *47*, 110–118. [CrossRef]
172. Lehr, H.A.; Weyrich, A.S.; Saetzler, R.K.; Jurek, A.; Arfors, K.E.; Zimmerman, G.A.; Prescott, S.M.; McIntyre, T.M. Vitamin C blocks inflammatory platelet-activating factor mimetics created by cigarette smoking. *J. Clin. Investig.* **1997**, *99*, 2358–2364. [CrossRef]
173. Lloberas, N.; Torras, J.; Herrero-Fresneda, I.; Cruzado, J.M.; Riera, M.; Hurtado, I.; Grinyo, J.M. Postischemic renal oxidative stress induces inflammatory response through PAF and oxidized phospholipids. Prevention by antioxidant treatment. *FASEB J.* **2002**, *16*, 908–910. [CrossRef]
174. Cao, Y.Z.; Cohen, Z.S.; Weaver, J.A.; Sordillo, L.M. Selenium modulates 1-O-alkyl-2-acetyl-sn-glycero-3-phosphocholine (PAF) biosynthesis in bovine aortic endothelial cells. *Antioxid. Redox Signal.* **2001**, *3*, 1147–1152. [CrossRef]
175. Kiran Gotru, S.; van Geffen, J.P.; Nagy, M.; Mammadova-Bach, E.; Eilenberger, J.; Volz, J.; Manukjan, G.; Schulze, H.; Wagner, L.; Eber, S.; et al. Defective Zn(2+) homeostasis in mouse and human platelets with alpha- and delta-storage pool diseases. *Sci. Rep.* **2019**, *9*, 8333. [CrossRef]
176. Lominadze, D.G.; Saari, J.T.; Miller, F.N.; Catalfamo, J.L.; Justus, D.E.; Schuschke, D.A. Platelet aggregation and adhesion during dietary copper deficiency in rats. *Thromb Haemost* **1996**, *75*, 630–634. [CrossRef]
177. Tako, E. Dietary Trace Minerals. *Nutrients* **2019**, *11*, 2823. [CrossRef]
178. Hampel, G.; Watanabe, K.; Weksler, B.B.; Jaffe, E.A. Selenium deficiency inhibits prostacyclin release and enhances production of platelet activating factor by human endothelial cells. *Biochim. Biophys. Acta* **1989**, *1006*, 151–158. [CrossRef]

179. Bao, B.; Prasad, A.S.; Beck, F.W.; Fitzgerald, J.T.; Snell, D.; Bao, G.W.; Singh, T.; Cardozo, L.J. Zinc decreases C-reactive protein, lipid peroxidation, and inflammatory cytokines in elderly subjects: A potential implication of zinc as an atheroprotective agent. *Am. J. Clin. Nutr.* **2010**, *91*, 1634–1641. [CrossRef]
180. Marx, G.; Krugliak, J.; Shaklai, M. Nutritional zinc increases platelet reactivity. *Am. J. Hematol* **1991**, *38*, 161–165. [CrossRef]
181. Watson, B.R.; White, N.A.; Taylor, K.A.; Howes, J.M.; Malcor, J.D.; Bihan, D.; Sage, S.O.; Farndale, R.W.; Pugh, N. Zinc is a transmembrane agonist that induces platelet activation in a tyrosine phosphorylation-dependent manner. *Metallomics* **2016**, *8*, 91–100. [CrossRef]
182. Hughes, S.; Samman, S. The effect of zinc supplementation in humans on plasma lipids, antioxidant status and thrombogenesis. *J. Am. Coll. Nutr.* **2006**, *25*, 285–291. [CrossRef]
183. Nunez, D.; Kumar, R.; Hanahan, D.J. Inhibition of [3H]platelet activating factor (PAF) binding by Zn^{2+}: A possible explanation for its specific PAF antiaggregating effects in human platelets. *Arch. Biochem. Biophys.* **1989**, *272*, 466–475. [CrossRef]
184. Huo, Y.; Ekholm, J.; Hanahan, D.J. A preferential inhibition by Zn^{2+} on platelet activating factor- and thrombin-induced serotonin secretion from washed rabbit platelets. *Arch. Biochem. Biophys.* **1988**, *260*, 841–846. [CrossRef]
185. Binder, H.; Arnold, K.; Ulrich, A.S.; Zschörnig, O. Interaction of Zn^{2+} with phospholipid membranes. *Biophys. Chem.* **2001**, *90*, 57–74. [CrossRef]
186. Vetlenyi, E.; Racz, G. The physiological function of copper, the etiological role of copper excess and deficiency. *Orv. Hetil.* **2020**, *161*, 1488–1496. [CrossRef]
187. Lynch, S. Effects of a dietary copper deficiency on plasma coagulation factor activities in male and female mice. *J. Nutr. Biochem.* **1992**, *3*, 387–391. [CrossRef]
188. Van Rensburg, M.J.; van Rooy, M.; Bester, M.J.; Serem, J.C.; Venter, C.; Oberholzer, H.M. Oxidative and haemostatic effects of copper, manganese and mercury, alone and in combination at physiologically relevant levels: An ex vivo study. *Hum. Exp. Toxicol.* **2019**, *38*, 419–433. [CrossRef]
189. Akinloye, O.; Oyewale, O.O.; Oguntibeju, O.O. Evaluation of trace elements in pregnant women with pre-eclampsia. *Afr. J. Biotechnol.* **2010**, *9*, 5196–5202.
190. Tesfa, E.; Nibret, E.; Munshea, A. Maternal Serum Zinc Level and Pre-eclampsia Risk in African Women: A Systematic Review and Meta-analysis. *Biol. Trace Elem. Res.* **2021**, *199*, 4564–4571. [CrossRef] [PubMed]
191. Bassiouni, B.A.; Foda, A.I.; Rafei, A.A. Maternal and fetal plasma zinc in pre-eclampsia. *Eur. J. Obstet. Gynecol. Reprod. Biol.* **1979**, *9*, 75–80. [CrossRef]
192. Jin, J.; Gao, L.; Zou, X.; Zhang, Y.; Zheng, Z.; Zhang, X.; Li, J.; Tian, Z.; Wang, X.; Gu, J.; et al. Gut Dysbiosis Promotes Preeclampsia by Regulating Macrophages and Trophoblasts. *Circ. Res.* **2022**, *131*, 492–506. [CrossRef] [PubMed]
193. Ma, Y.; Shen, X.; Zhang, D. The Relationship between Serum Zinc Level and Preeclampsia: A Meta-Analysis. *Nutrients* **2015**, *7*, 7806–7820. [CrossRef]
194. Martadiansyah, A.; Maulina, P.; Mirani, P.; Kaprianti, T. Zinc Serum Maternal Levels as a Risk Factor for Preeclampsia. *Biosci. Med. J. Biomed. Transl. Res.* **2021**, *5*, 693–701.
195. Shah, N.; Sethi, R.; Shah, S.; Jafri, K.; Duran, J.; Chang, Y.; Soni, C.; Wollocko, H. The Roles of Platelet-Activating Factor and Magnesium in Pathophysiology of Hypertension, Atherogenesis, Cardiovascular Disease, Stroke and Aging. *Cardiogenetics* **2022**, *12*, 49–62. [CrossRef]

nutrients

MDPI

Article

Alternative Healthy Eating Index-2010 and Incident Non-Communicable Diseases: Findings from a 15-Year Follow Up of Women from the 1973–78 Cohort of the Australian Longitudinal Study on Women's Health

Hlaing Hlaing-Hlaing [1,2,*], Xenia Dolja-Gore [1,2], Meredith Tavener [1,2], Erica L. James [1,2] and Alexis J. Hure [1,2]

1 School of Medicine and Public Health, University of Newcastle, Callaghan, Newcastle, NSW 2308, Australia
2 Hunter Medical Research Institute, New Lambton Heights, Newcastle, NSW 2305, Australia
* Correspondence: hlainghlainghlaing@uon.edu.au

Abstract: Non-communicable diseases (NCDs) and multimorbidity (≥two chronic conditions), are increasing globally. Diet is a risk factor for some NCDs. We aimed to investigate the association between diet quality (DQ) and incident NCDs. Participants were from the Australian Longitudinal Study on Women's Health 1973–78 cohort with no NCD and completed dietary data at survey 3 (2003, aged 25–30 years) who responded to at least one survey between survey 4 (2006) and survey 8 (2018). DQ was measured by the Alternative Healthy Eating Index-2010 (AHEI-2010). Outcomes included coronary heart disease (CHD), hypertension (HT), asthma, cancer (excluding skin cancer), diabetes mellitus (DM), depression and/or anxiety, multimorbidity, and all-cause mortality. Repeated cross-sectional multivariate logistic regressions were performed to investigate the association between baseline DQ and NCDs over 15 years. The AHEI-2010 mean (±sd) for participants (n = 8017) was 51.6 ± 11.0 (range: 19–91). There was an inverse association between AHEI-2010 and incident asthma at survey 4 ($OR_{Q5–Q1}$: 0.75, 95% CI: 0.57, 0.99). Baseline DQ did not predict the occurrence of any NCDs or multimorbidity between the ages of 25–45 years. Further well-planned, large prospective studies conducted in young women are needed to explore dietary risk factors before the establishment of NCDs.

Keywords: Alternative Healthy Eating Index-2010; non-communicable diseases; multimorbidity; childbearing age; women

Citation: Hlaing-Hlaing, H.; Dolja-Gore, X.; Tavener, M.; James, E.L.; Hure, A.J. Alternative Healthy Eating Index-2010 and Incident Non-Communicable Diseases: Findings from a 15-Year Follow Up of Women from the 1973–78 Cohort of the Australian Longitudinal Study on Women's Health. *Nutrients* **2022**, *14*, 4403. https://doi.org/10.3390/nu14204403

Academic Editors: William B. Grant and Ronan Lordan

Received: 23 August 2022
Accepted: 15 October 2022
Published: 20 October 2022

1. Introduction

Globally non-communicable diseases (NCDs) were the leading cause of mortality in women in 2019 [1,2]. Coupled with epidemiological transition, changing disease patterns from communicable diseases to NCDs, and demographic transition, resulting in population ageing and growth, NCD occurrence is expected to increase [3]. Furthermore, though NCDs are not restricted to a particular age group or sex, women are more likely to experience their socio-economic impacts compared with men [4]. Of concern, women of childbearing age are susceptible to NCDs, and the most common diseases are cardiovascular disease (CVD), hypertension (HT), cancer and diabetes mellitus (DM) [5]. Some health disorders during pregnancy are associated with NCDs in later life. For instance, hypertension during pregnancy (HTpreg) can affect the vascular health of the mother, thereby leading to the occurrence of CVD and stroke in later life [6,7]. In addition, gestational diabetes mellitus (GDM), one of the hyperglycemic disorders in pregnancy, increases the risk of developing DM in later life [8].

NCD and multimorbidity (the concurrence of two or more chronic diseases in the same person) [9,10] have become public health challenges. Australia is experiencing a catastrophic level of NCDs [11], and estimates from the 2017-18 National Health Survey indicated that almost half of Australian females were experiencing one or more NCDs, including asthma, cancer, DM, and mental and behavioural problems [12]. Notably, 43% of Australian women aged 45 years and younger reported as having at least one NCD [12].

With regard to disease aetiology, a differential distribution of NCD risk factors among men and women was observed in a European study [13]. It was accepted that a set of risk factors such as lifestyle or behavioural factors, environmental factors, underlying metabolic/physiological factors and infections are multifactorial causes that are related to NCD mortality and morbidity [14]. Of these, diet has been identified as one lifestyle or bahavioural risk for CVD, some cancers and DM [15]. Moreover, dietary risk factors are globally ranked second (to high blood pressure) for attributable deaths for women [16]. Diet, the combination of foods and nutrients, can be condensed into a simple measure or summative score [17–19] by constructing diet quality indices (DQIs) based on dietary guidelines or specific dietary pattern recommendations [20–22]. While this method also has limitations and measurement errors [23], it goes further than the single nutrient approach in recognising the synergistic nature of micro- and macronutrients [24].

A higher score on a DQI generally reflects healthier or optimal diet quality (DQ) or closer compliance to dietary recommendations [25]. The original DQIs such as the Healthy Eating Index (HEI), Diet Quality Index, and Diet Quality Index-Revised were developed in alignment with the Dietary Guidelines for Americans [26–28], whereas the Mediterranean Diet Score was based on the Mediterranean dietary pattern [29]. More recently, DQIs have been developed in alignment with country-specific dietary guidelines (e.g., the Dietary Guideline Index [30], China Dietary Guideline Index [31], and Healthy Dietary Habits Index [32]), or modified from earlier indices (e.g., Recommended Food Score [33], Australian Recommended Food Score [34], and alternative Mediterranean Diet [35]). Additionally, the Alternative Healthy Eating Index (AHEI), modelled on the original HEI, was constructed according to extensive epidemiological evidence for NCD prevention [33]. Although both HEI and AHEI are measurements of DQ, there are differences in scoring (e.g., the original HEI included total fat and protein (meat and beans) for fat and protein components [26], whereas the AHEI included trans-fat and the ratio of polyunsaturated fatty acid (PUFA) to saturated fatty acid (SFA) in the fat component; and the ratio of white to red meat, nuts and soy protein in the protein component [33]. Both the HEI and AHEI have been regularly updated (HEI-2005 [36], HEI-2010 [37], HEI-2015 [38], and HEI-2020 [39] for HEI; AHEI-2010 [40] for AHEI).

DQIs have been extensively applied in diet-health outcome relationship studies, including those considering NCDs [17,19,20,41–47]. The most commonly assessed health outcomes have been CVD or coronary heart disease (CHD) [41,44], HT [45], cancer [41,42], DM [41,43], depression [46,48], and all-cause mortality [41,47]. Previous evidence has shown that a high level of DQ assessed by the AHEI and/or AHEI-2010 is associated with a reduced risk of CVD, cancer, DM, neurodegenerative disease, and all-cause mortality [41]. However, in term of the clustering of NCDs, evidence on the relationship between overall diet and NCD multimorbidity is very limited [49]. Previously, we examined the relationship between DQ (measured as Healthy Eating Index for Australian Adults-2013 (HEIFA-2013), Mediterranean Diet Score (MDS), and Alternative Healthy Eating Index-2010 (AHEI-2010)) and the incidence of NCDs, including multimorbidity among a cohort of women born between 1946 and 1951 that was drawn from the Australian Longitudinal Study on Women's Health (ALSWH) [50]. Data showed that mid-aged women with the highest DQI quintiles had reduced odds of NCDs in later surveys (9 to 15 years later), and AHEI-2010 was the most sensitive DQI for prediction of NCDs [50].

Of note, taking a life course approach in NCD prevention and control has been introduced in recent years [51]. Despite slow progression and having a long latency in the development of NCDs [14], it is important to tackle risk factors in earlier life stages, transitional stages and critical life stages [52]. Partnering and parenting are major transitional stages and turning points for young adults [53], as they are associated with changes in dietary habits and physical activity [54]. In a 6-year follow-up of a nationally representative sample, young Australian women who were living with a partner at baseline survey and those who become partnered during follow-up or remained partnered consumed a relatively healthier diet compared to singles [55]. Parenting or living with children has been reported as having favourable and unfavourable effects on dietary intakes or quality [53,55–58].

Some risk factors for NCDs are prevalent in young women of childbearing age, for example, overweight or obesity [59]. Overweight, obesity, or weight gain in childbearing women is partly associated with marriage, pregnancy and motherhood [59], which could be related to changes in dietary behaviours and/or physical activity during such a transition period [59]. Overweight or obesity is linked to the occurrence of CVD [60], breast cancer and endometrial cancer [61], and DM [62].

The tracking of lifestyle or behavioural factors, especially unhealthy diet, in young women could be a useful strategy for NCD prevention. Given that the prevalence of NCDs is increasing with age, with women disproportionately affected [11,63], and almost half of Australian women who reported NCDs being of childbearing age [11,12], an investigation of DQ as a key modifiable predictor of NCDs among younger women is warranted.

It would be beneficial to make comparisons across generations in investigating the relationship between DQ and NCDs. A longitudinal analysis investigating the temporal associations between DQ and NCDs is needed. However, prior to the initiation of longitudinal analysis, initial cross-sectional analyses using baseline variables would provide context. This study was aimed to determine if DQ, assessed by AHEI-2010, is predictive of NCD outcomes during a 15-year follow-up of women born 1973–78 from the age of 25–30 years. The NCD outcome of interests included CHD, HT, asthma, cancer (except skin cancer), DM, depression and/or anxiety, and multimorbidity. We hypothesized that women with a high DQ assessed by AHEI-2010 would have reduced odds of NCD outcomes and multimorbidity during the 15-year follow-up compared to those with a low DQ.

2. Materials and Methods

2.1. Study Population

Data for this study were drawn from the Australian Longitudinal Study on Women's Health (ALSWH), a national population-based study funded by the Commonwealth Department of Health and Ageing. The ALSWH study commenced in 1996 and approximately 45,000 women born between 1973 and 1978, 1946 and 1951, and 1921 and1926 were selected from the Medicare health insurance database, which included all Australian citizens and permanent residents [64]. A new cohort of women born between 1989 and 1995 were enrolled by means of in-person, internet and social media contact methods [65]. Ethical clearance for all participants was granted from the Human Ethics committees of the University of Newcastle and the University of Queensland. Further details about the ALSWH can be found elsewhere [66].

The present study included data from a cohort of women born between 1973 and 1978. A total of 14,247 women aged 18–23 years responded to survey 1 (S1, 1996), then followed up every 3 years (apart from the 4 years between S1, 1996 and S2, 2000) until 2018. Self-administered questionnaires were sent to collect information on women's physical and mental health, health service use, and socio-demographic and behavioural characteristics. At S3 (aged 25–30 years in 2003), the dietary intakes of women were assessed by the Dietary Questionnaire for Epidemiological Studies version 2 (DQES-v2) [67]. The DQES-v2 has been validated among young Australian women and demonstrated its ability to assess habitual intake (energy-adjusted correlation coefficients: 0.28 to 0.70) [68].

This analysis comprised women at S3 (2003, aged 25–30 years), when dietary data were first assessed. Women were included if they responded to S3 with complete FFQ data and responded at least once between S4 (2006, aged 28–33 years) and S8 (2018, aged 40–45 years). To ensure that the dietary predictor was measured before the onset of the NCD (i.e., incident cases), women were excluded if they reported pre-existing NCDs prior to S3. An exception was made for asthma and depression and/or anxiety, where histories of these conditions were adjusted for and a recent episode was deemed an incident occurrence. Exclusions comprised the self-reported diagnosis of: CHD (heart disease) at S1, S2 and S3; HT (HT (high blood pressure) at S1, HT (high blood pressure) other than during pregnancy at S2 and S3); cancer at S1, S2 and S3; and DM (diabetes (high blood sugar) at S1, non-insulin dependent (type II) diabetes at S2 and S3). Women who had no or missing S3-FFQ data were excluded. Figure 1 shows a simple illustration of the ALSWH 1973–78 data for analysis. In the present analysis, 8017 women from S3 (2003) and afterwards were included (Figure 2). The number of respondents in five consecutive surveys (from S4, 2006 to S8, 2018) can be seen in Supplementary Table S1.

Survey (S)	S1	S2	S3	S4	S5	S6	S7	S8
Year	1996	2000	2003	2006	2009	2012	2015	2018
1973–78 cohort			FFQ					
Age (years)	18–23	22–27	25–30	28–33	31–36	34–39	37–42	40–45
Key Data			Baseline; Diet quality, Covariates	NCD outcomes	NCD outcomes	NCD outcomes	NCD outcomes	NCD outcomes
Exclusions	History of NCD	History of NCD	History of NCD					

Figure 1. Simplified diagram showing women born between 1973 and 1978 in the Australian Longitudinal Study on Women's Health (ALSWH).

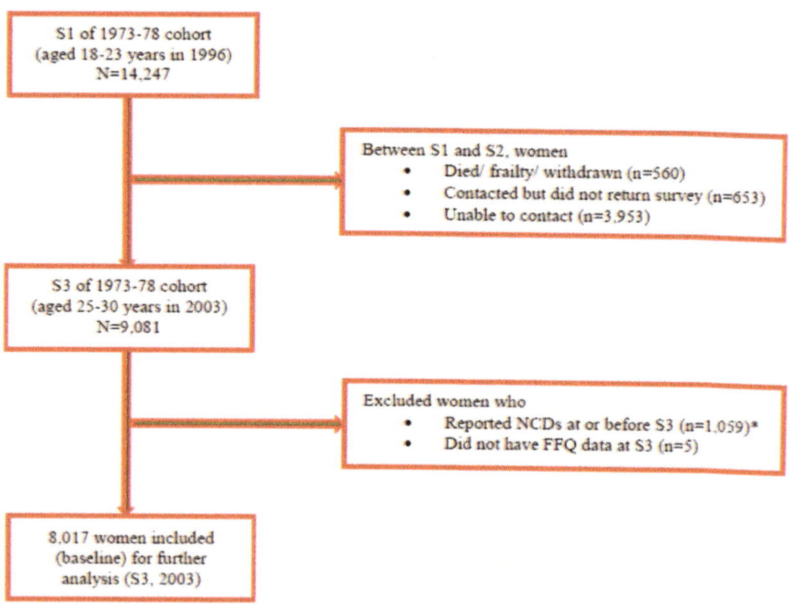

Figure 2. Diagram showing participants' sampling from the Australian Longitudinal Study on Women's Health (ALSWH), born in 1973–78. * NCDs at or before S3 were coronary heart disease, hypertension, cancer and diabetes mellitus.

2.2. Dietary Intake Assessment

The diet of the ALSWH 1973–78 cohort was assessed by using the DQES-v2 [67] at S3 (in 2003). The FFQ included questions for reporting the usual frequency of 74 food items and six alcoholic beverages in the last 12 months. The responses ranged from 1 to 10 points; "Never" to "$\geq$ 3 times per day" for food items and "Never" to "Every day" for beverages. Portion-size photographs were used to adjust respective servings. Further questions assessing the daily servings of fruit, vegetable, bread, dairy products, egg, fat spread, and sugar were also asked. The partaking of food items (in grams/day) and nutrients was calculated using the national food composition database of Australian foods, the NUTTAB95 [69].

2.3. Exposure Variable

The exposure variable of the ALSWH 1973–78 cohort was DQ assessed by the Alternative Healthy Eating Index-2010 (AHEI-2010) calculated from the DQES-v2 at S3. Dietary data at S3, assuming constancy throughout the 15-year period, were used in this cross-sectional study to investigate the association between the baseline DQ of the ALSWH 1973–78 cohort and NCDs during follow-ups (from S4 in 2006 to S8 in 2018). Previously, it was shown that the DQ of women in the ALSWH 1946–51 cohort was relatively stable over a 12-year period (from S3, 2001 to S7, 2013) [70]. Therefore, in our ALSWH 1973–78 cohort, DQ at S3 was used to examine the association between DQ and incident NCDs during a 15-year follow up (from S4 to S8). The AHEI-2010 was selected based on its relevance to current dietary recommendations [71], having performed well when critically appraised compared with other DQIs based on a specific dietary pattern [72], and because it allowed us to make comparisons across generations based on our previous ALSWH 1946–51 cohort analyses [50]. The detailed scoring of AHEI-2010 is provided in Supplementary Table S2.

The AHEI-2010 was constructed using foods and nutrients for NCD prevention including CVD, some cancers, and DM based on clinical and epidemiological investigations [33,40]. The components of this index were modelled on the original HEI [26] and the original AHEI [33]. Each component ranged from 0 (suboptimal diet) to 10 (optimal diet), and intermediate values were correspondingly assigned. The total score was set as the sum of 11 components so that the AHEI score had a possible range from 0 to 110. The computations of scoring for 11 components have been previously reported in detail [40]. The positive components were vegetables, fruit, whole grains, nuts and legumes, long chain omega-3 fats, and PUFAs. Intakes of sugar sweetened beverages (SSBs) and fruit juice, red and processed meat, trans-fat, and sodium were coded as negative components. Moderate alcohol consumption contributed to a higher score (i.e., higher diet quality).

2.4. Outcome Variables

The outcomes of the ALSWH 1973–78 cohort at five surveys (from S4, 2006 to S8, 2018) included the cumulative incidence of NCDs [CHD, HT, asthma, cancer (except skin cancer), DM, and depression and/or anxiety], multimorbidity (the concurrence of two or more of NCDs of interest), and all-cause mortality.

The incidence of each disease was based on the ALSHW survey when it was first reported. The occurrence of common NCDs in the ALSWH 1973–78 cohort was self-reported. At S1, participants were asked if they had a known diagnosis of heart disease, HT (high blood pressure), asthma, cancer, and/or diabetes (high blood sugar) by using the question "Have you ever been told by a doctor that you have ... ? (circle one number on each line)?" At S2, their known status of heart disease, HT (high blood pressure) other than during pregnancy, asthma, cancer (specify type), non-insulin dependent (type 2) diabetes, and depression (not postnatal) and anxiety disorder was assessed using the question "Have you ever been told by a doctor that you have ... ? (Mark all that apply)." The responses were "yes, in the last four years" and "yes, more than four years ago". At S3 and S4, the self-reported diagnosis of heart disease, HT other than during pregnancy, asthma, cancer (specify type), non-insulin dependent (type 2) diabetes, and depression (not postnatal) and anxiety disorder was assessed using the question "In the last three years, have you been diagnosed or treated for ... ? (Mark all that apply)". For S5 to S8, the self-reported status of HT, cancer and depression was assessed in terms of HT, cancer (skin cancer and other cancer) and depression by using the same format of S3 and S4. Skin cancer could not be excluded at S4, so cancer (except skin cancer) was used for S5 to S8 only.

The self-reported status of NCDs was considered in terms of enduring conditions, meaning participants who had reported an NCD (except for asthma and depression and/or anxiety) in any survey were counted as having that NCD throughout subsequent surveys (Supplementary Table S3). The self-reported diagnosis of NCDs among the ALSWH study participants was considered reliable and valid based on previous evaluations against administrative data [73,74]. The incidence of multimorbidity at five subsequent surveys (from S4, 2006 to S8, 2018) was calculated as the presence of two or more NCDs of interest in any combination. The deaths of ALSWH participants were reported through data linkage with the Australian National Death Index [75]. The Cause of Death file, containing the primary or underlying cause and other additional causes, is available for assessing mortality [75]. For the all-cause mortality of our sample, any death was counted and reported as descriptive data.

2.5. Covariates

To obtain unbiased estimates of the association between exposure and outcome, potential confounding variables are adjusted for in statistical analyses [76]. Generally, covariates are selected from possible candidates using statistical prerequisites, criteria for selecting variables (the Directed Acyclic Graph (DAG) based on background knowledge is one approach), and variable selection algorithms [77]. In this study, theoretical models in the form of DAGs [78–80] were developed based on background knowledge and literature to identify the relationship between exposure, outcome, and covariates across five surveys (from S4, 2006 to S8, 2018) (Supplementary Figures S1–S7). Covariates at S3 that were associated with both the exposure and outcomes were considered as confounders and adjusted for in this study. They were variables measuring socioeconomic status (residence status, marital status, education, occupation, and income stress), behaviour (physical activity) and childbearing (history of breastfeeding, history of GDM for DM, and history of HTpreg for HT). The role of prescribed medication in the diet-NCDs pathway differs by condition. For instance, taking cholesterol-lowering agents may prevent CHD, but for DM, participants were more likely to take prescribed medications for diabetes following diagnosis. Therefore, prescribed medicine was not included in DAGs but was included in the sensitivity analysis, testing the effect estimates with and without this variable. Being pregnant at the time of the survey was treated as an indicator variable in the regression models. The body mass index, which can be affected by DQ [81] and can influence NCD outcomes [82–88], was considered a mediator and not adjusted for in the models.

In this study, participants' responses for residence status were classified as "major cities, inner regional, and outer regional/rural". For marital status, responses were classified as "never married, married/de facto, and separated/divorced/widowed". Education status was classified as "no formal education, high school level, diploma, and university degree". Occupation was coded as "no paid employment and paid employment". Participants' income stress was assessed via questions asking how they were able to manage on available income and classified as "easy and difficult". Physical activity was measured using the Active Australia Survey items [89] incorporated into the ALSWH surveys by asking two questions on frequency and duration of walking (recreation and transport) and moderate- and vigorous-intensity activity over the previous 7 days. By using responses from these two questions, the metabolic equivalent per minute (MET.mins) was calculated as (3*minutes walking) + (4*minutes moderate activity) + (7.5* minutes vigorous activity). Then, 0–39 MET min/week was classified as "none/sedentary", 40–599 MET min/week as "low", 600–1199 MET min/week as "moderate", and $\geq$1200 MET min/week as "high" [90]. Childbearing variables were current pregnancy status (no or yes), parity (none or one and above), history of breastfeeding (no or yes), history of GDM (no or yes), and history of HTpreg (no or yes).

The participants' status of taking prescribed medicine and over-the-counter (OTC) medicine over the previous 4 weeks was reported as the quantity of various medications (S1 and S2) and a binary response (S3 and S4). From S5 onwards, they were asked to record the names of any medication taken over the previous 4 weeks without specifying as prescribed or OTC medicines [91]. We accessed the original medication data for S5 to S8 and created three new variables: taking both prescribed and OTC, prescribed medicine only, and OTC medicine only. Taking prescribed medicine, coded as a binary variable (no or yes), was used in this analysis.

At each ALSWH survey, women were asked "Are you currently pregnant?". The responses were "yes, no, or don't know" for S1 to S3 and "no, less than 3 months, 3 to 6 months, more than 6 months, or don't know" for the remaining surveys. We generated a binary variable for pregnancy status. Moreover, the women who had a child's date of birth recorded in the child dataset were included as pregnant if they returned a survey 0–9 months before a child's date of birth.

In the ALSWH child dataset, women who ever had a child were identified based on the date of birth and the breastfeeding status of children. Parity was categorised as "no" and "one and above". Women who ever had breastfed were recorded in the child dataset. Histories of breastfeeding across the surveys (from S2 to S8) were generated and assigned as a binary variable (no or yes).

In ALSWH main surveys, at S2, women were asked whether they had ever been told that they had GDM and/or HTpreg in the last four years or more than four years ago. At S3 and S4, women were asked to report their diagnosis or treatment status of GDM and HTpreg. From S5 to S7, they were asked to recall the diagnosis or treatment for GDM and HTpreg for each child during their pregnancy. Moreover, in the child dataset, the status of women's GDM and HTpreg was also recorded. From these two data, binary variables measuring the history of GDM and HTpreg were generated from S2 to S8.

There were missing values in covariates such as the area of residence, marital status, education, occupation, income stress, physical activity, and taking prescribed medicine. The carry-forward approach [92] from the subsequent survey was used from S4 to S7. For example, if the value at S4 was missing, values at adjacent surveys (S3 and S5) were checked and replaced the corresponding value labels. For a missing item at S8, the value at S7 was checked and replaced. After filling, no variable had missing values more than 5% of the total data. However, variables related to childbearing such as current pregnancy status, parity, history of breastfeeding, history of GDM, and history of HTpreg had missing values less than 5%, and the carry-forward approach for missing values was not deemed necessary.

2.6. Statistical Analysis

The statistical software package Stata version 15 was used in all analyses. A descriptive summary of AHEI-2010 was reported as a continuous measure (mean ± sd) and categorical measure (quintiles). The baseline characteristics of women (at S3) with respect to the AHEI-2010 quintiles were described as mean± standard deviation (sd) or n (%), and they were compared by using an analysis of variance (ANOVA) or chi-squared test. The occurrences of all-cause mortality and NCDs (each NCD and multimorbidity) were reported. The baseline characteristics of women who had been excluded and included in the current study were compared (Supplementary Table S4).

The univariate analyses of AHEI-2010 at S3 and NCD outcomes (each NCD and multimorbidity) over 15 years (from S4 to S8) were performed. The number of CHD events was low compared with other NCDs, so multivariate models were not fitted for the association between AHEI-2010 and CHD. For others, multivariate models (adjusted for covariates at S3 including age, residence, marital status, education, occupation, income stress, physical activity, current pregnancy status, parity, history of breastfeeding, history of GDM for DM, and history of HTpreg for HT) were fitted to investigate the short-term and long-term effects of AHEI-2010 on NCD outcomes. Multivariate logistic regression models [93] were used to investigate the effect of AHEI-2010 on the NCD outcomes. The DQI data from S3, 2003 were used for the prediction of NCD outcomes at five consecutive surveys (from S4, 2006 to S8, 2018). The odds ratios (ORs) and 95% confidence intervals (95% CIs) for NCD outcomes relative to AHEI-2010 were computed using the quintile 1 as the reference category, and they are presented as the main results.

To test the robustness of our results, sensitivity analyses were performed. We performed multiple tests, applying the Bonferroni correction in the logistic regression models to account for the covariates whose 95% CIs were near 1. Analyses that used taking prescribed medicine as a covariate (Supplementary Table S5) and changes of childbearing variables (Supplementary Table S6) were also performed.

3. Results

A total of 8017 women were included at baseline (S3, 2003), with the mean AHEI-2010 score at S3 being 51.6 (sd 11.0; range: 19–91). Women with the AHEI-2010 quintile 5 (Q5 AHEI-2010) were reported as married/de facto, living in major cities, having a university degree, having paid employment, easily managing income, and being more physically active than those in quintile 1 (Q1 AHEI-2010) (Table 1).

Table 1. Baseline characteristics of the sampled women (n = 8017; survey 3 in 2003) by quintiles of Alternative Healthy Eating Index-2010.

	AHEI-2010 Quintiles					
Characteristics	Q1 (n = 1635)	Q2 (n = 1572)	Q3 (n = 1582)	Q4 (n = 1636)	Q5 (n = 1592)	*p*-Value §
Age (years) [mean (sd)]	27.5 (1.5)	27.6 (1.5)	27.5 (1.5)	27.6 (1.5)	27.6 (1.4)	0.02 **
Marital status [n (%)]						<0.001 **
Never married	415 (25.5)	487 (31.1)	557 (35.3)	630 (38.6)	734 (46.3)	
Married/de facto	1141 (70.1)	1028 (65.6)	974 (61.7)	939 (57.5)	810 (51.1)	
Separated/divorced/widowed	72 (4.4)	52 (3.3)	47 (3.0)	64 (3.9)	42 (2.6)	
Area of residence [n (%)]						<0.001 **
Major cities	810 (49.6)	835 (53.2)	868 (55.0)	992 (60.7)	1002 (63.2)	
Inner regional	499 (30.5)	435 (27.7)	430 (27.2)	388 (23.7)	388 (24.4)	
Outer regional/rural	325 (19.9)	300 (19.1)	281 (17.8)	254 (15.6)	197 (12.4)	
Education [n (%)]						<0.001 **
No formal education	21 (1.3)	22 (1.4)	21 (1.3)	11 (0.7)	9 (0.6)	
High school level	582 (36.1)	507 (32.7)	452 (29.0)	375 (23.3)	289 (18.4)	
Diploma	449 (27.9)	414 (26.7)	398 (25.6)	402 (25.0)	340 (21.7)	
University degree	558 (34.7)	607 (39.2)	685 (44.1)	821 (51.0)	929 (59.3)	
Occupation [n (%)]						<0.001 **
No paid employment	393 (24.2)	354 (22.6)	269 (17.1)	257 (15.9)	195 (12.3)	
Paid employment	1228 (75.8)	1210 (77.4)	1304 (82.9)	1364 (84.1)	1385 (87.7)	
Income stress [n (%)]						<0.001 **
Easy	885 (54.3)	884 (56.4)	904 (57.4)	1031 (63.2)	1036 (65.2)	
Difficult	746 (45.7)	683 (43.6)	670 (42.6)	600 (36.8)	552 (34.8)	
Physical activity [n (%)]						<0.001 **
None/sedentary	204 (12.7)	167 (10.8)	129 (8.3)	101 (6.3)	56 (3.6)	
Low	625 (38.9)	578 (37.3)	510 (32.7)	482 (29.8)	375 (23.8)	
Moderate	359 (22.3)	355 (22.9)	380 (24.2)	414 (25.6)	381 (24.2)	
High	420 (26.1)	450 (29.0)	543 (34.8)	618 (38.3)	761 (48.4)	
Taking prescribed medicine [n (%)]						0.22
No	1170 (72.7)	1107 (71.5)	1134 (72.7)	1206 (74.6)	1171 (74.5)	
Yes	440 (27.3)	442 (28.5)	425 (27.3)	411 (25.4)	401 (25.5)	

The number of participants in each Alternative Healthy Eating Index-2010 (AHEI-2010) quintile varied because of missing data in the covariates. ** p-value < 0.05. § p-values were obtained from analysis of variance (ANOVA) for continuous variables and chi-squared test for categorical variables.

The relationship between AHEI-2010 and the risk of common NCDs (each NCD and multimorbidity) are presented in Table 2. Multivariate logistic regression models with and without a history of asthma were fitted. In the models without a history of asthma, women in the Q5 AHEI-2010 had lower odds of asthma in S4 (2006) when compared to those in the Q1 AHEI-2010 (OR: 0.75, 95% CI: 0.57, 0.99). After the Bonferroni correction was applied to the univariate model (asthma at S4) and multivariate model (asthma at S4 without a history of asthma variable) (Table 2), the odds of asthma in S4 (2006) when compared with the highest vs. lowest quintile of AHEI-2010 resulted in a non-significant association (Table 2, footnote: p-value = 0.41 and 0.42, respectively). In the model with a history of asthma, the association was attenuated and non-significant. Univariate inverse associations between AHEI-2010 and HT at S7 and S8 were found, though these associations became insignificant in the multivariate models. DQ did not predict the occurrence of DM, cancer, depression and/or anxiety, and multimorbidity (Table 2).

When performing the sensitivity analyses using prescribed medicine as a covariate and changes in childbearing variables, the odds of NCDs and multimorbidity remained consistent (Supplementary Tables S5 and S6).

Table 2. Odds of common non-communicable diseases (including multimorbidity) over 15 years of follow-up (from survey 4, S4 to survey 8, S8) based on quintiles of Alternative Healthy Eating Index-2010: 1973–78 Australian Longitudinal Study on Women's Health cohort.

Survey	S4	S5	S6	S7	S8
NCD	**OR (95% CI)**	**OR (95% CI)**	**OR (95% CI)**	**OR (95% CI)**	**OR (95% CI)**
CHD	**n = 11**	**n = 17**	**n = 29**	**n = 42**	**n = 69**
	S4 (n = 6871) [a]	S5 (n = 6127) [a]	S6 (n = 6017) [a]	S7 (n = 5452) [a]	S8 (n = 5394) [a]
Univariate	1.0 (0.1–7.0)	3.9 (0.4–34.6)	1.2 (0.3–4.3)	2.7 (0.6–13.6)	1.1 (0.5–2.5)
HT	**n = 77**	**n = 231**	**n = 346**	**n = 433**	**n = 556**
	S4 (n = 6871) [a]	S5 (n = 6127) [a]	S6 (n = 6017) [a]	S7 (n = 5452) [a]	S8 (n = 5394) [a]
Univariate	0.9 (0.4–1.8)	0.7 (0.4–1.1)	0.7 (0.5–1.0)	0.6 (0.4–0.9) *	0.7 (0.5–0.9) *
	S4 (n = 6608) [b]	S5 (n = 5905) [b]	S6 (n = 5814) [b]	S7 (n = 5268) [b]	S8 (n = 5214) [b]
Multivariate [c]	1.0 (0.5–2.3)	0.7 (0.4–1.2)	0.7 (0.5–1.0)	0.7 (0.5–1.1)	0.8 (0.6–1.1)
Asthma	**n = 662**	**n = 559**	**n = 558**	**n = 464**	**n = 478**
	S4 (n = 6871) [a]	S5 (n = 6127) [a]	S6 (n = 6017) [a]	S7 (n = 5452) [a]	S8 (n = 5394) [a]
Univariate	0.76 (0.59–0.99) * § ¥	0.8 (0.6–1.0)	0.8 (0.6–1.1)	0.9 (0.6–1.2)	0.9 (0.6–1.1)
	S4 (n = 6621) [b]	S5 (n = 5914) [b]	S6 (n = 5824) [b]	S7 (n = 5279) [b]	S8 (n = 5226) [b]
Multivariate [d]	0.75 (0.57–0.99) * § ¥	0.8 (0.6–1.0)	0.8 (0.6–1.1)	0.8 (0.6–1.1)	0.8 (0.6–1.1)
Multivariate [d+]	0.8 (0.6–1.1)	0.8 (0.6–1.1)	0.9 (0.7–1.2)	0.9 (0.6–1.2)	0.8 (0.6–1.2)
Cancer (excludes skin cancer)	**n = 68**	**n = 100**	**n = 148**	**n = 200**	**n = 274**
	S4 (n = 6871) [a]	S5 (n = 6127) [a]	S6 (n = 6017) [a]	S7 (n = 5452) [a]	S8 (n = 5394) [a]
Univariate	1.3 (0.6–2.8)	1.5 (0.7–2.9)	1.1 (0.6–1.8)	1.0 (0.6–1.6)	0.9 (0.6–1.3)
	S4 (n = 6621) [b]	S5 (n = 5866) [b]	S6 (n = 5785) [b]	S7 (n = 5279) [b]	S8 (n = 5226) [b]
Multivariate	1.4 (0.6–3.1)	1.4 (0.7–3.0)	1.1 (0.6–2.0)	0.9 (0.5–1.5)	0.85 (0.6–1.3)
DM	**n = 24**	**n = 62**	**n = 110**	**n = 136**	**n = 167**
	S4 (n = 6871) [a]	S5 (n = 6127) [a]	S6 (n = 6017) [a]	S7 (n = 5452) [a]	S8 (n = 5394) [a]
Univariate	0.8 (0.2–3.0)	1.4 (0.5–3.6)	0.8 (0.4–1.6)	0.7 (0.4–1.4)	0.6 (0.3–1.1)
	S4 (n = 6560) [b]	S5 (n = 5905) [b]	S6 (n = 5814) [b]	S7 (n = 5268) [b]	S8 (n = 5214) [b]
Multivariate [e]	0.9 (0.2–4.0)	1.5 (0.5–4.5)	0.8 (0.4–1.6)	0.7 (0.3–1.4)	0.6 (0.3–1.3)
Depression and/or anxiety	**n = 999**	**n = 1106**	**n = 1199**	**n = 1058**	**n = 1024**
	S4 (n = 6871) [a]	S5 (n = 6127) [a]	S6 (n = 6017) [a]	S7 (n = 5452) [a]	S8 (n = 5394) [a]
Univariate	1.0 (0.8–1.2)	1.0 (0.8–1.2)	1.0 (0.8–1.2)	0.9 (0.8–1.2)	0.9 (0.8–1.2)
	S4 (n = 6621) [b]	S5 (n = 5914) [b]	S6 (n = 5824) [b]	S7 (n = 5279) [b]	S8 (n = 5226) [b]
Multivariate [f]	1.0 (0.8–1.3)	1.0 (0.8–1.2)	1.1 (0.9–1.4)	1.0 (0.8–1.3)	1.0 (0.8–1.3)
Multimorbidity	**n = 198**	**n = 253**	**n = 360**	**n = 346**	**n = 413**
	S4 (n = 6871) [a]	S5 (n = 6127) [a]	S6 (n = 6017) [a]	S7 (n = 5452) [a]	S8 (n = 5394) [a]
Univariate	0.9 (0.6–1.5)	0.9 (0.6–1.4)	0.9 (0.6–1.2)	1.0 (0.7–1.4)	0.9 (0.6–1.2)
	S4 (n = 6621) [b]	S5 (n = 5914) [b]	S6 (n = 5824) [b]	S7 (n = 5279) [b]	S8 (n = 5226) [b]
Multivariate [f]	1.1 (0.7–1.8)	0.9 (0.6–1.4)	0.9 (0.6–1.3)	1.0 (0.7–1.5)	0.8 (0.6–1.2)

CHD: coronary heart disease; CI: confidence interval; DM: diabetes mellitus; HT: hypertension; NCD: non-communicable disease; OR: odds ratio; S: survey. OR (95% CI) expressed in the table is the comparison of odds of having NCDs (each disease and multimorbidity) in the highest quintile to the lowest quintile of AHEI-2010. The main presented results were ORs (95% CI) obtained from respective models. ORs (95% CIs) in respective models were rounded off to 1 decimal place. ¥ ORs (95% CIs) were expressed in 2 decimal places when the CI was close to 1 and rounding would have altered the interpretation. The Bonferroni correction was applied to logistic regression models where 95% CIs were near 1. § Inverse association became non-significant after Bonferroni correction was applied (p-value = 0.41 for univariate model and p-value = 0.42 for multivariate model without a history of asthma variable). Age, socioeconomic status (marital status, residence, education, occupation, and income stress), and the behavioural variable (physical activity) were adjusted for in multivariate models of all NCD outcomes. [a] Number in bracket shows the number of participants in respective surveys for univariate regression; [b] Number in bracket shows the number of participants in respective surveys for multivariate regression; [c] A history of hypertension during pregnancy at S3 was included as a covariate; [d] Model without a history of asthma at S3; [d+] Model with a history of asthma at S3; [e] A history of gestational diabetes mellitus at S3 was included as a covariate; [f] A history of depression and/or anxiety at S3 was included as a covariate. Texts in bold and italic represent accumulative figures of NCD cases (except asthma, depression and/or anxiety) in every survey. * p-value < 0.05.

4. Discussion

In this large population-based study of Australian women aged 25–30 years at baseline, greater adherence to the AHEI-2010 was only associated with the occurrence of asthma at 3 years after the measure of diet. Overall, there was no association between the AHEI-2010 and NCDs and multimorbidity. This suggests that a specific dietary pattern, based on clinical and epidemiological evidence and representing dietary components that have been associated with a lower risk of chronic diseases, is not immediately and obviously associated with NCDs in this group of women.

In the present study, we found no association between AHEI-2010 and CHD in univariate analysis. This finding is inconsistent with the previous studies. In an analysis conducted among Caucasian nurses aged 38–63 years at baseline, those in the Q5 AHEI-2010 had a reduced risk of CHD compared to those in the Q1 AHEI-2010 (HR: 0.66, 95% CI: 0.58, 0.78) over 24 years [40]. The inverse association between AHEI-2010 and incident CVD over 22–26 years, including CHD has been supported in US community-based Atherosclerosis Risk in Communities (ARIC) study participants (aged 45–64 years, 56% women at baseline) [94,95]. Furthermore, in a 10-year follow-up of the NutriNet-Sante (NNS) study cohort (mean age = 41 years, 79% women), participants in quartile 4 of AHEI-2010 had a 22% reduced risk of CVD compared to those in quartile 1 [96]. One reason the results may be inconsistent is that the samples in previous research have generally been older than the sample in the current study. DQ quintiles and its distributions reported in previous studies were higher than those of [95] or similar to our study [40,94]. Many components included in AHEI-2010 are foods and nutrients that have been shown to prevent CVD, including CHD [40]; however, we could not investigate the relationship between AHEI-2010 and CHD because of the very low incidence in this cohort. The self-reported prevalences of CHD in Australian women at various age groups were 0.3% at 18–44 years, 0.7% at 45–54 years, 3.6% at 55–64 years, 6.0% at 65–74 years, and 8.1% at 75 years and older [97].

There was no association between AHEI-2010 and HT in multivariate models. HT, as a risk factor of CVD and investigated as raised systolic blood pressure (SBP) or diastolic blood pressure (DBP), has been investigated in NCD-related research [98–100]. A significant reduction in SBP, not DBP, was demonstrated among people in quartile 4 of AHEI-2010 compared to those in quartile 1 of the 2007–2010 National Health and Nutrition Examination Survey (NHANES) (n = 4097, aged $\geq$ 20 years) [100]. Furthermore, there is now considerable evidence on food and nutrients beneficial for blood pressure (BP) or HT such as fruits [101], vegetables [101], legumes [102], omega-3 fatty acids and PUFAs [103]. Nevertheless, there was no association between AHEI-2010 and HT or BP in a nested case-control study conducted within the Singapore Chinese Health Study (SCHS) (n = 1994, aged 45–75 years, 35% women) [99].

In the present study, women in the Q5 AHEI-2010 had a 25% reduction in the odds of asthma at S4 (2006) compared to those in the Q1 AHEI-2010. However, the association was attenuated and was not found to be statistically significant after applying a Bonferroni correction in the multivariate logistic regression models. Current evidence on DQ and asthma outcome is inconclusive, and studies investigating the role of diet in adults have been limited [104]. The preventive roles of fruits, vegetables, and vitamin E [105,106], as well as fiber [107], and the deleterious roles of red processed meat [108,109] and SSB [110] on asthma have been documented in observational studies. A favourable effect of DQ on asthma was observed in the French prospective Epidemiological study on the Genetics and Environment and Asthma Study (n = 969, mean age = 43 years, 51% women), indicating that a 10-unit increase in AHEI-2010 was shown to improve asthma symptoms, measured as the frequency of respiratory symptoms during the last year [111]. In another French cohort study (n = 26,197 women, aged ≥ 18 years), a 21% reduced odds of asthma symptoms in women and AHEI-2010 was also observed [112]. Opposite findings were observed among 73,228 women from NHS [113] and among 12,687 adult participants (60.2% women) from the Hispanic Community Health Study/Study of Latinos [114]. The conflicting findings between studies could be related to different methodological approaches of measurements for asthma such as asthma symptom score [111,112], self-reported diagnosis [113], current asthma status [114] and for dietary assessments such as FFQ [111,113] or 24-hour dietary recalls [112,114].

With regard to overall cancer and DQ, an inverse association between AHEI-2010 and overall cancer was observed in 71,495 women aged 38–63 years from the NHS over 24 years [40] and 41,543 participants from the NNS study over 9 years (aged ≥40 years, 73.5% women) [115]. In our study, we did not find any association between DQ and cancer (excluding skin cancer). The previous analysis conducted among the ALSWH cohort showed similar results [50]. The discrepancies between studies might be partly explained by shorter follow-up durations since cancer cannot be detected within a far shorter period [116]. Another reason may be the components measured in AHEI-2010 were not cancer-specific but based on dietary factors related to chronic disease risk in general [40]. Given that specific types of fruits and vegetables have an impact on cancer, their effect could not be found when all fruits and vegetables are combined as components in a DQI [117]. Furthermore, the endpoint for total cancer is heterogeneous compared with other NCDs such as CHD, asthma, or DM [40].

Previous studies have demonstrated a significant inverse association between AHEI-2010 and the risk of DM among participants from the NHS over 24 years of follow-up (n = 71,495, aged 38–63 years) [40], among white women from the MEC study (aged 45–74 years, 53% women) [118], among participants from the Women's Health Initiative Observational Study (aged 50–79 years) [119], among women from the Singapore Chinese Health Study (aged 45–74 years) [120], and among the ALSWH 1946–51 cohort (n = 5350, aged 50–55 years) [50]. However, these findings were not replicated in the present study or in a recent analysis of the ARIC Study over a median follow-up of 22 years (n = 10,808, aged 45–64 years, 56.1% women) [95]. Many components in AHEI-2010 had been constructed based on the optimal dietary factors for the prevention of DM [40]. Notwithstanding AHEI-2010 being the most sensitive DQI for incident DM among the three DQIs tested in the previous ALSWH cohort analysis [50], these results cannot extend to the younger age ALSWH cohort.

Previous evidence showed that healthy dietary patterns characterized by vegetables, fruits and whole grains was inversely associated with depression [121]. The underlying preventive effects of vegetables and fruits for depression may be related to (1) the antioxidant actions of vitamin C, vitamin E, and carotenoid compounds; (2) the balancing of neurotransmitter levels, for instance, by reducing homocysteine concentrations by folates [122]. The anti-inflammatory properties of long-chain omega-3 PUFA also contribute to neurotransmission and are beneficial for depression [123]. The deleterious effects of red/processed meat and SSBs have been shown in previous studies [124,125]. In the scoring of AHEI-2010, higher points are given to high intakes of beneficial components and low intakes of deleterious components. The preventive potential of DQ assessed by AHEI-2010 on depression and/or anxiety was observed among Iranian adults (n = 3363, mean age = 36 years, 58.3% women) [126] and Spanish adults (n = 15,093, aged ≥18 years,

59% women) [127]. However, there was no association between AHEI-2010 and depression and/or anxiety in the present study, nor in a French cohort (n = 26,225, aged 18–86 years, 76% women) [128]. Compared with our results, reported prevalence or incidence of depression were higher in the Iranian study (30% in total participants, 35% in women) [126] but lower in the Spanish (~7%) [127] and French (~8.3%) [128] studies. Given that the incidence of depression and/or anxiety in our participants (~15%) was lower than Iranian women (35%), it might have been underpowered to detect the difference.

Regarding the relationship between dietary factors and NCD multimorbidty, the literature is limited and inconclusive [50,129–136]. Inverse associations between vegetable and fruits consumption and NCD multimorbidity have been documented in cross-sectional [130,131] and longitudinal [132–134] studies. The harmful effect of soft drink consumption on multimorbidity was reported in an Australian study (n = 36,663, aged $\geq$ 16 years, 51.1% women) [129]. In exploring the relationship between overall diet assessed by DQIs and multimorbidity, decreased relative odds were observed among European adults who adhered to the Mediterranean diet (n = 291,778, aged 43–58 years, 64% women vs. n = 1140, aged 18 years, 56% women) [135,136] and Australian women who adhered to the Australian Dietary Guideline and AHEI-2010 (n = 5,350, aged 50–55 years) [50] when compared to those who did not. In contrast to the previous findings [50], there was no association between AHEI-2010 and NCD multimorbidity in this study. A potential explanation for the lack of association could be the differences between the AHEI-2010 scores amongst the ALSWH 1946–51 cohort (mean ($\pm$sd): 56.0 $\pm$ 10.3, range: 26–93.8) and this study cohort (mean ($\pm$sd): 51.6 $\pm$ 11.0, range: 19–91).

Descriptive summaries of AHEI-2010 have been reported in previous studies: similar to [40,94,119], higher than [50,95,98,118,127], or smaller than [113] the present study. Comparisons cannot made with some studies since they did not include trans-fat [99,100,111,112,115,120] or alcohol [126] because those data were not available.

The application of the most sensitive DQI, AHEI-2010, in predicting NCD outcomes in the previous study [50] is one of the strengths of this study. AHEI-2010 is the latest version based on the clinical and epidemiological evidence, and it is widely used in diet-health relationship studies [41]. However, most of these studies have been conducted in the United States [41]. A few Australian studies have used AHEI-2010 as a measurement of DQ [137–141], and only two studies were related to NCDs such as urothelial cell carcinoma [141] and ovarian cancer survival [137]. Hence, this study adds to the body of evidence on DQ. The analysis of healthy childbearing women from a nationally representative population [142] is another strength of this study.

Several limitations should be acknowledged. Firstly, in comparison with the originally recruited young women at S1 (n = 14,247), those excluded from the study for the reasons of having NCDs and missing FFQ at S3 (n = 1,064) were more likely to be married, originate from inner regional and outer regional/rural areas, have no formal education, have no paid employment, have difficulty in income management, have lower physical activity, have poor/fair self-rated health, and have taken prescription medications (Supplementary Table S4). Women excluded from the present study were distributed in lower DQ quintiles such as quintiles 1 and 2 (data not shown). This affected the examination of DQ differences between the groups, making it difficult to detect small variations that could have been biased towards the null. Secondly, the exposure and variables at S3 were used in logistic regression models. Cross-sectional studies cannot determine whether there is temporal relationship between exposure and outcome of interests. A longitudinal analysis that allows for the adjustment of time-varying covariates should be performed. However, the effects of change in childbearing variables such as current pregnancy status, parity, history of breastfeeding, history of GDM, and history of HTpreg were investigated in sensitivity analysis (Supplementary Table S6). Thirdly, the insufficient NCD cases at each survey, particularly CHD and DM, could have introduced high uncertainty in our effect sizes. Fourthly, the number of cancer cases (except skin cancer) could have been affected by the inability to exclude skin cancer at earlier surveys (from S1 to S4). However, when the proportions of cancer cases (not specified) at S1 to S4 were checked amongst skin cancer cases across S5 to S8, these data comprised no more than 5% (data not shown). The selected NCDs are characterised by increasing incidence with age, which makes it difficult to obtain results based on clinical endpoints such as CHD and DM. However, looking at modifiable risk factors early, even prior to disease onset, is important given that Australian reports have shown that 43% of women aged 45 and younger are affected by one or more NCDs [11,12]. Exploring NCD risk factors and considering which NCDs are the most prevalent at different life stages [143,144] are important for developing appropriate primary prevention strategies. In assessing dietary intake, FFQs, such as the DQES-v2, are more useful in providing relative ranking than absolute values of intake [145]. Calculating a DQI from FFQs introduces further uncertainty in the dietary measure and by no means represents a complete picture of diet [146]. Instead, it provides a simple summary that performs well across large-scale national and international studies [23]. Lastly, residual confounding is still possible even though appropriate covariates adjustments were made with multivariate analysis and selected based on DAGs.

5. Conclusions

A high DQ measured by AHEI-2010 was only associated with the occurrence of asthma at S4. A cross-sectional analysis examining the association between overall DQ and NCDs could not determine the causal relationship. A longitudinal analysis is therefore necessary to investigate the temporal associations. While most research examining diet-health outcome relationship has suggested the preventive effect of overall diet on NCDs, the present findings did not demonstrate early evidence of this between the ages of 25–45 years. Further well-planned, large prospective studies with sensitive indicators of NCD risk factors and incident NCDs are indispensable.

Supplementary Materials: The supplementary figures and tables can be downloaded at: https://www.mdpi.com/article/10.3390/nu14204403/s1.

Author Contributions: H.H.-H., X.D.-G., A.J.H., E.L.J. and M.T. contributed to conceptualization and methodology; H.H.-H., A.J.H. and X.D.-G. conducted formal analysis; H.H.-H. drafted the paper; H.H.-H., A.J.H., E.L.J., M.T. and X.D.-G. contributed to a critical review and editing of the manuscript. All authors have read and agreed to the published version of the manuscript.

Funding: This research received no external funding.

Institutional Review Board Statement: Ethical approval for the ALSWH study was granted by the Human Ethics Committees of the University of Newcastle (reference: h-076-0795) and University of Queensland (reference: 200400224). All procedures in this study were in accordance with the guidelines laid down in the Declaration of Helsinski.

Informed Consent Statement: Written informed consent of the ALSWH study was obtained fromall participants prior to inclusion to each survey.

Data Availability Statement: ALSWH survey data are owned by the Australian Government Department of Health and due to the personal nature of the data collected, release by ALSWH is subject to strict contractual and ethical restrictions. Ethical review of ALSWH is by the Human Research Ethics Committees at The University of Queensland and The University of Newcastle. De-identified data are available to collaborating researchers where a formal request to make use of the material has been approved by the ALSWH Data Access Committee. The committee is receptive of requests for datasets required to replicate results. Information on applying for ALSWH data is available from https://alswh.org.au/for-data-users/applying-for-data/ (accessed on 7 March 2021).

Acknowledgments: HH-H acknowledges support from the University of Newcastle PhD scholarship. The research on which this paper is based was conducted as part of the Australian Longitudinal Study on Women's Health (ALSWH) by the University of Queensland and the University of Newcastle. We are grateful to the Australian Government Department of Health and Aged Care for funding and to the women who provided the survey data. The authors thank Professor Graham Giles and Professor Roger Milne of the Cancer Epidemiology Centre of Cancer Council Victoria, for permission to use the Dietary Questionnaire for Epidemiological Studies-Version 2 (DQES-v2), Melbourne: Cancer Council

Victoria, 1996. We would like to thank Professor Gita Mishra (ALSWH Liaison) and Professor Allison Hodge (DQES Representative). The authors acknowledge the assistance of the Data Linkage Unit at the Australian Institute of Health and Welfare (AIHW) for undertaking the data linkage to the National Death Index (NDI).

Conflicts of Interest: The authors declare no conflict of interest.

Abbreviations

AHEI: Alternative Healthy Eating Index, AHEI-2010: Alternative Healthy Eating Index-2010, ALSWH: Australian Longitudinal Study on Women's Health, ANOVA: analysis of variance, ARIC: Atherosclerosis Risk in Communities, BP: blood pressure, CHD: coronary heart disease, CI: confidence interval, CVD: cardiovascular disease; DAG: Directed Acyclic Graph, DBP: diastolic blood pressure, DM: diabetes mellitus, DQES-v2: Dietary Questionnaire for Epidemiological Studies version 2, DQ: diet quality; DQI: diet quality index, DQIs: diet quality indices, FFQ: Food Frequency Questionnaire, GDM: gestational diabetes mellitus; HEI: Healthy Eating Index, HEIFA-2013: Healthy Eating Index for Australian Adult-2013, HPFS: Health Professional Follow-up Study; HR: hazard ratio; HT: hypertension, HTpreg: Hypertension during pregnancy; MDS: Mediterranean Diet Scale, MEC: Multiethnic Cohort; MET: metabolic equivalent of task, NCD: non-communicable disease, NHANES: National Health and Nutrition Examination Survey; NHS: Nurses' Health Study, NNS: NutriNet-Santè, OTC: over-the-counter, OR: odds ratio; PUFA: polyunsaturated fatty acid, S: survey; SBP: systolic blood pressure; SCHS: Singapore Chinese Health Study; SES: Socioeconomic status, SSBs: sugar-sweetened beverages.

References

1. World Health Organization. Global Health Estimates 2020: Deaths by Cause, Age, Sex, by Country and by Region, 2000–2019. Available online: https://www.who.int/data/gho/data/themes/mortality-and-global-health-estimates/ghe-leading-causes-of-death (accessed on 9 August 2021).
2. World Health Organization. *World Health Statistics 2021: Monitoring Health for the SDGs, Sustainable Development Goals*; World Health Organization: Geneva, Switzerland, 2021; ISBN 978-92-4-002705-3.
3. Afshar, S.; Roderick, P.J.; Kowal, P.; Dimitrov, B.D.; Hill, A.G. Multimorbidity and the inequalities of global ageing: A cross-sectional study of 28 countries using the World Health Surveys. *BMC Public Health* **2015**, *15*, 776. [CrossRef] [PubMed]
4. NCD Alliance. *Non-Communicable Diseases: A Priortity for Women's Health and Development*; The NCD Alliance: Geneva, Switzerland, 2011.
5. Taskforce on Women and NCDs. Women and Non-Communicable Diseases: A Call to Action. Available online: https://ncdalliance.org/sites/default/files/resource_files/Women%20and%20NCDs%20infographic_WEB_fv.pdf (accessed on 11 November 2021).
6. Canoy, D.; Cairns, B.J.; Balkwill, A.; Wright, F.L.; Khalil, A.; Beral, V.; Green, J.; Reeves, G.; Collaborators, M.W.S. Hypertension in pregnancy and risk of coronary heart disease and stroke: A prospective study in a large UK cohort. *Int. J. Cardiol.* **2016**, *222*, 1012–1018. [CrossRef] [PubMed]
7. Veerbeek, J.H.; Hermes, W.; Breimer, A.Y.; Van Rijn, B.B.; Koenen, S.V.; Mol, B.W.; Franx, A.; De Groot, C.J.; Koster, M.P. Cardiovascular disease risk factors after early-onset preeclampsia, late-onset preeclampsia, and pregnancy-induced hypertension. *Hypertension* **2015**, *65*, 600–606. [CrossRef]
8. Li, G.; Wei, T.; Ni, W.; Zhang, A.; Zhang, J.; Xing, Y.; Xing, Q. Incidence and risk factors of gestational diabetes mellitus: A prospective cohort study in Qingdao, China. *Front. Endocrinol.* **2020**, *11*, 636. [CrossRef] [PubMed]
9. Johnston, M.C.; Crilly, M.; Black, C.; Prescott, G.J.; Mercer, S.W. Defining and measuring multimorbidity: A systematic review of systematic reviews. *Eur. J. Public Health* **2019**, *29*, 182–189. [CrossRef] [PubMed]
10. Le Reste, J.Y.; Nabbe, P.; Manceau, B.; Lygidakis, C.; Doerr, C.; Lingner, H.; Czachowski, S.; Munoz, M.; Argyriadou, S.; Claveria, A. The European General Practice Research Network presents a comprehensive definition of multimorbidity in family medicine and long term care, following a systematic review of relevant literature. *J. Am. Med. Dir. Assoc.* **2013**, *14*, 319–325. [CrossRef]
11. Harris, M.L.; Egan, N.; Forder, P.M.; Loxton, D. Increased chronic disease prevalence among the younger generation: Findings from a population-based data linkage study to inform chronic disease ascertainment among reproductive-aged Australian women. *PLoS ONE* **2021**, *16*, e0254668. [CrossRef]
12. Australian Institute of Health and Welfare. The Health of Australia's Females. Available online: https://www.aihw.gov.au/reports/men-women/female-health/contents/who-are (accessed on 31 August 2021).

13. World Health Organization. *Gender and Noncommunicable Diseases in Europe: Analysis of STEPS Data*; World Health Organization, Regional Office for Europe: Geneva, Switzerland, 2020.
14. Park, K. *Park's Textbook of Preventive and Social Medicine*, 23rd ed.; Banarsidas Bhanot: Madhya Pradesh, India, 2015.
15. World Cancer Research Fund / American Institute for Cancer Research. *Continuous Update Project Expert Report 2018. Diet, Nutrition, Physical Activity and Colorectal Cancer*; World Cancer Research Fund International: London, UK, 2018; ISBN 978-1-912259-00-7.
16. Murray, C.J.; Aravkin, A.Y.; Zheng, P.; Abbafati, C.; Abbas, K.M.; Abbasi-Kangevari, M.; Abd-Allah, F.; Abdelalim, A.; Abdollahi, M.; Abdollahpour, I. Global burden of 87 risk factors in 204 countries and territories, 1990–2019: A systematic analysis for the Global Burden of Disease Study 2019. *Lancet* **2020**, *396*, 1223–1249. [CrossRef]
17. Waijers, P.M.C.M.; Feskens, E.J.M.; Ocke, M.C. A critical review of predefined diet quality scores. *Br. J. Nutr.* **2007**, *97*, 219–231. [CrossRef]
18. Arvaniti, F.; Panagiotakos, D.B. Healthy indexes in public health practice and research: A review. *Crit. Rev. Food Sci. Nutr.* **2008**, *48*, 317–327. [CrossRef]
19. Wirt, A.; Collins, C.E. Diet quality—What is it and does it matter? *Public Health Nutr.* **2009**, *12*, 2473–2492. [CrossRef] [PubMed]
20. Kant, A.K. Dietary patterns and health outcomes. *J. Am. Diet. Assoc.* **2004**, *104*, 615–635. [CrossRef] [PubMed]
21. Newby, P.K.; Tucker, K.L. Empirically derived eating patterns using factor or cluster analysis: A review. *Nutr. Rev.* **2004**, *62*, 177–203. [CrossRef] [PubMed]
22. Kourlaba, G.; Panagiotakos, D.B. Dietary quality indices and human health: A review. *Maturitas* **2009**, *62*, 1–8. [CrossRef]
23. Hodge, A.; Bassett, J. What can we learn from dietary pattern analysis? *Public Health Nutr.* **2016**, *19*, 191–194. [CrossRef]
24. Gerber, M. The comprehensive approach to diet: A critical review. *J. Nutr.* **2001**, *131*, 3051S–3055S. [CrossRef]
25. Thorpe, M.G.; Kestin, M.; Riddell, L.J.; Keast, R.S.; McNaughton, S.A. Diet quality in young adults and its association with food-related behaviours. *Public Health Nutr.* **2013**, *17*, 1767–1775. [CrossRef]
26. Kennedy, T.E.; Ohls, J.; Carlson, S.; Fleming, K. The Healthy Eating Index: Design and applications. *J. Am. Diet. Assoc.* **1995**, *95*, 1103–1108. [CrossRef]
27. Patterson, R.E.; Haines, P.S.; Popkin, B.M. Diet quality index: Capturing a multidimensional behavior. *J. Am. Diet. Assoc.* **1994**, *94*, 57–64. [CrossRef]
28. Haines, P.S.; Siega-Riz, A.M.; Popkin, B.M. The Diet Quality Index Revised: A measurement instrument for populations. *J. Am. Diet. Assoc.* **1999**, *99*, 697–704. [CrossRef]
29. Trichopoulou, A.; Kouris-Blazos, A.; Wahlqvist, M.L.; Gnardellis, C.; Lagiou, P.; Polychronopoulos, E.; Vassilakou, T.; Lipworth, L.; Trichopoulos, D. Diet and overall survival in elderly people. *BMJ* **1995**, *311*, 1457–1460. [CrossRef] [PubMed]
30. McNaughton, S.A.; Ball, K.; Crawford, D.; Mishra, G.D. An index of diet and eating patterns is a valid measure of diet quality in an Australian population. *J. Nutr.* **2008**, *138*, 86–93. [CrossRef] [PubMed]
31. Wang, Z.; Siega-Riz, A.M.; Gordon-Larsen, P.; Cai, J.; Adair, L.S.; Zhang, B.; Popkin, B.M. Diet quality and its association with type 2 diabetes and major cardiometabolic risk factors among adults in China. *Nutr. Metab. Cardiovasc. Dis.* **2018**, *28*, 987–1001. [CrossRef]
32. Wong, J.E.; Haszard, J.J.; Howe, A.S.; Parnell, W.R.; Skidmore, P.M.L. Development of a Healthy Dietary Habits Index for New Zealand adults. *Nutrients* **2017**, *9*, 454. [CrossRef] [PubMed]
33. McCullough, M.L.; Feskanich, D.; Stampfer, M.J.; Giovannucci, E.L.; Rimm, E.B.; Hu, F.B.; Spiegelman, D.; Hunter, D.J.; Colditz, G.A.; Willett, W.C. Diet quality and major chronic disease risk in men and women: Moving toward improved dietary guidance. *Am. J. Clin. Nutr.* **2002**, *76*, 1261–1271. [CrossRef] [PubMed]
34. Collins, C.E.; Young, A.F.; Hodge, A. Diet quality is associated with higher nutrient intake and self-rated health in mid-aged women. *J. Am. Coll. Nutr.* **2008**, *27*, 146–157. [CrossRef] [PubMed]
35. Wu, J.; Song, X.; Chen, G.-C.; Neelakantan, N.; van Dam, R.M.; Feng, L.; Yuan, J.-M.; Pan, A.; Koh, W.-P. Dietary pattern in midlife and cognitive impairment in late life: A prospective study in Chinese adults. *Am. J. Clin. Nutr.* **2019**, *110*, 912–920. [CrossRef]
36. Guenther, P.M.; Reedy, J.; Krebs-Smith, S.M. Development of the Healthy Eating Index-2005. *J. Am. Diet. Assoc.* **2008**, *108*, 1896–1901. [CrossRef]
37. Guenther, P.M.; Casavale, K.O.; Reedy, J.; Kirkpatrick, S.I.; Hiza, H.A.; Kuczynski, K.J.; Kahle, L.L.; Krebs-Smith, S.M. Update of the Healthy Eating Index: HEI-2010. *J. Acad. Nutr. Diet.* **2013**, *113*, 569–580. [CrossRef]
38. Krebs-Smith, S.M.; Pannucci, T.E.; Subar, A.F.; Kirkpatrick, S.I.; Lerman, J.L.; Tooze, J.A.; Wilson, M.M.; Reedy, J. Update of the Healthy Eating Index: HEI-2015. *J. Acad. Nutr. Diet.* **2018**, *118*, 1591–1602. [CrossRef]
39. Reedy, J.; Lerman, J.; Herrick, K.; Shams-White, M.; Zimmer, M.; Mathieu, K.M.; Dodd, K.; Kahle, L.; Stoody, E.; Pannucci, T. Measuring Diet Quality Across the Lifespan: Introducing the New Healthy Eating Index-Toddlers-2020 and Healthy Eating Index-2020. *Curr. Dev. Nutr.* **2022**, *6*, 392. [CrossRef]
40. Chiuve, S.E.; Fung, T.T.; Rimm, E.B.; Hu, F.B.; McCullough, M.L.; Wang, M.; Stampfer, M.J.; Willett, W.C. Alternative dietary indices both strongly predict risk of chronic disease. *J. Nutr.* **2012**, *142*, 1009–1018. [CrossRef] [PubMed]
41. Morze, J.; Danielewicz, A.; Hoffmann, G.; Schwingshackl, L. Diet quality as assessed by the Healthy Eating Index, Alternate Healthy Eating Index, Dietary Approaches to Stop Hypertension Score, and health outcomes: A second update of a systematic review and meta-analysis of cohort studies. *J. Acad. Nutr. Diet.* **2020**, *120*, 1998–2031.e1915. [CrossRef] [PubMed]
42. Morze, J.; Danielewicz, A.; Przybyłowicz, K.; Zeng, H.; Hoffmann, G.; Schwingshackl, L. An updated systematic review and meta-analysis on adherence to mediterranean diet and risk of cancer. *Eur. J. Nutr.* **2021**, *60*, 1561–1586. [CrossRef] [PubMed]

43. Martín-Peláez, S.; Fito, M.; Castaner, O. Mediterranean diet effects on type 2 diabetes prevention, disease progression, and related mechanisms. a review. *Nutrients* **2020**, *12*, 2236. [CrossRef]
44. Martínez-González, M.A.; Gea, A.; Ruiz-Canela, M. The Mediterranean diet and cardiovascular health: A critical review. *Circ. Res.* **2019**, *124*, 779–798. [CrossRef]
45. Cowell, O.R.; Mistry, N.; Deighton, K.; Matu, J.; Griffiths, A.; Minihane, A.M.; Mathers, J.C.; Shannon, O.M.; Siervo, M. Effects of a Mediterranean diet on blood pressure: A systematic review and meta-analysis of randomized controlled trials and observational studies. *J. Hypertens.* **2021**, *39*, 729–739. [CrossRef]
46. Shafiei, F.; Salari-Moghaddam, A.; Larijani, B.; Esmaillzadeh, A. Adherence to the Mediterranean diet and risk of depression: A systematic review and updated meta-analysis of observational studies. *Nutr. Rev.* **2019**, *77*, 230–239. [CrossRef]
47. Soltani, S.; Jayedi, A.; Shab-Bidar, S.; Becerra-Tomás, N.; Salas-Salvadó, J. Adherence to the Mediterranean diet in relation to all-cause mortality: A systematic review and dose-response meta-analysis of prospective cohort studies. *Adv. Nutr.* **2019**, *10*, 1029–1039. [CrossRef]
48. Altun, A.; Brown, H.; Szoeke, C.; Goodwill, A.M. The Mediterranean dietary pattern and depression risk: A systematic review. *Neurol. Psychiatry Brain Res.* **2019**, *33*, 1–10. [CrossRef]
49. Singh-Manoux, A.; Fayosse, A.; Sabia, S.; Tabak, A.; Shipley, M.; Dugravot, A.; Kivimäki, M. Clinical, socioeconomic, and behavioural factors at age 50 years and risk of cardiometabolic multimorbidity and mortality: A cohort study. *PLoS Med.* **2018**, *15*, e1002571. [CrossRef] [PubMed]
50. Hlaing-Hlaing, H.; Dolja-Gore, X.; Tavener, M.; James, E.L.; Hodge, A.M.; Hure, A.J. Diet quality and Incident non-communicable disease in the 1946–1951 cohort of the Australian Longitudinal Study on Women's Health. *Int. J. Environ. Res. Public Health* **2021**, *18*, 11375. [CrossRef] [PubMed]
51. Mikkelsen, B.; Williams, J.; Rakovac, I.; Wickramasinghe, K.; Hennis, A.; Shin, H.-R.; Farmer, M.; Weber, M.; Berdzuli, N.; Borges, C. Life course approach to prevention and control of non-communicable diseases. *BMJ* **2019**, *364*, l257. [CrossRef] [PubMed]
52. World Health Organization. *The Minsk Declaration: The Life-Course Approach in the Context of Health 2020*; World Health Organization, Regional Office for Europe: Geneva, Switzerland, 2015.
53. Smith, K.J.; McNaughton, S.A.; Gall, S.L.; Otahal, P.; Dwyer, T.; Venn, A.J. Associations between partnering and parenting transitions and dietary habits in young adults. *J. Acad. Nutr. Diet.* **2017**, *117*, 1210–1221. [CrossRef]
54. Ball, K.; Jeffery, R.W.; Abbott, G.; McNaughton, S.A.; Crawford, D. Is healthy behavior contagious: Associations of social norms with physical activity and healthy eating. *Int. J. Behav. Nutr. Phys. Act* **2010**, *7*, 86. [CrossRef]
55. Elstgeest, L.E.; Mishra, G.D.; Dobson, A.J. Transitions in living arrangements are associated with changes in dietary patterns in young women. *J. Nutr.* **2012**, *142*, 1561–1567. [CrossRef]
56. Olson, C.M. Tracking of food choices across the transition to motherhood. *J. Nutr. Educ. Behav.* **2005**, *37*, 129–136. [CrossRef]
57. Laroche, H.H.; Wallace, R.B.; Snetselaar, L.; Hillis, S.L.; Steffen, L.M. Changes in diet behavior when adults become parents. *J. Acad. Nutr. Diet.* **2012**, *112*, 832–839. [CrossRef]
58. Berge, J.M.; Larson, N.; Bauer, K.W.; Neumark-Sztainer, D. Are parents of young children practicing healthy nutrition and physical activity behaviors? *Pediatrics* **2011**, *127*, 881–887. [CrossRef]
59. Pegington, M.; French, D.P.; Harvie, M.N. Why young women gain weight: A narrative review of influencing factors and possible solutions. *Obes. Rev.* **2020**, *21*, e13002. [CrossRef]
60. Powell-Wiley, T.M.; Poirier, P.; Burke, L.E.; Després, J.-P.; Gordon-Larsen, P.; Lavie, C.J.; Lear, S.A.; Ndumele, C.E.; Neeland, I.J.; Sanders, P. Obesity and cardiovascular disease: A scientific statement from the American Heart Association. *Circulation* **2021**, *143*, e984–e1010. [CrossRef] [PubMed]
61. Lauby-Secretan, B.; Scoccianti, C.; Loomis, D.; Grosse, Y.; Bianchini, F.; Straif, K. Body fatness and cancer—viewpoint of the IARC Working Group. *N. Engl. J. Med.* **2016**, *375*, 794–798. [CrossRef]
62. Rowley, W.R.; Bezold, C.; Arikan, Y.; Byrne, E.; Krohe, S. Diabetes 2030: Insights from yesterday, today, and future trends. *Popul. Health Manag.* **2017**, *20*, 6–12. [CrossRef] [PubMed]
63. Australian Institute of Health and Welfare. *Australia's Health 2020: In Brief*; Australian Institute of Health and Welfare: Canberra, Australia, 2020; ISBN 978-1-76054-691-5.
64. Dobson, A.J.; Hockey, R.; Brown, W.J.; Byles, J.E.; Loxton, D.J.; McLaughlin, D.P.; Tooth, L.R.; Mishr, G.D. Cohort profile update: Australian Longitudinal Study on Women's Health. *Int. J. Epidemiol.* **2015**, *44*, 1547. [CrossRef] [PubMed]
65. Loxton, D.; Tooth, L.; Harris, M.L.; Forder, P.M.; Dobson, A.; Powers, J.; Brown, W.; Byles, J.; Mishra, G. Cohort profile: The Australian longitudinal study on Women's health (ALSWH) 1989–1995 cohort. *Int. J. Epidemiol.* **2018**, *47*, 391–392e. [CrossRef] [PubMed]
66. Brown, W.J.; Dobson, A.J.; Team, A.R. The Australian Longitudinal Study on Women's Health: Study design and sample. *N. S. W. Public Health Bull.* **2000**, *11*, 3–4. [CrossRef] [PubMed]
67. Ireland, P.; Jolley, D.; Giles, G.; O'Dea, K.; Powles, J.; Rutishauser, I.; Wahlqvist, M.L.; Williams, J. Development of the Melbourne FFQ: A food frequency questionnaire for use in an Australian prospective study involving an ethnically diverse cohort. *Asia Pac. J. Clin. Nutr.* **1994**, *3*, 19–31. [PubMed]
68. Hodge, A.; Patterson, A.J.; Brown, W.J.; Ireland, P.; Giles, G. The Anti Cancer Council of Victoria FFQ: Relative validity of nutrient intakes compared with weighed food records in young to middle-aged women in a study of iron supplementation. *Aust. N. Z. J. Public Health* **2000**, *24*, 576–583. [CrossRef]

69. Lewis, J.; Milligan, G.C.; Hunt, A. *Nuttab95: Nutrient Data Table for Use in Australia*; Food Standards Australia New Zealand: Barton, Australia, 1995.
70. Baldwin, J.N.; Forder, P.M.; Haslam, R.L.; Hure, A.J.; Loxton, D.J.; Patterson, A.J.; Collins, C.E. Change in diet quality over 12 years in the 1946–1951 cohort of the Australian Longitudinal Study on Women's Health. *Nutrients* **2020**, *12*, 147. [CrossRef] [PubMed]
71. Shivappa, N.; Hebert, J.R.; Kivimaki, M.; Akbaraly, T. Alternative Healthy Eating Index 2010, Dietary Inflammatory Index and risk of mortality: Results from the Whitehall II cohort study and meta-analysis of previous Dietary Inflammatory Index and mortality studies. *Br. J. Nutr.* **2017**, *118*, 210–221. [CrossRef] [PubMed]
72. Hlaing-Hlaing, H.; Pezdirc, K.; Tavener, M.; James, E.L.; Hure, A. Diet quality indices used in Australian and New Zealand adults: A systematic review and critical appraisal. *Nutrients* **2020**, *12*, 3777. [CrossRef] [PubMed]
73. Cristina, T.J.N.; Williams, J.A.S.; Parkinson, L.; Sibbritt, D.W.; Byles, J.E. Identification of diabetes, heart disease, hypertension and stroke in mid- and older-aged women: Comparing self-report and administrative hospital data records. *Geriatr. Gerontol. Int.* **2016**, *16*, 95–102. [CrossRef]
74. Stavrou, E.; Vajdic, C.M.; Loxton, D.; Pearson, S.-A. The validity of self-reported cancer diagnoses and factors associated with accurate reporting in a cohort of older Australian women. *Cancer Epidemiol.* **2011**, *35*, e75–e80. [CrossRef] [PubMed]
75. Australian Longitudinal Study on Women's Health. ALSWH: Linked Data Overview. Available online: https://alswh.org.au/for-data-users/linked-data-overview/national-data/ (accessed on 26 February 2022).
76. Hernán, M.A. The C-word: Scientific euphemisms do not improve causal inference from observational data. *Am. J. Public Health* **2018**, *108*, 616–619. [CrossRef] [PubMed]
77. Heinze, G.; Wallisch, C.; Dunkler, D. Variable selection–a review and recommendations for the practicing statistician. *Biom. J.* **2018**, *60*, 431–449. [CrossRef] [PubMed]
78. Greenland, S.; Pearl, J.; Robins, J.M. Causal diagrams for epidemiologic research. *Epidemiology* **1999**, *10*, 37–48. [CrossRef] [PubMed]
79. Textor, J.; Hardt, J.; Knüppel, S. DAGitty: A graphical tool for analyzing causal diagrams. *Epidemiology* **2011**, *22*, 745. [CrossRef] [PubMed]
80. Tennant, P.W.; Murray, E.J.; Arnold, K.F.; Berrie, L.; Fox, M.P.; Gadd, S.C.; Harrison, W.J.; Keeble, C.; Ranker, L.R.; Textor, J. Use of directed acyclic graphs (DAGs) to identify confounders in applied health research: Review and recommendations. *Int. J. Epidemiol.* **2021**, *50*, 620–632. [CrossRef]
81. Asghari, G.; Mirmiran, P.; Yuzbashian, E.; Azizi, F. A systematic review of diet quality indices in relation to obesity. *Br. J. Nutr.* **2017**, *117*, 1055–1065. [CrossRef]
82. Peacock, A.S.; Bogossian, F.; McIntyre, H.D.; Wilkinson, S. A review of interventions to prevent type 2 diabetes after gestational diabetes. *Women Birth* **2014**, *27*, e7–e15. [CrossRef]
83. Mongraw-Chaffin, M.L.; Peters, S.A.; Huxley, R.R.; Woodward, M. The sex-specific association between BMI and coronary heart disease: A systematic review and meta-analysis of 95 cohorts with 1 × 2 million participants. *Lancet Diabetes Endocrinol.* **2015**, *3*, 437–449. [CrossRef]
84. Jackson, C.; Herber-Gast, G.-C.; Brown, W. Joint effects of physical activity and BMI on risk of hypertension in women: A longitudinal study. *J. Obes.* **2014**, *2014*, 271532. [CrossRef]
85. Peters, U.; Dixon, A.E.; Forno, E. Obesity and asthma. *J. Allergy Clin. Immunol.* **2018**, *141*, 1169–1179. [CrossRef] [PubMed]
86. Tyrrell, J.; Mulugeta, A.; Wood, A.R.; Zhou, A.; Beaumont, R.N.; Tuke, M.A.; Jones, S.E.; Ruth, K.S.; Yaghootkar, H.; Sharp, S. Using genetics to understand the causal influence of higher BMI on depression. *Int. J. Epidemiol.* **2019**, *48*, 834–848. [CrossRef] [PubMed]
87. Bhaskaran, K.; Douglas, I.; Forbes, H.; dos-Santos-Silva, I.; Leon, D.A.; Smeeth, L. Body-mass index and risk of 22 specific cancers: A population-based cohort study of 5.24 million UK adults. *Lancet* **2014**, *384*, 755–765. [CrossRef]
88. Gondek, D.; Bann, D.; Brown, M.; Hamer, M.; Sullivan, A.; Ploubidis, G.B. Prevalence and early-life determinants of mid-life multimorbidity: Evidence from the 1970 British birth cohort. *BMC Public Health* **2021**, *21*, 1319. [CrossRef] [PubMed]
89. Australian Institute of Health and Welfare. *The Active Australia Survey: A Guide and Manual for Implementation, Analysis and Reporting*; Australian Institute of Health and Welfare: Canberra, Australia, 2003; ISBN 1740242580.
90. Brown, W.J.; Burton, N.W.; Marshall, A.L.; Miller, Y.D. Reliability and validity of a modified self-administered version of the Active Australia physical activity survey in a sample of mid-age women. *Aust. N. Z. J. Public Health* **2008**, *32*, 535–541. [CrossRef] [PubMed]
91. Dolja-Gore, X.; Graves, A.; Gibson, R.; Wheway, V. ALSWH Data Dictionary Supplement: Section 6 Self-Reported Medications. Available online: https://alswh.org.au/wp-content/uploads/2020/08/DSSSection10Medications.pdf (accessed on 27 January 2022).
92. Engels, J.M.; Diehr, P. Imputation of missing longitudinal data: A comparison of methods. *J. Clin. Epidemiol.* **2003**, *56*, 968–976. [CrossRef]
93. Lalonde, T.L.; Wilson, J.R.; Yin, J. GMM logistic regression models for longitudinal data with time-dependent covariates and extended classifications. *Stat. Med.* **2014**, *33*, 4756–4769. [CrossRef] [PubMed]
94. Hu, E.A.; Steffen, L.M.; Coresh, J.; Appel, L.J.; Rebholz, C.M. Adherence to the healthy eating index–2015 and other dietary patterns may reduce risk of cardiovascular disease, cardiovascular mortality, and all-cause mortality. *J. Nutr.* **2020**, *150*, 312–321. [CrossRef]

95. Xu, Z.; Steffen, L.M.; Selvin, E.; Rebholz, C.M. Diet quality, change in diet quality and risk of incident CVD and diabetes. *Public Health Nutr.* **2020**, *23*, 329–338. [CrossRef] [PubMed]
96. Trebuchet, A.; Julia, C.; Fezeu, L.; Touvier, M.; Chaltiel, D.; Hercberg, S.; Galan, P.; Adjibade, M.; Kesse-Guyot, E. Prospective association between several dietary scores and risk of cardiovascular diseases: Is the Mediterranean diet equally associated to cardiovascular diseases compared to National Nutritional Scores? *Am. Heart J.* **2019**, *217*, 1–12. [CrossRef] [PubMed]
97. Australian Institute of Health and Welfare. Cardiovascular Disease. Available online: https://www.aihw.gov.au/reports/heart-stroke-vascular-diseases/cardiovascular-health-compendium/contents/what-is-cardiovascular-disease (accessed on 24 July 2021).
98. Wu, P.Y.; Huang, C.L.; Lei, W.S.; Yang, S.H. Alternative health eating index and the Dietary Guidelines from American Diabetes Association both may reduce the risk of cardiovascular disease in type 2 diabetes patients. *J. Hum. Nutr. Diet.* **2016**, *29*, 363–373. [CrossRef] [PubMed]
99. Neelakantan, N.; Naidoo, N.; Koh, W.-P.; Yuan, J.-M.; van Dam, R.M. The Alternative Healthy Eating Index is associated with a lower risk of fatal and nonfatal acute myocardial infarction in a Chinese adult population. *J. Nutr.* **2016**, *146*, 1379–1386. [CrossRef]
100. Al-Ibrahim, A.A.; Jackson, R.T. Healthy Eating Index versus Alternate Healthy Index in relation to diabetes status and health markers in US adults: NHANES 2007–2010. *Nutr. J.* **2019**, *18*, 26. [CrossRef]
101. Gibbs, J.; Gaskin, E.; Ji, C.; Miller, M.A.; Cappuccio, F.P. The effect of plant-based dietary patterns on blood pressure: A systematic review and meta-analysis of controlled intervention trials. *J. Hypertens.* **2021**, *39*, 23–37. [CrossRef]
102. Martini, D.; Godos, J.; Marventano, S.; Tieri, M.; Ghelfi, F.; Titta, L.; Lafranconi, A.; Trigueiro, H.; Gambera, A.; Alonzo, E. Nut and legume consumption and human health: An umbrella review of observational studies. *Int. J. Food Sci. Nutr.* **2021**, *72*, 871–878. [CrossRef]
103. Colussi, G.; Catena, C.; Novello, M.; Bertin, N.; Sechi, L. Impact of omega-3 polyunsaturated fatty acids on vascular function and blood pressure: Relevance for cardiovascular outcomes. *Nutr. Metab. Cardiovasc. Dis.* **2017**, *27*, 191–200. [CrossRef] [PubMed]
104. Bédard, A.; Li, Z.; Ait-Hadad, W.; Camargo Jr, C.A.; Leynaert, B.; Pison, C.; Dumas, O.; Varraso, R. The role of nutritional factors in asthma: Challenges and opportunities for epidemiological research. *Int. J. Environ. Res. Public Health* **2021**, *18*, 3013. [CrossRef]
105. Guilleminault, L.; Williams, E.J.; Scott, H.A.; Berthon, B.S.; Jensen, M.; Wood, L.G. Diet and asthma: Is it time to adapt our message? *Nutrients* **2017**, *9*, 1227. [CrossRef] [PubMed]
106. Leynaert, B.; Le Moual, N.; Neukirch, C.; Siroux, V.; Varraso, R. Environmental risk factors for asthma developement. *Presse Med.* **2019**, *48*, 262–273. [CrossRef] [PubMed]
107. Andrianasolo, R.M.; Hercberg, S.; Kesse-Guyot, E.; Druesne-Pecollo, N.; Touvier, M.; Galan, P.; Varraso, R. Association between dietary fibre intake and asthma (symptoms and control): Results from the French national e-cohort NutriNet-Santé. *Br. J. Nutr.* **2019**, *122*, 1040–1051. [CrossRef] [PubMed]
108. Andrianasolo, R.M.; Hercberg, S.; Touvier, M.; Druesne-Pecollo, N.; Adjibade, M.; Kesse-Guyot, E.; Galan, P.; Varraso, R. Association between processed meat intake and asthma symptoms in the French NutriNet-Santé cohort. *Eur. J. Nutr.* **2020**, *59*, 1553–1562. [CrossRef] [PubMed]
109. Li, Z.; Rava, M.; Bédard, A.; Dumas, O.; Garcia-Aymerich, J.; Leynaert, B.; Pison, C.; Le Moual, N.; Romieu, I.; Siroux, V. Cured meat intake is associated with worsening asthma symptoms. *Thorax* **2017**, *72*, 206–212. [CrossRef] [PubMed]
110. DeChristopher, L.R.; Tucker, K.L. Excess free fructose, high-fructose corn syrup and adult asthma: The Framingham Offspring Cohort. *Br. J. Nutr.* **2018**, *119*, 1157–1167. [CrossRef]
111. Li, Z.; Kesse-Guyot, E.; Dumas, O.; Garcia-Aymerich, J.; Leynaert, B.; Pison, C.; Moual, N.L.; Romieu, I.; Siroux, V.; Camargo Jr, C.A.; et al. Longitudinal study of diet quality and change in asthma symptoms in adults, according to smoking status. *Br. J. Nutr.* **2017**, *117*, 562–571. [CrossRef]
112. Andrianasolo, R.M.; Kesse-Guyot, E.; Adjibade, M.; Hercberg, S.; Galan, P.; Varraso, R. Associations between dietary scores with asthma symptoms and asthma control in adults. *Eur. Respir. J.* **2018**, *52*, 1702572. [CrossRef]
113. Varraso, R.; Chiuve, S.E.; Fung, T.T.; Barr, R.G.; Hu, F.B.; Willett, W.C.; Camargo, C.A. Alternate Healthy Eating Index 2010 and risk of chronic obstructive pulmonary disease among US women and men: Prospective study. *BMJ* **2015**, *350*, h286. [CrossRef]
114. Han, Y.-Y.; Jerschow, E.; Forno, E.; Hua, S.; Mossavar-Rahmani, Y.; Perreira, K.M.; Sotres-Alvarez, D.; Afshar, M.; Punjabi, N.M.; Thyagarajan, B. Dietary patterns, asthma, and lung function in the Hispanic Community Health Study/Study of Latinos. *Ann. Am. Thorac. Soc.* **2020**, *17*, 293–301. [CrossRef]
115. Lavalette, C.; Adjibade, M.; Srour, B.; Sellem, L.; Fiolet, T.; Hercberg, S.; Latino-Martel, P.; Fassier, P.; Deschasaux, M.; Kesse-Guyot, E. Cancer-specific and general nutritional scores and cancer risk: Results from the prospective NutriNet-Sante cohort. *Cancer Res.* **2018**, *78*, 4427–4435. [CrossRef]
116. Wirfält, E.; Drake, I.; Wallström, P. What do review papers conclude about food and dietary patterns? *Food Nutr. Res.* **2013**, *57*, 20523. [CrossRef] [PubMed]
117. McCullough, M.L.; Willett, W.C. Evaluating adherence to recommended diets in adults: The Alternate Healthy Eating Index. *Public Health Nutr.* **2006**, *9*, 152–157. [CrossRef]
118. Jacobs, S.; Harmon, B.E.; Boushey, C.J.; Morimoto, Y.; Wilkens, L.R.; Le Marchand, L.; Kröger, J.; Schulze, M.B.; Kolonel, L.N.; Maskarinec, G. A priori-defined diet quality indexes and risk of type 2 diabetes: The Multiethnic Cohort. *Diabetologia* **2015**, *58*, 98–112. [CrossRef]

119. Cespedes, E.M.; Hu, F.B.; Tinker, L.; Rosner, B.; Redline, S.; Garcia, L.; Hingle, M.; Van Horn, L.; Howard, B.V.; Levitan, E.B. Multiple healthful dietary patterns and type 2 diabetes in the Women's Health Initiative. *Am. J. Epidemiol.* **2016**, *183*, 622–633. [CrossRef] [PubMed]
120. Chen, G.-C.; Koh, W.-P.; Neelakantan, N.; Yuan, J.-M.; Qin, L.-Q.; van Dam, R.M. Diet quality indices and risk of type 2 diabetes mellitus: The Singapore Chinese Health Study. *Am. J. Epidemiol.* **2018**, *187*, 2651–2661. [CrossRef]
121. Rahe, C.; Unrath, M.; Berger, K. Dietary patterns and the risk of depression in adults: A systematic review of observational studies. *Eur. J. Nutr.* **2014**, *53*, 997–1013. [CrossRef] [PubMed]
122. Li, Y.; Lv, M.-R.; Wei, Y.-J.; Sun, L.; Zhang, J.-X.; Zhang, H.-G.; Li, B. Dietary patterns and depression risk: A meta-analysis. *Psychiatry Res.* **2017**, *253*, 373–382. [CrossRef]
123. Kiecolt-Glaser, J.K. Stress, food, and inflammation: Psychoneuroimmunology and nutrition at the cutting edge. *Psychosom. Med.* **2010**, *72*, 365. [CrossRef]
124. Hu, D.; Cheng, L.; Jiang, W. Sugar-sweetened beverages consumption and the risk of depression: A meta-analysis of observational studies. *J. Affect. Disord.* **2019**, *245*, 348–355. [CrossRef]
125. Nucci, D.; Fatigoni, C.; Amerio, A.; Odone, A.; Gianfredi, V. Red and processed meat consumption and risk of depression: A systematic review and meta-analysis. *Int. J. Environ. Res. Public Health* **2020**, *17*, 6686. [CrossRef]
126. Saneei, P.; Hajishafiee, M.; Keshteli, A.H.; Afshar, H.; Esmaillzadeh, A.; Adibi, P. Adherence to Alternative Healthy Eating Index in relation to depression and anxiety in Iranian adults. *Br. J. Nutr.* **2016**, *116*, 335–342. [CrossRef] [PubMed]
127. Sánchez-Villegas, A.; Henríquez-Sánchez, P.; Ruiz-Canela, M.; Lahortiga, F.; Molero, P.; Toledo, E.; Martínez-González, M.A. A longitudinal analysis of diet quality scores and the risk of incident depression in the SUN Project. *BMC Med.* **2015**, *13*, 197. [CrossRef] [PubMed]
128. Adjibade, M.; Lemogne, C.; Julia, C.; Hercberg, S.; Galan, P.; Assmann, K.E.; Kesse-Guyot, E. Prospective association between adherence to dietary recommendations and incident depressive symptoms in the French NutriNet-Santé cohort. *Br. J. Nutr.* **2018**, *120*, 290–300. [CrossRef] [PubMed]
129. Shi, Z.; Ruel, G.; Dal Grande, E.; Pilkington, R.; Taylor, A.W. Soft drink consumption and multimorbidity among adults. *Clin. Nutr. ESPEN* **2015**, *10*, e71–e76. [CrossRef]
130. Jeong, D.; Kim, J.; Lee, H.; Kim, D.-Y.; Lim, H. Association of cardiometabolic multimorbidity pattern with dietary factors among adults in South Korea. *Nutrients* **2020**, *12*, 2730. [CrossRef]
131. Pereira, B.P.; Bortolotto, C.C.; Tomasi, E.; Gonzalez, M.C.; Gomes, A.P.; Gonçalves, H.; Bielemann, R.M. Food consumption and multimorbidity among non-institutionalized elderly people in Pelotas, 2014: A cross-sectional study. *Epidemiol. E Serviços De Saúde* **2020**, *29*, e2019050. [CrossRef]
132. Ruel, G.; Shi, Z.; Zhen, S.; Zuo, H.; Kröger, E.; Sirois, C.; Lévesque, J.-F.; Taylor, A.W. Association between nutrition and the evolution of multimorbidity: The importance of fruits and vegetables and whole grain products. *Clin. Nutr.* **2014**, *33*, 513–520. [CrossRef]
133. Wikström, K.; Lindström, J.; Harald, K.; Peltonen, M.; Laatikainen, T. Clinical and lifestyle-related risk factors for incident multimorbidity: 10-year follow-up of Finnish population-based cohorts 1982–2012. *Eur. J. Intern. Med.* **2015**, *26*, 211–216. [CrossRef]
134. Dekker, L.H.; de Borst, M.H.; Meems, L.M.; de Boer, R.A.; Bakker, S.J.; Navis, G.J. The association of multimorbidity within cardio-metabolic disease domains with dietary patterns: A cross-sectional study in 129,369 men and women from the Lifelines cohort. *PLoS ONE* **2019**, *14*, e0220368. [CrossRef]
135. Freisling, H.; Viallon, V.; Lennon, H.; Bagnardi, V.; Ricci, C.; Butterworth, A.S.; Sweeting, M.; Muller, D.; Romieu, I.; Bazelle, P. Lifestyle factors and risk of multimorbidity of cancer and cardiometabolic diseases: A multinational cohort study. *BMC Med.* **2020**, *18*, 5. [CrossRef]
136. Kyprianidou, M.; Panagiotakos, D.; Faka, A.; Kambanaros, M.; Makris, K.C.; Christophi, C.A. Adherence to the Mediterranean diet in Cyprus and its relationship to multi-morbidity: An epidemiological study. *Public Health Nutr.* **2021**, *24*, 4546–4555. [CrossRef] [PubMed]
137. Al Ramadhani, R.M.; Nagle, C.M.; Ibiebele, T.I.; Grant, P.; Friedlander, M.; DeFazio, A.; Webb, P.M. Pre-and post-diagnosis diet quality and ovarian cancer survival. *Cancer Epidemiol. Biomark. Prev.* **2021**, *30*, 229–232. [CrossRef]
138. Shang, X.; Scott, D.; Hodge, A.; Khan, B.; Khan, N.; English, D.; Giles, G.G.; Ebeling, P.; Sanders, K. Dietary quality is associated with abdominal aortic calcification: A mean of 18-year longitudinal study in community-dwelling older adults. *J. Nutr. Health Aging* **2017**, *21*, 147–151. [CrossRef] [PubMed]
139. Chan, C.H.; Conley, M.; Reeves, M.M.; Campbell, K.L.; Kelly, J.T. Evaluating the impact of goal setting on improving diet quality in chronic kidney disease. *Front. Nutr.* **2021**, *8*, 627753. [CrossRef]
140. Hodge, A.M.; Karim, M.N.; Hébert, J.R.; Shivappa, N.; Milne, R.L.; De Courten, B. Diet scores and prediction of general and abdominal obesity in the Melbourne Collaborative Cohort study. *Public Health Nutr.* **2021**, *24*, 6157–6168. [CrossRef] [PubMed]
141. Dugué, P.A.; Hodge, A.M.; Brinkman, M.T.; Bassett, J.K.; Shivappa, N.; Hebert, J.R.; Hopper, J.L.; English, D.R.; Milne, R.L.; Giles, G.G. Association between selected dietary scores and the risk of urothelial cell carcinoma: A prospective cohort study. *Int. J. Cancer* **2016**, *139*, 1251–1260. [CrossRef]
142. Powers, J.; Loxton, D. The impact of attrition in an 11-year prospective longitudinal study of younger women. *Ann. Epidemiol.* **2010**, *20*, 318–321. [CrossRef]

143. Irani, F.; Coquoz, E.; von Wolff, M.; Bitterlich, N.; Stute, P. Awareness of non-communicable diseases in women: A cross-sectional study. *Arch. Gynecol. Obstet.* **2022**, *306*, 801–810. [CrossRef]
144. Bista, B.; Dhungana, R.R.; Chalise, B.; Pandey, A.R. Prevalence and determinants of non-communicable diseases risk factors among reproductive aged women of Nepal: Results from Nepal Demographic Health Survey 2016. *PLoS ONE* **2020**, *15*, e0218840. [CrossRef]
145. Shim, J.-S.; Oh, K.; Kim, H.C. Dietary assessment methods in epidemiologic studies. *Epidemiol. Health* **2014**, *36*, e2014009. [CrossRef]
146. Olendzki, B.; Procter-Gray, E.; Magee, M.F.; Youssef, G.; Kane, K.; Churchill, L.; Ockene, J.; Li, W. Racial differences in misclassification of healthy eating based on food frequency questionnaire and 24-hour dietary recalls. *J. Nutr. Health Aging* **2017**, *21*, 787–798. [CrossRef] [PubMed]

nutrients

MDPI

Article

Genetic Predisposition, Fruit Intake and Incident Stroke: A Prospective Chinese Cohort Study

Jun Wang [1,2,3,†], Jianxin Li [1,2,†], Fangchao Liu [1], Keyong Huang [1], Xueli Yang [4], Xiaoqing Liu [5], Jie Cao [1], Shufeng Chen [1], Chong Shen [6], Ling Yu [7], Fanghong Lu [8], Liancheng Zhao [1], Ying Li [1], Dongsheng Hu [9,10], Jianfeng Huang [1], Dongfeng Gu [1,2,11] and Xiangfeng Lu [1,2,*]

1 Department of Epidemiology, Fuwai Hospital, National Center for Cardiovascular Diseases, Chinese Academy of Medical Sciences and Peking Union Medical College, Beijing 100037, China
2 Key Laboratory of Cardiovascular Epidemiology, Chinese Academy of Medical Sciences, Beijing 100037, China
3 National Institute of Environmental Health, Chinese Center for Disease Control and Prevention, Beijing 100021, China
4 Tianjin Key Laboratory of Environment, Nutrition and Public Health, Department of Occupational and Environmental Health, School of Public Health, Tianjin Medical University, Tianjin 300070, China
5 Division of Epidemiology, Guangdong Provincial People's Hospital and Cardiovascular Institute, Guangzhou 510080, China
6 Department of Epidemiology and Biostatistics, School of Public Health, Nanjing Medical University, Nanjing 211166, China
7 Department of Cardiology, Fujian Provincial Hospital, Fuzhou 350014, China
8 Cardio-Cerebrovascular Control and Research Center, Institute of Basic Medicine, Shandong Academy of Medical Sciences, Jinan 250062, China
9 Department of Epidemiology and Health Statistics, College of Public Health, Zhengzhou University, Zhengzhou 450001, China
10 Department of Biostatistics and Epidemiology, School of Public Health, Shenzhen University Health Science Center, Shenzhen 518071, China
11 School of Medicine, Southern University of Science and Technology, Shenzhen 518055, China
* Correspondence: luxf@pumc.edu.cn; Tel./Fax: +86-10-60866599
† These authors contributed equally to this work.

Citation: Wang, J.; Li, J.; Liu, F.; Huang, K.; Yang, X.; Liu, X.; Cao, J.; Chen, S.; Shen, C.; Yu, L.; et al. Genetic Predisposition, Fruit Intake and Incident Stroke: A Prospective Chinese Cohort Study. *Nutrients* **2022**, *14*, 5056. https://doi.org/10.3390/nu14235056

Academic Editors: William B. Grant and Ronan Lordan

Received: 20 October 2022
Accepted: 24 November 2022
Published: 28 November 2022

Abstract: The aim of this study was to evaluate the association between fruit intake and stroke risk considering the genetic predisposition. We used data from 34,871 participants from the project of Prediction for Atherosclerotic Cardiovascular Disease Risk in China (China-PAR project) from 2007 to 2020. A polygenic risk score comprising 534 genetic variants associated with stroke and its related factors was constructed to categorize individuals into low, intermediate, and high genetic risk groups. The associations of genetic and fruit intake with incident stroke were assessed by the Cox proportional hazard regression. We documented 2586 incident strokes during a median follow-up of 11.2 years. Compared with fruit intake < 200 g/week, similar relative risk reductions in stroke with adherence to fruit intake > 100 g/day across the genetic risk categories were observed (28–32%), but the absolute risk reductions were relatively larger in the highest genetic risk group (p for trend = 0.03). In comparison to those with a fruit intake < 200 g/week, those with a fruit intake >100 g/day in the low, intermediate, and high genetic risk groups had an average of 1.45 (95% CI, 0.61–2.31), 2.12 (1.63–2.59), and 2.19 (1.13–3.22) additional stroke-free years at aged 35, respectively. Our findings suggest that individuals with a high genetic risk could gain more absolute risk reductions and stroke-free years than those with a low genetic risk from increasing fruit intake for the stroke primary prevention.

Keywords: polygenic risk score; fruit; stroke

1. Introduction

Stroke, caused by a combination of both a genetic predisposition and environmental risk factors, is the second-leading cause of death globally [1]. Notably, the stroke incidence is increasing over time in China, with a severe disease burden [2]. Fruit, including micronutrients and bioactive compounds, have been widely recommended in stroke prevention [3,4]. However, according to the China National Nutrition Surveys, only 4% of Chinese residents met the guideline-recommended level of fruit intake in 2012 [5]. Although there was a slight increase in average daily fruit intake among Chinese adults from 1982 to 2012, a low fruit intake still ranked as the second most common cause of stroke death [6].

Recent genome-wide association studies have identified a variety of genetic variants associated with stroke and its correlated traits, such as blood pressure (BP), type 2 diabetes, blood lipids, body mass index (BMI), etc. [7–11]. Subsequently, the polygenic risk score (PRS) was developed by combing multiple risk alleles to predict the risk of incident stroke [12–14]. Using PRS, some researchers estimated whether fruit intake could modify the effect of the genetic predisposition to stroke and its risk factors. Previous research in the USA found that increasing the fruit intake could attenuate the genetic predisposition on long-term weight gain [15]. However, the UK Biobank study failed to find an interaction between a genetic risk and a healthy diet, involving fruit intake, on stroke [12]. Further, it is still unclear whether the interaction on stroke existed among Chinese adults, and the extent to which an increased genetic predisposition to stroke can be offset by increasing the fruit intake, considering the different genetic backgrounds and dietary patterns from western populations.

Therefore, the aim of this study was to examine whether associations of fruit intake with the risk of stroke varied in different genetic risk groups among Chinese adults from the project of Prediction for Atherosclerotic Cardiovascular Disease Risk in China (China-PAR project).

2. Materials and Methods

2.1. Study Design and Participants

The study design and methods of the China-PAR project have been reported in detail, previously [16,17]. The study included three prospective cohorts with dietary records that were part of the China-PAR project, comprising the China Multi-Center Collaborative Study of Cardiovascular Epidemiology (China MUCA-1998), the International Collaborative Study of Cardiovascular Disease in Asia (InterASIA), and the Community Intervention of Metabolic Syndrome in China and Chinese Family Health Study (CIMIC). The baseline examination was 1998 for China MUCA-1998, 2000–2001 for InterASIA, and 2007–2008 for CIMIC, respectively. A follow-up survey was conducted for both the China MUCA-1998 and InterASIA cohorts during 2007 to 2008. Then all three cohorts were followed up in 2012–2015 and 2018–2020. The methods for the follow-up surveys were identical for all cohorts. The data collection was carried out after written informed consent was obtained from the participants, including an interviewer-administered questionnaire, physical examination, and blood sampling. This study was approved by the Institutional Review Board at Fuwai Hospital (Beijing, China). To avoid the influences of the economic development on fruit intake across the three cohorts enrolled at different periods [18], the survey in 2007–2008 with a unified questionnaire on dietary intake was used as the baseline in the present study.

Among a total of 41,006 participants from three cohorts with available genotypic data [19], we further excluded 4856 individuals with missing information on fruit intake, 1265 with prevalent cardiovascular disease or cancer at 2007–2008, and 14 with missing data on follow-up, leaving 34,871 eligible participants in the final analysis.

2.2. Fruit Intake and Covariate Assessment

Dietary intake, including fruit intake, was collected by a simple standardized food-frequency questionnaire (FFQ), indicating acceptable reproducibility and validity in previous study [20]. Consumption frequency (daily, weekly, monthly, yearly, or never) and average consumption amount per unit of time for each food group during the previous year were reported by the participants, which were further converted into average daily intake. Processed fruit was excluded. Fruit intake was classified into three categories, based on the tertiles (<200 g/week, 200 g/week–100 g/day, and >100 g/day). The dietary score was computed, according to the consumption of four commonly eaten food groups in cardiometabolic health, based on current dietary guideline recommendations and our previous study evidence, including red meat, legumes, fish, and tea. The ideal levels were red meat < 75 g/day, legumes ≥ 125 g/day, fish ≥ 200 g/week, and tea ≥ 3 times/week [21,22]. The participants were given one point for each dietary component that reached the ideal level, otherwise they were given zero points. Based on this, we assigned equal weight for each food type and summed all points together, ranging from 0 to 4, with a higher index indicating a healthier dietary lifestyle.

The following variables were self-reported via standardized questionnaires, such as age, sex, region, urbanization, educational level, current smoking status, alcohol drinking status, physical activity, and medical conditions (use of antihypertensive, antidiabetic and lipid-lowering medications). Anthropometric measurements (height, weight, blood pressure, fasting blood glucose, and serum lipid levels) were performed following standard techniques by trained staff. Current smoking status was classified as smoker or non-smoker by asking the participant whether he or she had smoked more than 100 cigarettes and whether he or she kept smoking up till then. Alcohol drinking was defined as alcohol consumption at least 12 times in the previous year. The physical activity level was evaluated by summing the time spent on moderate physical activity and on vigorous physical activity weighted by 2, and the ideal physical activity level was defined as at least 150 min/week of moderate physical activity or 75 min/week of vigorous physical activity or an equivalent combination. Body height and weight were measured twice by standardized anthropometric procedures with the participant wearing light garments and no shoes. The BMI was calculated as weight (kg)/height (m)^2. Hypertension was diagnosed as mean systolic BP ≥ 140 mmHg and/or diastolic BP ≥ 90 mmHg (averaging three times standardized measurements with 30 s intervals in a seated position), and/or antihypertensive treatment in last two weeks. Overnight fasting blood samples were drawn for the measurements of serum glucose and lipid concentrations. Diabetes mellitus was defined as a fasting glucose level ≥ 126 mg/dL, and/or previous diagnosed diabetes, and/or antidiabetic treatment within two weeks. Subjects with a total cholesterol ≥ 240 mg/dL, or triglycerides ≥ 200 mg/dL, or low-density lipoprotein cholesterol ≥ 160 mg/dL, or high-density lipoprotein cholesterol < 40 mg/dL, or taking lipid-lowing medicine within two weeks were classified as dyslipidemia.

2.3. Ascertainment of Incident Stroke Events

Information on stroke incidence was collected by well-trained staff to obtain medical records or death certificates from participants or their proxies. Local investigators initially recorded fatal and nonfatal stroke events. All medical and death records were reviewed by the central adjudication committee at Fuwai Hospital (Beijing, China) to determine the final diagnosis. Two adjudication committee members, blinded to the baseline information of the participants, verified the events independently, and any discrepancies were resolved by discussion with involvement of an additional committee member. Incident stroke was defined as a confirmed first ever fatal or nonfatal stroke event during the follow-up period, consisting of ischemic stroke (International Classification of Diseases, Tenth Revision I63), hemorrhagic stroke (I60–I62), and unspecified stroke (I64–I69) [19].

2.4. Polygenetic Risk Score for Stroke

The detailed information on single nucleotide polymorphisms (SNPs) selection and genotyping process used in the China-PAR study has been described, previously [19]. Briefly, a combined PRS was derived, based on 534 SNPs by using a meta-analytic approach and large genome-wide association results for stroke and stroke-related traits, including BP, type 2 diabetes, lipids, obesity, atrial fibrillation, and coronary artery disease, in East Asians. The detailed information on the selected SNPs is provided in Table S1. Genotyping for all participants was performed by multiplex PCR targeted amplicon sequencing technology. We designed multiplexed primers targeting each SNP and amplified the target regions for a high-throughput sequencing on a Hiseq X10 sequencer (Illumina, San Diego, CA, USA). Individual SNPs were coded as 0, 1, and 2 according to the number of risk alleles. The weight coefficient for each SNP was calculated in a previous publication. The PRS was formulated as the sum of the number of risk alleles at each locus multiplied by the respective weight coefficient, which was developed in a training set and validated in a validation set. The participants were categorized into low (lowest quintile of PRS), intermediate (quintiles 2 to 4 of PRS), and high (highest quintile of PRS) genetic risk groups, as described, previously [19].

2.5. Statistical Analysis

The baseline characteristics of the participants were described as mean (standard deviation) for continuous variables and numbers (percentages) for categorical variables. Age- and sex-adjusted incidence rates of stroke were calculated using the Poisson regression [23].

Person-years of follow-up were calculated from the date of the baseline examination to the date of the incident stroke, death, or follow-up interview for each study participant, whichever occurred first. To address the potential effect variation among the sub-cohorts, cohort-stratified Cox proportional hazards models with the duration of the follow-up as the time metric were applied to estimate the hazard ratios (HRs) and their 95% confidence intervals (CIs) of incident stroke, with the lowest category of genetic risk score or fruit intake as reference. Proportional hazards assumptions were not violated when assessed using the Schoenfeld residuals ($p > 0.05$). The selection process of the potential confounders was based on an extensive literature search and then tested by a univariate Cox regression analysis. Two models were constructed to account for the potential confounders. The base model included age and sex for the genetic risk association analysis, as previously described [19], and included age, sex, region, urbanization, and educational level, for the fruit intake association analysis. The full model further included the current smoking status, alcohol drinking status, physical activity, BMI, diet score, and vegetables intake. The tests of trend were conducted by modeling the median of the PRS or fruit intake of each category, as a continuous variable. Potential nonlinear relationships of incident stroke, associated with the PRS and fruit intake were assessed by restricted cubic splines with three knots at the 25th, 50th (reference), and 75th percentiles of the distribution.

Moreover, we investigated the combined effects of the PRS and fruit intake on stroke, with the lowest genetic risk and fruit intake > 100 g/day, as reference. We further explored their potential interaction between the PRS and fruit intake. To test for the multiplicative interaction, we included a multiplicative interaction term in the full model. An additive interaction was also tested by calculating the relative excess risk due to the interaction (RERI) and the attributable proportion due to the interaction (AP) [24].

We calculated the cumulative incidence of stroke for three categories of fruit intake within each genetic risk stratum. The absolute risk reductions (ARRs) in 10-year stroke incidence between low and high fruit intake groups were computed, and their trends across the genetic risk categories were tested using the weighted least-squares model [19].

The gained stroke-free years related to fruit intake were estimated as the differences of the areas under the survival curves, based on the fully-adjusted Cox proportional hazard model with age as the underlying timescale, and 95% CIs were derived by drawing 500 bootstrap samples from the estimation dataset [25].

To examine the robustness of our results, we performed several sensitivity analyses by (1) using the chronological age as the primary timescale instead of the time-on-study; (2) accounting for competing the risk of a non-stroke death using Fine and Gray's approach [26]; (3) adjusting for the smoking packs per year and daily alcohol intake as continuous variables instead of categorical variables; (4) further adjusting for the family history of stroke or per-capita household income; (5) separately investigating the interaction between each SNP in the PRS and fruit intake.

We additionally estimated the association of the frequent intake of fruit with stroke risk stratified by the genetic risk category. All statistical analyses were performed using SAS software, version 9.4 (SAS Institute, Cary, NC, USA) and R software, version 4.2.1 (R Foundation for Statistical Computing, Vienna, Austria). The R package "forestplot", "cmprsk", and "survival" were used. A two-sided p value < 0.05 indicated statistical significance.

3. Results

During a median of 11.2 years of follow-up, we documented 2586 first-ever incident stroke events, including 1834 ischemic stroke, 439 hemorrhagic stroke, 55 both ischemic and hemorrhagic stroke, and 258 strokes with an unknown subtype, with the incidence of 7.7 per 1000 person-years. The baseline characteristics of the study population, stratified according to the presence of incident stroke are shown in Table 1, and the characteristics according to the genetic risk category and fruit intake are presented in Table S2. The mean age of the participants eligible for the analysis was 55 ± 10 years, and 58% were female. The average fruit intake was 83 g/day. Compared to the non-cases, the stroke cases were more likely to be older, had a higher BMI and waist circumference, a higher baseline prevalence of hypertension, diabetes mellitus and dyslipidemia, a poorer ideal level of diet score, and a lower fruit and vegetable intake.

Table 1. Baseline characteristics of the participants.

Characteristic	No Incident Stroke (n = 32,285)	Incident Stroke (n = 2586)	p Value
Age, year	55 ± 10	60 ± 9	<0.001
Female, n (%)	18,701 (58)	1427 (55)	0.007
Northern, n (%)	15,389 (48)	1437 (56)	<0.001
Urban residence, n (%)	5462 (17)	151 (5.8)	<0.001
High-school or above, n (%)	15,544 (48)	779 (30)	<0.001
Current smoker, n (%)	7560 (23)	633 (25)	0.22
Alcohol drinker, n (%)	7571 (23)	521 (20)	<0.001
Family history of stroke, n (%)	3188 (10)	303 (12)	0.003
Ideal physical activity, n (%)	20,244 (63)	1662 (64)	0.11
Ideal diet score, n (%)	17,019 (53)	1086 (42)	<0.001
Body mass index, kg/m^2	24 ± 4	24 ± 4	<0.001
Waist circumference, cm	82 ± 10	84 ± 10	<0.001
Hypertension, n (%)	11,945 (37)	1435 (56)	<0.001
Diabetes mellitus, n (%)	2596 (8.3)	329 (13)	<0.001
Dyslipidemia, n (%)	10,047 (32)	1025 (41)	<0.001
Fruit intake, g/day	85 ± 86	66 ± 78	<0.001
Vegetable intake, g/day	335 ± 157	327 ± 158	0.02

Data are mean ± standard deviation for the continuous variables and numbers (percentages) for the dichotomous variables. Ideal physical activity level was defined as at least 150 min/week of moderate physical activity or 75 min/week of vigorous physical activity or an equivalent combination. The ideal diet was defined as a healthy diet score ≥ 2 components: red meat < 75 g/day, legumes ≥ 125 g/day, fish ≥ 200 g/week, and tea ≥ 3 times/week.

The univariate Cox regression analysis results of potential confounders were shown in Table S3. The genetic risk score showed a normal distribution (Figure S1). The stroke risk increased monotonically across the range of the genetic risk score (p overall < 0.001, p nonlinear = 0.37). Compared with those with a low genetic risk score, the age- and sex-adjusted HRs (95% CIs) for the intermediate and high genetic risk groups were 1.19 (1.06–1.32) and 1.51 (1.33–1.71) when only adjusted for age and sex, respectively (Table S4). The results were basically unchanged by the adjustment for other factors. Conversely, a higher fruit intake was associated with a lower risk of stroke (p overall < 0.001, p nonlinear = 0.02). Based on the fully adjusted model, the participants with a fruit intake of 200 g/week–100 g/day and >100 g/day had 19% (HR, 0.81; 95% CI, 0.74–0.89) and 31% (HR, 0.69; 95% CI, 0.62–0.77) lower risk of stroke, respectively, compared with those consuming less than 200 g/week.

When the genetic risk and fruit intake were combined, the overall risk of incident stroke increased as both the genetic risk and unfavorable fruit intake increased (Figure 1). Specifically, the age- and sex-adjusted incidence rates of stroke per 1000 person-years ranged from 5.7 (95% CI, 4.6–7) for participants with a low genetic risk and a high fruit intake to 9.3 (95% CI, 8.2–10.5) for participants with a high genetic risk and a low fruit intake. Participants with a high genetic risk and a low fruit intake had the highest risk of incident stroke, with a HR of 1.87 (95% CI, 1.46–2.41). No significant multiplicative or additive interactions between the PRS and fruit intake on stroke were recorded.

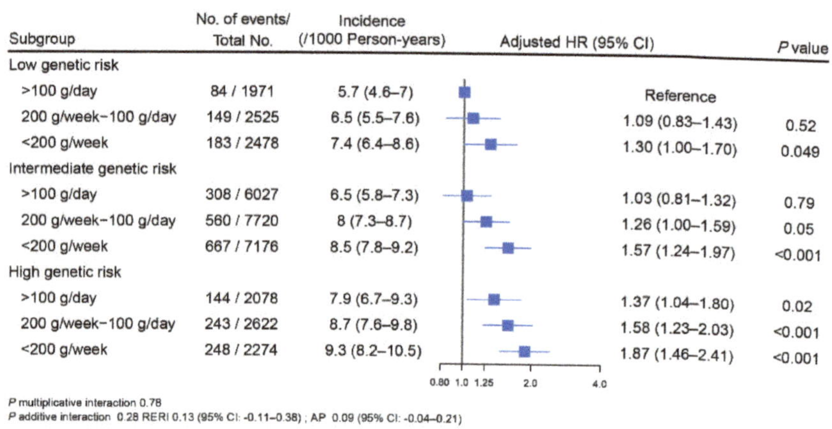

Figure 1. The joint association of genetic risk and fruit intake with incident stroke. Age- and sex-adjusted incidence rates per 1000 person-years of stroke were calculated using the Poisson regression. The overall genetic risk for stroke was defined as high (quintile 5 of PRS), intermediate (quintile 2–4 of PRS), and low (quintile 1 of PRS). In these comparisons, the participants with a low genetic risk and a fruit intake > 100 g/day served as the reference group. HRs were derived from the Cox proportional hazards models stratified by cohort and adjusted for age (continuous), sex (male, female), region (North/South China), urbanization (urban, rural), education level (less than high school, high school, or above), current smoking status (yes, no), alcohol drinking status (yes, no), physical activity (continuous), body mass index (continuous), diet score (continuous), and vegetables intake (daily intake $\geq$ 500 g/day or not). Multiplicative interaction was evaluated using the hazard ratios for the product term between the fruit intake (<200 g/week vs. >100 g/day) and the PRS (low vs. high). p multiplicative interaction was 0.78. To estimate the additive interaction, the participants with a fruit intake of $\geq$200 g/week and the low genetic risk (quintile 1 of PRS) were used as the reference with consideration of the practical interpretations. Two indexes were calculated: the relative excess risk due to the interaction (RERI) and the attributable proportion due to the interaction (AP). p additive interaction was 0.28. RERI was 0.13 (95% CI: −0.11–0.38). AP was 0.09 (95% CI: −0.04–0.21). Abbreviations: CI, confidence interval; HR, hazard ratio; PRS, polygenic risk score.

Further analyses stratified by the genetic risk category with a fruit intake less than 200 g/week as the reference group confirmed that a higher fruit intake was associated with a lower risk of stroke within each category of genetic risk, with adjusted HRs (95% CIs) of 0.72 (0.54–0.95), 0.68 (0.59–0.78), and 0.70 (0.56–0.87) among the participants in the low, intermediate, and high genetic risk groups, respectively (Figure S2). However, in terms of the absolute benefit of increasing the fruit intake, there was a significant gradient of ARRs across the low, intermediate, and high genetic risk categories (1.3%, 1.8%, and 2.3%, p for trend = 0.03, Figure 2).

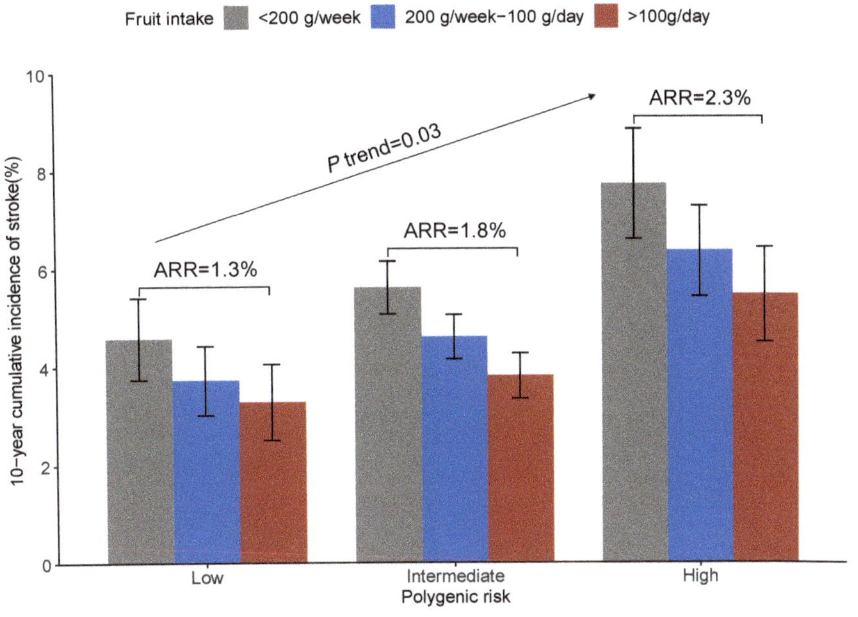

Figure 2. Adjusted 10-year cumulative incidence of stroke according to the genetic risk and fruit intake. Standardized to the means of cohort, age, sex, region, urbanization, education level, current smoking status, alcohol drinking status, physical activity, body mass index, diet score, and vegetables intake within the study population. Abbreviations: ARR, absolute risk reduction.

We also estimated the gains of the stroke-free years of increment in the fruit intake under the overall and the three genetic risk groups. In participants aged 35 years, a fruit intake of 200 g/week–100 g/day and >100 g/day were associated with an average of 0.64 (95% CI, 0.38–0.89) and 2 (95% CI, 1.57–2.4) additional stroke-free years, respectively, compared to a fruit intake of less than 200 g/week (Figure 3). The additional stroke-free years of a fruit intake >100 g/day were 1.45 (95% CI, 0.61–2.31), 2.12 (95% CI, 1.63–2.59), and 2.19 (95% CI, 1.13–3.22) for the participants in the low, intermediate, and high genetic risk groups, respectively, as compared with a fruit intake of less than 200 g/week at the age of 35.

The results were consistent in a series of sensitivity analyses (Table S5). For the interactions between the individual SNPs and fruit intake, two SNPs (rs12999907 at *AC092684.1* and rs2531995 at *ADCY9*) showed significant additive interactions with $p < 0.001$, however, they would not reach significance after a Bonferroni adjustment ($p < 0.05/534$) (Table S1). Moreover, a significant gradient of the absolute benefit from the daily consumption of fruit across the low, intermediate, and high polygenic risk categories was also observed (p trend = 0.03, Table S6).

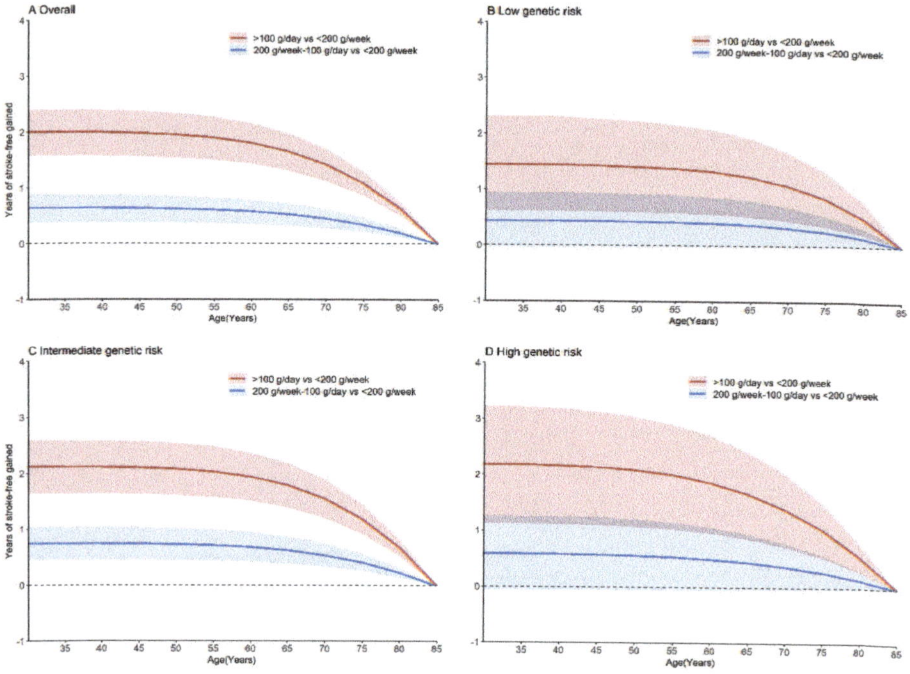

Figure 3. Stroke-free years gained associated with the fruit intake. The overall genetic risk for stroke was defined as high (quintile 5 of PRS), intermediate (quintile 2–4 of PRS), and low (quintile 1 of PRS). Models were adjusted for the cohort, age, sex, region, urbanization, education level, current smoking status, alcohol drinking status, physical activity, body mass index, diet score, and vegetables intake. Abbreviations: PRS, polygenic risk score.

4. Discussion

Using a large-scale Chinese prospective cohort study, an increased PRS was positively associated with a stroke risk. Adherence to a higher fruit intake was associated with a reduced risk of developing stroke and the absolute risk reduction was most evident among those at the highest genetic risk. It was also estimated that increasing the fruit intake from less than 200 g/week to above 100 g/day resulted in longer stroke-free years among those with a high genetic risk, compared with those with a low genetic risk at the age of 35 years.

Our findings of the positive influence of adherence to an increased fruit intake to prevent stroke are generally comparable with the previous studies [3,4,27–29]. However, the findings from the Prospective Urban Rural Epidemiology (PURE) did not find an association between fruit intake and incident stroke [30]. The difference in the types of fruit consumed may partly explain the negative result in the PURE study, which were also found in an early study [31]. Moreover, the association between the fruit intake and stroke risk in individuals with different genetic burdens for stroke were not evaluated in these studies. Although two prior studies showed that the increased cardiovascular disease risk associated with the genetic variants in the chromosome 9p21 region, such as rs2383206 and rs4977574, appeared to be modified by the adherence to a higher fruit and vegetable intake [32,33], population-based studies that qualify the potential interactions of the aggregated genetic susceptibility and fruit intake with stroke risk were rather limited.

With the advances in the field of nutritional genetics, some researchers cast doubt on the "one size fits all" public Dietary Reference Intakes for individuals. The reality of nutrition and dietetic practice is complicated. Accordingly, it is time to unravel the degree to which the diet influences the disease development, and whether it may depend more on a person's genetic background [34,35]. The UK Biobank study reported that a healthy diet including a fruit intake could not decrease the stroke risk across all genetic risk categories, using 11 SNPs [36]. However, a significant relative risk of incident stroke associated with an unhealthy diet was merely observed among the high genetic risk group when extending to 90 SNPs [12]. That might be explained by the high predictive power due to the inclusion of more SNPs. Consistent with our study, no evidence was identified for the interaction between the genetic risk and adherence to a healthy diet, including the fruit intake [12]. Nonetheless, it is worth noting that individuals with the highest burden of genetic risk derived the greatest absolute risk reductions and largest gain in stroke-free years with a fruit intake >100 g/day, in the present study, which may provide evidence for the beneficial effects of increasing the fruit intake on stroke prevention among Chinese populations, especially for those with a high genetic risk.

The potential mechanisms that an increased fruit intake might be beneficial to cardiovascular health have been suggested. Fruit provides excellent sources of fiber, potassium, micronutrients, and bioactive compounds which are necessary for human health. The involved biological mechanisms included protecting the vascular endothelial function, regulating lipids metabolism, modulating blood pressure, alleviating ischemia/reperfusion injury, suppressing thrombosis, reducing oxidative stress, and attenuating inflammation [37].

The strengths of this study included its large sample size, long-term follow-up period, and the strict and comprehensive data collection. We first estimated the interaction of the genetic risk and fruit intake on incident stroke, as well as stroke-free years by increasing the fruit intake among the Chinese population. Our results were robust to a variety of sensitivity analyses. Nevertheless, the present study has several limitations. First, the fruit intake was assessed by a simple FFQ. A recall bias of self-reported dietary information is of concern, which might lead to misclassification. However, the FFQ has been well validated among the Chinese population [20]. Second, the total fruit were analyzed with a lack of the information on different types of fruit available in the present study. Hence, it is still not clear which types of fruit may be the most relevant in gene-diet interactions, which need further studies. Third, participants with increased fruit consumption are likely to have a relatively healthier lifestyle. Although we adjusted for several major lifestyle and dietary factors in the analysis, some potential confounders, such as salt or sugar intake (not measured in this study), may modulate the genetic association with stroke. Fourth, a change in fruit intake over the follow-up period was not evaluated, and future studies would be warranted. Fifth, due to the limited number of hemorrhagic stroke cases and the heterogeneity in the influence of the genetic variants on different stroke subtypes, the associations of fruit intake and subtype-specific stroke genetic risk scores are not evaluated [14]. Finally, our study was restricted to the Chinese population, and further study is needed to evaluate the generalizability of our stroke PRS to other East Asian populations.

5. Conclusions

In summary, increasing the fruit intake was associated with a significantly decreased risk of stroke, and individuals with a high genetic risk derived greater benefit from increasing their fruit intake. Our study supports the recommendations of the adherence to fruit intake for the reduction of strokes in the Chinese population, particularly in those at a high genetic risk.

Supplementary Materials: The following supporting information can be downloaded at: https://www.mdpi.com/article/10.3390/nu14235056/s1, Table S1. Information about the selected SNPs and the interplay between a single SNP and fruit intake. Table S2. Baseline characteristics by the genetic risk score and the daily fruit intake. Table S3. Association of the stroke risk with the sociodemographic and behavioral factors. Table S4. Multivariable-adjusted HRs (95% CIs) for genetic risk, fruit intake, and incident stroke. Table S5. Sensitivity analysis of the fruit intake associated with incident stroke, according to the genetic risk category. Table S6. Risk of incident stroke, according to the fruit intake frequency and genetic risk. Figure S1. Associations between the genetic risk score and incident stroke. Figure S2. Cumulative incidence of stroke, according to the categories of the genetic risk and fruit intake.

Author Contributions: Conceptualization: D.G. and X.L. (Xiangfeng Lu); methodology, formal analysis, writing—original draft preparation: J.W. and J.L.; investigation, writing—review and editing: all authors. All authors have read and agreed to the published version of the manuscript.

Funding: This work was funded by the Chinese Academy of Medical Sciences (CAMS) Innovation Fund for Medical Sciences (2021-I2M-1-010, 2019-I2M-2-003, 2017-I2M-1-004), Research Unit of Prospective Cohort of Cardiovascular Diseases and Cancers, CAMS (2019RU038), the National Key Research and Development Program of China (2018YFE0115300, 2017YFC0211700), the National Natural Science Foundation of China (82030102, 12126602, 91857118), the Taikang Yicai Public Health and Epidemic Control Fund (TKYC-GW-2020), the National High Level Hospital Clinical Research Funding (2022-GSP-GG-1, 2022-GSP-GG-2), and the National Clinical Research Center for Cardiovascular Diseases, Fuwai Hospital, CAMS (NCRC2020006). The sources of funding had no role in the study design, data collection, analyses, interpretation, and decision to submit the article for publication.

Institutional Review Board Statement: The Institutional Review Board at Fuwai Hospital in Beijing approved all three sub-cohorts (approval code 2018-1062 and date of approval 17 July 2018).

Informed Consent Statement: Each participant signed the written informed consent before the baseline survey and each follow-up visit.

Data Availability Statement: All data supporting the findings of this study are available within the article and its supplementary information files. The datasets used and analyzed during the current study are available from the corresponding author upon reasonable request.

Acknowledgments: The authors thank the staff and participants of the China-PAR project for their important participation and contribution.

Conflicts of Interest: The authors declare that they have no known competing financial interests or personal relationships that could have appeared to influence the work reported in this paper.

References

1. Feigin, V.L.; Stark, B.A.; Johnson, C.O.; Roth, G.A.; Bisignano, C.; Abady, G.G.; Abbasifard, M.; Abbasi-Kangevari, M.; Abd-Allah, F.; Abedi, V.; et al. Global, regional, and national burden of stroke and its risk factors, 1990–2019: A systematic analysis for the Global Burden of Disease Study 2019. *Lancet Neurol.* **2021**, *20*, 795–820. [CrossRef]
2. Ma, Q.; Li, R.; Wang, L.; Yin, P.; Wang, Y.; Yan, C.; Ren, Y.; Qian, Z.; Vaughn, M.G.; McMillin, S.E.; et al. Temporal trend and attributable risk factors of stroke burden in China, 1990–2019: An analysis for the Global Burden of Disease Study 2019. *Lancet Public Health* **2021**, *6*, e897–e906. [CrossRef] [PubMed]
3. Hu, D.; Huang, J.; Wang, Y.; Zhang, D.; Qu, Y. Fruits and vegetables consumption and risk of stroke: A meta-analysis of prospective cohort studies. *Stroke* **2014**, *45*, 1613–1619. [CrossRef]
4. Aune, D.; Giovannucci, E.; Boffetta, P.; Fadnes, L.T.; Keum, N.; Norat, T.; Greenwood, D.C.; Riboli, E.; Vatten, L.J.; Tonstad, S. Fruit and vegetable intake and the risk of cardiovascular disease, total cancer and all-cause mortality-a systematic review and dose-response meta-analysis of prospective studies. *Int. J. Epidemiol.* **2017**, *46*, 1029–1056. [CrossRef] [PubMed]
5. Zhao, L.Y.; He, Y.N. *China National Nutrition and Health Surveillance Report: 2010–2013 Food and Nutrient Intake*, 1st ed.; People's Medical Publishing House: Beijing, China, 2018; pp. 18–24. (In Chinese)
6. He, Y.; Li, Y.; Yang, X.; Hemler, E.C.; Fang, Y.; Zhao, L.; Zhang, J.; Yang, Z.; Wang, Z.; He, L.; et al. The dietary transition and its association with cardiometabolic mortality among Chinese adults, 1982-2012: A cross-sectional population-based study. *Lancet Diabetes Endocrinol.* **2019**, *7*, 540–548. [CrossRef] [PubMed]
7. Dichgans, M.; Pulit, S.L.; Rosand, J. Stroke genetics: Discovery, biology, and clinical applications. *Lancet Neurol.* **2019**, *18*, 587–599. [CrossRef] [PubMed]
8. Padmanabhan, S.; Dominiczak, A.F. Genomics of hypertension: The road to precision medicine. *Nat. Rev. Cardiol.* **2020**, *18*, 235–250. [CrossRef] [PubMed]
9. Laakso, M. Biomarkers for type 2 diabetes. *Mol. Metab.* **2019**, *27S*, S139–S146. [CrossRef]
10. Lu, X.; Peloso, G.M.; Liu, D.J.; Wu, Y.; Zhang, H.; Zhou, W.; Li, J.; Tang, C.S.; Dorajoo, R.; Li, H.; et al. Exome chip meta-analysis identifies novel loci and East Asian-specific coding variants that contribute to lipid levels and coronary artery disease. *Nat. Genet* **2017**, *49*, 1722–1730. [CrossRef]
11. Rohde, K.; Keller, M.; la Cour Poulsen, L.; Bluher, M.; Kovacs, P.; Bottcher, Y. Genetics and epigenetics in obesity. *Metabolism* **2019**, *92*, 37–50. [CrossRef]

12. Rutten-Jacobs, L.C.; Larsson, S.C.; Malik, R.; Rannikmae, K.; Sudlow, C.L.; Dichgans, M.; Markus, H.S.; Traylor, M. Genetic risk, incident stroke, and the benefits of adhering to a healthy lifestyle: Cohort study of 306 473 UK Biobank participants. *BMJ* **2018**, *363*, k4168. [CrossRef]
13. Abraham, G.; Malik, R.; Yonova-Doing, E.; Salim, A.; Wang, T.; Danesh, J.; Butterworth, A.S.; Howson, J.M.M.; Inouye, M.; Dichgans, M. Genomic risk score offers predictive performance comparable to clinical risk factors for ischaemic stroke. *Nat. Commun.* **2019**, *10*, 5819. [CrossRef] [PubMed]
14. Abraham, G.; Rutten-Jacobs, L.; Inouye, M. Risk Prediction Using Polygenic Risk Scores for Prevention of Stroke and Other Cardiovascular Diseases. *Stroke* **2021**, *52*, 2983–2991. [CrossRef]
15. Wang, T.; Heianza, Y.; Sun, D.; Zheng, Y.; Huang, T.; Ma, W.; Rimm, E.B.; Manson, J.E.; Hu, F.B.; Willett, W.C.; et al. Improving fruit and vegetable intake attenuates the genetic association with long-term weight gain. *Am. J. Clin. Nutr.* **2019**, *110*, 759–768. [CrossRef] [PubMed]
16. Lu, X.; Liu, Z.; Cui, Q.; Liu, F.; Li, J.; Niu, X.; Shen, C.; Hu, D.; Huang, K.; Chen, J.; et al. A polygenic risk score improves risk stratification of coronary artery disease: A large-scale prospective Chinese cohort study. *Eur. Heart J.* **2022**, *43*, 1702–1711. [CrossRef] [PubMed]
17. Yang, X.; Li, J.; Hu, D.; Chen, J.; Li, Y.; Huang, J.; Liu, X.; Liu, F.; Cao, J.; Shen, C.; et al. Predicting the 10-year risks of atherosclerotic cardiovascular disease in Chinese population: The China-PAR project (Prediction for ASCVD Risk in China). *Circulation* **2016**, *134*, 1430–1440. [CrossRef] [PubMed]
18. Miller, V.; Yusuf, S.; Chow, C.K.; Dehghan, M.; Corsi, D.J.; Lock, K.; Popkin, B.; Rangarajan, S.; Khatib, R.; Lear, S.A.; et al. Availability, affordability, and consumption of fruits and vegetables in 18 countries across income levels: Findings from the Prospective Urban Rural Epidemiology (PURE) study. *Lancet Glob. Health* **2016**, *4*, e695–e703. [CrossRef] [PubMed]
19. Lu, X.; Niu, X.; Shen, C.; Liu, F.; Liu, Z.; Huang, K.; Wang, L.; Li, J.; Hu, D.; Zhao, Y.; et al. Development and Validation of a Polygenic Risk Score for Stroke in the Chinese Population. *Neurology* **2021**, *97*, e619–e628. [CrossRef] [PubMed]
20. Zhao, W.-H.; Huang, Z.-P.; Zhang, X.; He, L.; Willett, W.; Wang, J.-L.; Hasegawa, K.; Chen, J.-S. Reproducibility and Validity of a Chinese Food Frequency Questionnaire. *Biomed. Environ. Sci.* **2010**, *23*, 1–38. [CrossRef]
21. Han, C.; Liu, F.; Yang, X.; Chen, J.; Li, J.; Cao, J.; Li, Y.; Shen, C.; Yu, L.; Liu, Z.; et al. Ideal cardiovascular health and incidence of atherosclerotic cardiovascular disease among Chinese adults: The China-PAR project. *Sci. China Life Sci.* **2018**, *61*, 504–514. [CrossRef] [PubMed]
22. Wang, X.; Liu, F.; Li, J.; Yang, X.; Chen, J.; Cao, J.; Wu, X.; Lu, X.; Huang, J.; Li, Y.; et al. Tea consumption and the risk of atherosclerotic cardiovascular disease and all-cause mortality: The China-PAR project. *Eur. J. Prev. Cardiol.* **2020**, *27*, 1956–1963. [CrossRef] [PubMed]
23. Zhao, D. Poisson regression adjustment of event rates and its macro procedure ADJ_POIS. In Proceedings of the SAS Conference Proceedings: SAS Users Group International 24, Miami Beach, FL, USA, 11–14 April 1999.
24. Knol, M.J.; VanderWeele, T.J.; Groenwold, R.H.; Klungel, O.H.; Rovers, M.M.; Grobbee, D.E. Estimating measures of interaction on an additive scale for preventive exposures. *Eur. J. Epidemiol.* **2011**, *26*, 433–438. [CrossRef] [PubMed]
25. Lin, Y.; Yang, X.; Liang, F.; Huang, K.; Liu, F.; Li, J.; Xiao, Q.; Chen, J.; Liu, X.; Cao, J.; et al. Benefits of active commuting on cardiovascular health modified by ambient fine particulate matter in China: A prospective cohort study. *Ecotoxicol. Environ. Saf.* **2021**, *224*, 112641. [CrossRef] [PubMed]
26. Lau, B.; Cole, S.R.; Gange, S.J. Competing risk regression models for epidemiologic data. *Am. J. Epidemiol.* **2009**, *170*, 244–256. [CrossRef] [PubMed]
27. He, F.J.; Nowson, C.A.; MacGregor, G.A. Fruit and vegetable consumption and stroke: Meta-analysis of cohort studies. *Lancet* **2006**, *367*, 320–326. [CrossRef] [PubMed]
28. Tong, T.Y.N.; Appleby, P.N.; Key, T.J.; Dahm, C.C.; Overvad, K.; Olsen, A.; Tjonneland, A.; Katzke, V.; Kuhn, T.; Boeing, H.; et al. The associations of major foods and fibre with risks of ischaemic and haemorrhagic stroke: A prospective study of 418 329 participants in the EPIC cohort across nine European countries. *Eur. Heart J.* **2020**, *41*, 2632–2640. [CrossRef]
29. Du, H.; Li, L.; Bennett, D.; Guo, Y.; Key, T.J.; Bian, Z.; Sherliker, P.; Gao, H.; Chen, Y.; Yang, L.; et al. Fresh fruit consumption and major cardiovascular disease in China. *N. Engl. J. Med.* **2016**, *374*, 1332–1343. [CrossRef]
30. Miller, V.; Mente, A.; Dehghan, M.; Rangarajan, S.; Zhang, X.; Swaminathan, S.; Dagenais, G.; Gupta, R.; Mohan, V.; Lear, S.; et al. Fruit, vegetable, and legume intake, and cardiovascular disease and deaths in 18 countries (PURE): A prospective cohort study. *Lancet* **2017**, *390*, 2037–2049. [CrossRef]
31. Oude Griep, L.M.; Verschuren, W.M.; Kromhout, D.; Ocke, M.C.; Geleijnse, J.M. Colors of fruit and vegetables and 10-year incidence of stroke. *Stroke* **2011**, *42*, 3190–3195. [CrossRef]
32. Do, R.; Xie, C.; Zhang, X.; Mannisto, S.; Harald, K.; Islam, S.; Bailey, S.D.; Rangarajan, S.; McQueen, M.J.; Diaz, R.; et al. The effect of chromosome 9p21 variants on cardiovascular disease may be modified by dietary intake: Evidence from a case/control and a prospective study. *PLoS Med.* **2011**, *8*, e1001106. [CrossRef] [PubMed]
33. Hindy, G.; Ericson, U.; Hamrefors, V.; Drake, I.; Wirfalt, E.; Melander, O.; Orho-Melander, M. The chromosome 9p21 variant interacts with vegetable and wine intake to influence the risk of cardiovascular disease: A population based cohort study. *BMC Med. Genet* **2014**, *15*, 1220. [CrossRef] [PubMed]
34. Murgia, C.; Adamski, M.M. Translation of Nutritional Genomics into Nutrition Practice: The Next Step. *Nutrients* **2017**, *9*, 366. [CrossRef]

35. Mullins, V.A.; Bresette, W.; Johnstone, L.; Hallmark, B.; Chilton, F.H. Genomics in Personalized Nutrition: Can You "Eat for Your Genes"? *Nutrients* **2020**, *12*, 3118. [CrossRef] [PubMed]
36. Said, M.A.; Verweij, N.; van der Harst, P. Associations of combined genetic and lifestyle risks with incident cardiovascular disease and diabetes in the UK Biobank study. *JAMA Cardiol.* **2018**, *3*, 693–702. [CrossRef] [PubMed]
37. Zhao, C.N.; Meng, X.; Li, Y.; Li, S.; Liu, Q.; Tang, G.Y.; Li, H.B. Fruits for prevention and treatment of cardiovascular diseases. *Nutrients* **2017**, *9*, 598. [CrossRef] [PubMed]

MDPI
St. Alban-Anlage 66
4052 Basel
Switzerland
www.mdpi.com

Nutrients Editorial Office
E-mail: nutrients@mdpi.com
www.mdpi.com/journal/nutrients

www.ingramcontent.com/pod-product-compliance
Lightning Source LLC
Chambersburg PA
CBHW061508210926
51850CB00008B/1849
* 9 7 8 3 0 3 6 5 9 5 5 0 4 *